Thermodynamics of Natural Systems

Theory and Applications in Geochemistry and Environmental Science

Third Edition

Thermodynamics deals with energy levels and energy transfers between states of matter, and is therefore fundamental to all branches of science. This new edition provides an accessible introduction to the subject, specifically tailored to the interests of Earth and environmental science students. Beginning at an elementary level, the first four chapters explain all necessary concepts via a simple graphical approach. Throughout the rest of the book the author emphasizes the importance of field observations and demonstrates that, despite being derived from idealized circumstances, thermodynamics is crucial to understanding ore formation, acid mine drainage, and other real-world geochemical and geophysical problems. Exercises now follow each chapter, with answers provided at the end of the book. An associated website includes extra chapters and password-protected answers to additional problems. This textbook is ideal for undergraduate and graduate students studying geochemistry and environmental science.

Greg Anderson is Professor Emeritus in the Department of Earth Sciences at the University of Toronto. He began his career as a mining engineer and exploration geologist before becoming interested in geochemistry. After completing his Ph.D. at Toronto he spent five years at Pennsylvania State University doing experimental work on mineral solubilities. Afterward, he returned to Toronto, where he has since divided his time between studies of mineral systems, lead-zinc ore deposits, and theoretical geochemistry. He is the co-author of two other textbooks for Earth scientists, and in 2000 was awarded the Past Presidents' Medal (now named the Peacock Medal) by the Mineralogical Association of Canada for contributions to geochemistry.

Thermodynamics of Natural Systems

Theory and Applications in Geochemistry and Environmental Science

Third Edition

G.M. ANDERSON

CAMBRIDGE
UNIVERSITY PRESS

University Printing House, Cambridge CB2 8BS, United Kingdom

One Liberty Plaza, 20th Floor, New York, NY 10006, USA

477 Williamstown Road, Port Melbourne, VIC 3207, Australia

314-321, 3rd Floor, Plot 3, Splendor Forum, Jasola District Centre, New Delhi - 110025, India

79 Anson Road, #06-04/06, Singapore 079906

Cambridge University Press is part of the University of Cambridge.

It furthers the University's mission by disseminating knowledge in the pursuit of education, learning and research at the highest international levels of excellence.

www.cambridge.org
Information on this title: www.cambridge.org/9781107175211
10.1017/9781316796856

First published 2017

A catalogue record for this publication is available from the British Library

Library of Congress Cataloging in Publication data
Names: Anderson, G. M. (Gregor Munro), 1932– author.
Title: Thermodynamics of natural systems : theory and applications in geochemistry and environmental science / G.M. Anderson.
Description: Third edition. | Cambridge, United Kingdom ; New York, NY : Cambridge University Press, 2017. | Includes bibliographical references and index.
Identifiers: LCCN 2016041118 | ISBN 9781107175211 (Hardback ; alk. paper) | ISBN 1107175216 (Hardback ; alk. paper)
Subjects: LCSH: Thermodynamics–Textbooks. | Geochemistry–Textbooks.
Classification: LCC QE515.5.T46 A53 2017 | DDC 541/.369–dc23
LC record available at https://lccn.loc.gov/2016041118

ISBN 978-1-107-17521-1 Hardback

Additional resources for this publication at www.cambridge.org/thermodynamics

To my wife, Khodjasteh Hedjran Anderson

CONTENTS

ONLINE MATERIALS: ADDITIONAL CHAPTERS FROM THE SECOND EDITION
Available at www.cambridge.org/thermodynamics

PREFACE

Why another book on thermodynamics?

I wrote the first edition of this book because *Thermodynamics in Geochemistry* (Anderson and Crerar, 1993) was not suitable for the teaching I was doing at the time, a second-year geochemistry course for geology specialists and geological engineers. I wanted something shorter and less detailed. Later I wanted to write something more suitable for my students and other graduate students, resulting in the second edition.

But then the first edition went out of print, and several people told me that it was better for teaching than the second edition, which was seen as more of a reference book despite the fact that it includes just about all of the first edition. So the objective with this edition is to make a book more like the first edition available again, that is, one more suitable for a first introduction to thermodynamics, though with many improvements. Several chapters and appendices of the second edition have been cut, and others made shorter and more concise. Because thermodynamic potentials are the central concept in chemical thermodynamics, the first four chapters extend the ball-in-valley analogy to introduce these, as well as exact and inexact differentials, considerably improving the introduction to thermodynamic theory. There is also an increased emphasis on the importance of field observations in the application of thermodynamics to natural systems.

The three editions of this book and *Thermodynamics in Geochemistry* have several features in common, besides the fact that they have an Earth science point of view, and regarding which they differ from other books on thermodynamics. These differences include the following points.

- Presentation of thermodynamics as a model or idealization of real systems.
- Emphasis on the significance of the fact that thermodynamics consists of single-valued continuous functions.
- An approach to entropy which avoids discussion of cycles and the entropy of the Universe: Gibbs, not Carnot/Clausius.
- Emphasis on the importance of metastable equilibrium states and thermodynamic constraints.
- Emphasis on the fact that time is not a thermodynamic variable.

The presentation is exclusively classical, meaning that statistical mechanics is barely mentioned; only equilibrium states and macroscopic variables are considered. Some consider this approach obsolete, believing that classical thermodynamics has largely been superseded by the greater explanatory power of statistical mechanics. Because of this, many books on thermodynamics, perhaps a majority, have a chapter or a section on statistical mechanics. Teachers of thermodynamics now commonly blend thermodynamics and statistical mechanics together into a course in "thermal physics," in which classical thermodynamics inevitably becomes subordinate (Callen, 1985, p. viii).

This is not the best approach for scientists and engineers dealing with real-world problems. Knowing that equilibrium requires that energy be distributed among the system's degrees of freedom in a manner described by a Boltzmann distribution adds nothing useful for a geochemist investigating a waste disposal site or a metallic ore deposit. Gibbs said that the equilibrium state has no gradients in temperature, pressure, or composition, and that is entirely sufficient. What might help is the knowledge that in thermodynamics this state is an idealization, so it won't exist at the disposal site and never existed at the ore deposit site, and that despite this thermodynamics will be essential in the study of such natural systems.

There is no doubt about the explanatory power of statistical mechanics, but, although a chapter or a section on statistical mechanics will give a general idea of the ideas involved, it is not sufficient for a good understanding; in fact, it is a distraction from a first attempt at learning thermodynamics. Statistical mechanics may well be in some sense more fundamental than classical thermodynamics, but more fundamental does not mean more useful. It is even possible to question whether thermodynamics can be reduced to statistical mechanics; whether concepts such as entropy in thermodynamics and statistical mechanics are the same or at least simply related (Sklar, 1993, p. 337), but such philosophical arguments are of even less concern to geochemists. In his book on statistical mechanics, Nash (1974) says

From a union of the entropy and energy concepts ... there was born a notably abstract science with innumerable concrete applications; a science of thermodynamics that combines magnificent generality with unfailing reliability to a degree unrivalled by any other science known to man.

It has not been superseded.

Geochemical Modeling

In common with other sciences, thermodynamics deals with idealized concepts, and there are always attempts to make it seem more realistic. It is better to teach and learn thermodynamics as it really is, with a clear distinction between real systems and the thermodynamic models of these systems. The connection between thermodynamics and reality is made, not by explaining entropy with statistical mechanics, but by using thermodynamics to elucidate the countless number of practical real-world problems in all aspects of Earth science – ore formation, chemical oceanography, the environment, most of igneous and metamorphic petrology, and so on. In this sense, doing exercises and problems is even more important than in other subjects. But it can be taken too far. There is a tendency to add realism by including topics such as fluid flow and reactive transport, the subjects of geochemical modeling programs such as the USGS program PHREEQC (Parkhurst and Appelo, 2013) or the programs in The Geochemist's Workbench (2015). These applications greatly enhance the power of thermodynamics in practical situations, and tech-savvy students love using them. The problem is that, much like the inclusion of statistical mechanics, including such programs dilutes the concentration on thermodynamics itself, and obscures the idea that thermodynamics itself is a model. An exception is the use of these programs to do speciation (Chapter 10), which greatly expands the ability of thermodynamics to deal with problems involving aqueous solutions. Having at least one course which deals only with classical equilibrium thermodynamics, with problem sets that must be done with a calculator or a spreadsheet, is the only way to really learn the subject.

Acknowledgments

There are many references in the text to data from Johnson *et al.* (1992), which in the second edition was referred to as SUPCRT92. These data have been superseded by data from THERMOCALC (Holland and Powell, 2011) but, as the data are used for illustrative purposes only, there was no point in recalculating all the results from SUPCRT92. I thank Roger Powell and Tim Holland for supplying data for those places where it seemed important.

It is a pleasure to acknowledge the encouragement and support of Dr. Norman Evensen in the writing of this book. His wide-ranging knowledge of science and mathematics provided valuable insights.

ONLINE RESOURCES AND EXERCISES

This book is intended as an introduction to thermodynamics. To give instructors the option of including other aspects of particular thermodynamic topics, several chapters and sections from the second edition (Anderson, 2005) are available on the Cambridge University Press website for this book, www.cambridge.org/thermodynamics.

These sections are as follows.

- Chapter 10 Real Solutions
- Chapter 11 The Phase Rule
- Chapter 13 Equations of State
- Chapter 14 Solid Solutions
- Chapter 15 Electrolyte Solutions
- The van 't Hoff Equilibrium Box
- Topics in Mathematics

At appropriate places in the text, reference to the additional information available online is noted in a box like this, on page 154:

Equations of State

The ideal gas equation $PV = nRT$ or $PV = RT$ is an equation of state; an equation relating the volume of an ideal gas to its pressure and temperature. For real gases there have been many modifications of this equation, and there are other equations of state which interrelate not only P, V, and T, but also thermodynamic parameters such as enthalpy, Gibbs energy, and entropy. See Chapter 13 Equations of State, in the online material.

Other boxes add worked examples or more details of the material in the text. For example on page 21 there is this:

Box 2.1 Scientific vs. Engineering Units

In science, *molar* properties, such as molar volumes and molar energies, are most commonly used. In engineering, on the other hand, *specific* properties are more common. Specific properties are mass-related rather than mole-related. Thus the specific volume of water at 25 °C is $1.0029 \, \text{cm}^3 \, \text{g}^{-1}$. Molar and specific properties are related by the molar mass (or so-called gram formula weight, gfw) of the substance. That for water is $18.0153 \, \text{g} \, \text{mol}^{-1}$, so $1.0029 \, \text{cm}^3 \, \text{g}^{-1} \times 18.0153 \, \text{g} \, \text{mol}^{-1} = 18.068 \, \text{cm}^3 \, \text{mol}^{-1}$.

Exercises and Additional Problems

Exercises

At the end of each chapter there is a set of exercises on the material in that chapter under the heading **Exercises**. These are intended to be done by students as an aid to understanding, and the answers are in Appendix C.

Additional Problems

Following that, in most chapters under the heading **Additional Problems** there are more problems, often a bit more difficult, the answers to which are available to instructors from Cambridge University Press.

1 What Is Thermodynamics?

1.1 Introduction

Thermodynamics is the branch of science that deals with relative energy levels and transfers of energy between systems and between different states of matter. Because these subjects arise in virtually every other branch of science, thermodynamics is one of the cornerstones of scientific training. Various scientific specialties place varying degrees of emphasis on the subject areas covered by thermodynamics – a text on thermodynamics for physicists can look quite different from one for chemists, or one for mechanical engineers. For chemists, biologists, geologists, and environmental scientists of various types, the thermodynamics of chemical reactions is of course a central concern, and that is the emphasis to be found in this book. Let us start by considering a few simple reactions and the questions that arise in doing this.

1.2 What Is the Problem?

1.2.1 Some Simple Chemical Reactions

A chemical reaction involves the rearrangement of atoms from one structure or configuration to another, normally accompanied by an energy change. Let's consider some simple examples.

- Take an ice cube from the freezer of your refrigerator and place it in a cup on the counter. After a few minutes, the ice begins to melt, and it soon is completely changed to water. When the water has warmed up to room temperature, no further change can be observed, even if you watch for hours. If you put the water back in the freezer, it changes back to ice within a few minutes, and again there is no further change. Evidently, this substance (H_2O) has at least two different forms, and it will change spontaneously from one to the other depending on its surroundings.

- Take an egg from the refrigerator and fry it on the stove, then cool to room temperature. Again, all change seems now to have stopped – the reaction is complete. However, putting the fried egg back in the refrigerator will not change it back into a raw egg. This change seems not to be reversible. What is different in this case?

- Put a teaspoonful of salt into a cup of water. The salt, which is made up of a great many tiny fragments of the mineral halite ($NaCl$), quickly disappears into the water. It is still there, of course, in some dissolved form, because the water now tastes salty, but why did it dissolve? And is there any way to reverse this reaction?

Eventually, of course, we run out of experiments that can be performed in the kitchen. Consider two more reactions.

- On a museum shelf, you see a beautiful clear diamond and a piece of black graphite side by side. You know that these two specimens have exactly the same chemical composition (pure carbon, C), and that experiments at very high pressures and temperatures have succeeded in changing graphite into diamond. But how is it that these two different forms of carbon can exist side by side for years, while the two different forms of H_2O cannot?

- When a stick of dynamite explodes, a spectacular chemical reaction takes place. The solid material of the dynamite changes very rapidly into a mixture of gases, plus some leftover solids, and the sudden expansion of the gases gives the dynamite its destructive power. The reaction would seem to be nonreversible, but the fact that energy is obviously released may furnish a clue to understanding our other examples, where energy changes were not obvious.

These reactions illustrate many of the problems addressed by chemical thermodynamics. You may have used ice in your drinks for years without realizing that there was a problem, but it is actually a profound and very difficult one. It can be stated in the following way. What controls the changes (reactions) that we observe taking place in substances? Why do they occur? And why can some reactions go in the forward and backward directions (i.e., ice → water or water → ice) while others can only go in one direction (i.e., raw egg → fried egg)? Scientists puzzled over these questions during most of the nineteenth century before the answers became clear. Having the answers is important; they furnish the ability to control the power of chemical reactions for human uses, and thus form one of the cornerstones of modern science.

1.3 A Mechanical Analogy

Wondering why things happen the way they do goes back much further than the last century and includes many things other than chemical reactions. Some of these things are much simpler than chemical reactions, and we might look to these for analogies, or hints, as to how to explain what is happening.

A simple mechanical analogy would be a ball rolling in a valley, as in Figure 1.1. Balls have always been observed to roll down hills. In physical terms, this is "explained" by saying that mechanical systems have a tendency to change so as to reduce their *potential energy* to a minimum. Strang (1991, p. 102) says

In mechanics, nature chooses minimum energy.

In the case of the ball on the surface, the potential energy (for a ball of given mass) is determined by the height of the ball above the lowest valley, or some other reference level. It follows that the ball will spontaneously roll downhill, losing potential energy as it goes, to the lowest point it can reach. Thus it will always come to rest (equilibrium) at the bottom

1.3 A Mechanical Analogy

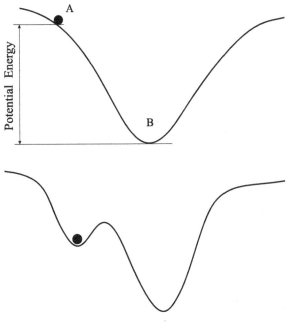

Figure 1.1 A mechanical analogy for a chemical system – a ball on a slope. The ball will spontaneously roll into the valley.

Figure 1.2 The ball has rolled into a valley, but there is a deeper valley.

of a valley. However, if there is more than one valley, it may get stuck in a valley that is not the lowest available, as shown in Figure 1.2.

It was discovered quite early that most chemical reactions are accompanied by a release or liberation of heat. This is most easily seen when you strike a match, but in fact the freezing of water is also a heat-liberating process. It was quite natural, then, by analogy with mechanical systems, to think that various substances contained various quantities of heat, which was thought to be a substance called caloric, and that reactions would occur if substances could rearrange themselves (react) so as to *lower* their heat or caloric content. These days we know that heat is a form of energy transfer, and we prefer to say that reactions lower some kind of *energy*, not just heat. According to this view, ice would have less of this energy (per gram, or per mole) than has water in the freezer, so water changes spontaneously to ice, and the salt in dissolved form would have less of this energy than solid salt, so salt dissolves in water. In the case of the diamond and graphite, perhaps the story is basically the same, but carbon is somehow "stuck" in the diamond structure.

Of course, chemical systems are not mechanical systems, and analogies can be misleading. You would be making a possibly fatal mistake if you believed that the energy of a stick of dynamite could be measured by how far above the ground it was. Nevertheless, the analogy is useful. Perhaps chemical systems will react so as to lower (in fact, minimize) their *chemical* energy, although sometimes, like diamond, they may get stuck in a valley (an energy level) higher than another nearby valley. We will see that this is in fact the case. The analogy *is* useful. We know that energy is manifested in different forms, so the problem lies in discovering just what kind of energy is being minimized. What is this *chemical energy*?

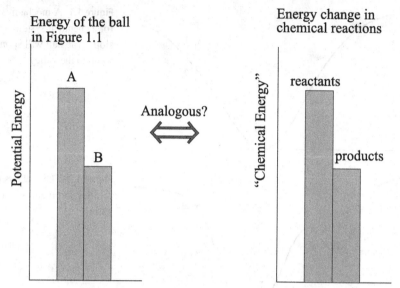

Figure 1.3 Mechanical processes always act so as to lower the potential energy content of the mechanical system. Perhaps, by analogy, chemical systems have some sort of "chemical energy" that is lowered during chemical reactions.

1.3.1   Chemical Energy

We mentioned that an early idea was that it is the heat (caloric) content of systems that is minimized in chemical systems; that is, reactions will occur if heat is liberated. This is another way of saying that the caloric content of the *products* is less than the caloric content of the *reactants* of a reaction, so that the reaction liberates heat (Figure 1.3). This view of things was common in the nineteenth century, and a great deal of effort was expended in measuring the flow of heat in chemical reactions. However, we don't even have to leave our kitchen to realize that this cannot be entirely correct. The melting of ice is obviously a process in which heat is *absorbed*, not liberated, which is why it is useful in cooling drinks. Therefore, despite the appealing simplicity of the "heat content" argument for explaining why chemical reactions occur, it cannot be the whole story. Nevertheless, the idea that some kind of "chemical energy" is liberated in reactions, or that "chemical energy" is minimized in systems at rest (equilibrium) is a powerful one. Perhaps heat is not the only factor involved. What other factors might there be? Not too many, we hope!

1.3.2   Plus Something Else?

Another important clue we must pay attention to is the fact that some chemical reactions are able to take place with no energy change at all. For example, when gases mix together at low pressures, virtually no heat energy is liberated *or* absorbed. The situation is similar for a drop of ink spreading in a glass of water. These are spontaneous processes. We are using the terms *reaction* and *process* more or less synonymously here. Strictly speaking there is a difference; a chemical reaction is one kind of process, but there are others. Some spontaneous processes are characterized by *mixing*, rather than by a reorganization of molecular structures like graphite → diamond, or raw egg → fried egg. In fact even just warming or cooling is a process, and some processes, like metamorphism, might include many reactions. We will give a more exact definition in Chapter 2, but clearly our "chemical energy" term will have to take account of all these observations.

At this point, we might become discouraged, and conclude that our idea that some sort of chemical energy is being reduced in all reactions must be wrong – there seem to be many complications. It certainly was a puzzle for a long time. But we have the benefit of hindsight, and, because we now know that this concept of decreasing chemical energy of some kind is in fact the correct answer, we will continue to pursue this line of thought.

1.4 Thermodynamic Potentials

The ball-in-valley analogy as introduced in Figures 1.1 and 1.2 is intended to show *not only* that chemical systems strive to achieve equilibrium by decreasing some kind of "chemical energy" but also that, when at equilibrium, i.e., at the bottom of some valley, they achieve an *extremum* in the mathematical sense. This is instructive because it turns out that thermodynamics has functions called *thermodynamic potentials* (Section 4.3), which exhibit a minimum or maximum for many systems in a state of equilibrium. To illustrate what this means, we can look more closely at the analogy.

Consider a ball which has achieved a minimum potential energy as in Figure 1.2. It doesn't have to be the lowest minimum available or a symmetrical valley, so, to be a little more general, we can choose a valley described by the function (Strang, 1991, p. 98)

$$y = x + \frac{1}{x} \tag{1.1}$$

shown in Figure 1.4, for which the derivative is

$$\frac{dy}{dx} = 1 - \frac{1}{x^2} \tag{1.2}$$

The minimum of the function occurs where the derivative, the slope of the curve, is zero, which in this case is where $x = 1$ and $y = 2$. Because the second derivative $d^2y/dx^2 > 0$ it is a minimum. In other words, when the derivative dy/dx equals zero, this tells us that the function has an extremum, and implies that the differential is also zero, or $dy = 0$.

1.4.1 An Exact Differential

Differentials are ubiquitous in thermodynamics. It is absolutely essential to have a clear understanding of what they mean. The differential dy is defined in terms of the derivative of the function; that is,

$$dy = \frac{dy}{dx} dx \tag{1.3}$$

$$= y'(x)dx \tag{1.4}$$

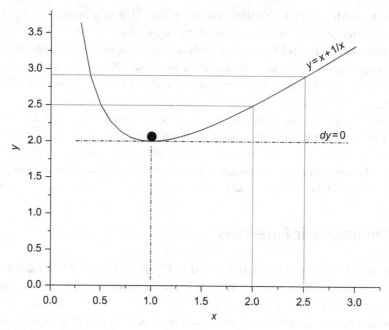

Figure 1.4 A valley described by the function $y = x + 1/x$. The minimum occurs where both the derivative dy/dx and the differential dy are zero.

where, in the case of the function in Equation (1.1), dy/dx is defined in Equation (1.2) and $y'(x)$ is another way of writing the derivative. This way is perhaps preferable because it emphasizes that the derivative is just another function of x. The derivative dy/dx is not, as it appears, a fraction composed of two quantities called dx and dy, but defining the differential separates them; it creates the separate quantities dx and dy. Each can now have any magnitude, and they are not necessarily infinitesimals. Furthermore, if the derivative is zero, as it is at an extremum in the function, then dy is also zero, by Equation (1.3) or (1.4).

If we integrate dy between limits, we get

$$\int_a^b dy = \int_a^b y'(x)dx \tag{1.5}$$

$$= y(b) - y(a) \tag{1.6}$$

$$= \Delta y \tag{1.7}$$

The significance of this in thermodynamics is that, if y is some physical parameter (say volume $\mathbf{V}$) which is a function of some other parameter x (say temperature), then the change in $\mathbf{V}$ because of some change in x from a to b can be calculated *no matter what the volume of the real system actually does* as x changes from a to b. The system may be completely out of equilibrium between the equilibrium states $\mathbf{V}_a$ and $\mathbf{V}_b$ so the system doesn't actually have a defined volume during the change, but we can calculate the change anyway if we have a function describing the change of $\mathbf{V}$ from $\mathbf{V}_a$ to $\mathbf{V}_b$. In this case $d\mathbf{V}$ is called an exact differential; it is derived from a continuous function and can be integrated. The reason for discussing this here is that we will soon come across other differentials (the differentials of work and heat) which are not derived from any function, and cannot be integrated in this way.

Just to complete this example, let's say $a = 2.0$ and $b = 2.5$. Then

$$\Delta y = y(2.5) - y(2.0)$$
$$= (2.5 + 1/2.5) - (2.0 + 1/2.0)$$
$$= 2.9 - 2.5$$
$$= 0.4$$

As x changes from 2.0 to 2.5, the change in y is 0.4. Perfectly simple when dealing with a mathematical function like $y = y'(x)$. Not quite as simple when the variables are thermodynamic parameters representing real things.

We will find that thermodynamics has functions that behave exactly like this; they are continuous, have exact differentials, and may have a maximum or minimum. We take a closer look at them in Chapter 4. Because these functions, called thermodynamic potentials, are a central feature of thermodynamics, the concept of differentials in calculus becomes important. If you have any doubts about what a differential is, or the difference between a differential and a derivative, see the discussion in the online resources or in one of the many calculus textbooks.

Topics in Mathematics

At several places in the text there is a reference to some mathematical subject in the online resources. Subjects like Legendre transforms, Euler's theorem, homogeneous functions, etc., as well as some of the more arcane aspects of Gibbs' fundamental writings on thermodynamics, are not necessary for an introduction to the subject, but they do offer a different perspective on thermodynamics which in some ways is simpler, though not to everyone's taste. There is also a bit more discussion of some essential topics in the text such as here, on the meaning of differentials.

1.5 Idealization in Thermodynamics

Despite its immense usefulness in real life, thermodynamics, like all physical theories, describes not real systems, but models of real systems. It is easy to confuse these two things. We have already illustrated the difference by referring to a valley as the function $y = x + 1/x$. No real valley is described exactly by this function, though some might come close. The function and the curve it describes constitute a two-dimensional or cross-sectional *model* of a real valley. As Cartwright (1983, p. 129) says,

...fundamental equations do not govern objects in reality; they govern only objects in models.

Feynman *et al.* (1963, p. 12–2) say

... in order to understand physical laws you must understand that they are all some kind of approximation. ... The trick is the idealizations.

Thermodynamic theory is a bunch of mathematical expressions such as simple linear differential equations. But Feynman *et al.* also say

... mathematical definitions can never work in the real world.

Engineering design problems are also based on models. Jones and Dugan (1996, pp. 17–18) say

In solving physical problems, we usually focus our attention not on the actual system but rather on some idealized system that is similar to, but simpler than, the actual system. ... For example, in calculating the mechanical advantage of a system of ropes and pulleys (*actual system*), we might start by considering a *model* composed of non-stretching weightless ropes and frictionless pulleys.

In this view classical chemical thermodynamics is a model of energy relationships in chemical systems in which many aspects of real systems are represented by equations which work only for idealized systems. Thermodynamics seeks to represent or to simulate reality, but, because it is made up of the differential and integral calculus, it can do this only in an idealized way. We will see that this idealization means that only equilibrium states can be represented by our equations, which is why the subject is often called *equilibrium* thermodynamics or thermostatics.

The reason for emphasizing this somewhat philosophical point is that many aspects of thermodynamics (e.g., ideal solutions, reversible processes) are physically unrealistic. It helps to remember that we are using mathematics to simulate real systems.

The Equilibrium Paradox

The fact that our thermodynamic functions represent *only* equilibrium states, plus that natural systems are rarely, if ever, at equilibrium, obviously presents a paradox. How is it that a bunch of equations derived for idealized, non-existent equilibrium conditions can be useful in trying to understand real, not even close to equilibrium, processes like ore deposit formation and groundwater contamination? This is an important question to keep in mind as you progress through the text.

1.6   Limitations of the Thermodynamic Model

This book outlines the essential elements of a first understanding of chemical thermodynamics, especially as applied to natural systems. However, it is useful at the start to have some idea of the scope of our objective – just how useful is this subject, and what are its limitations? It is at the same time very powerful and very limited. With the concepts described here, you can predict the equilibrium state for most chemical systems, and therefore the direction and amount of reaction that should occur, including the composition of all phases when reaction has stopped. The operative word here is "should." Our model consists of comparing equilibrium states with one another, and determining which is more stable under the circumstances. We will not consider how fast the reaction will proceed, or how to tell whether it will proceed at all. Many reactions that "should" occur do not occur, for various reasons. We will also say very little about what "actually" happens during these reactions – the specific interactions of ions and molecules that result in the new arrangements or structures that are more stable. In other words, our model will say virtually nothing about *why* one arrangement is more stable than another or has less "chemical energy," just that it does, and how to determine that it does.

These are serious limitations. Obviously, we will often need to know not only that a reaction *should* occur but also *whether* it occurs, and at what rate. A great deal of effort has also been directed toward understanding the structures of crystals and solutions, and what happens during reactions, shedding much light on why things happen the way they do. However, these fields of study are not completely independent. The subject of this book is really a prerequisite for any more advanced understanding of chemical reactions, which is why every chemist, environmental scientist, biochemist, geochemist, soil scientist, and the like must be familiar with it.

But, in a sense, the limitations of our subject are also a source of its strength. The concepts and procedures described here are so firmly established partly because they are independent of our understanding of *why* they work. The laws of thermodynamics are distillations from our experience, and that goes for all the deductions from these laws, such as are described in this book. As a scientist dealing with problems in the real world, you need to know the subject described here. You need to know other things as well, but this subject is so fundamental that virtually every scientist has it in some form in his tool kit.

1.7   Summary

The fundamental problem addressed here is why things (specifically, chemical reactions) happen the way they do. Why does ice melt and water freeze? Why does graphite change into diamond, or vice versa? Or why does it *not* change? Taking a cue from the study of simple mechanical systems, such as a ball rolling in a valley, we propose that these reactions happen if some kind of energy is being reduced, much as the ball rolls down in order to reduce its potential energy. However, we quickly find that this cannot be the whole story – some reactions occur with *no* decrease in energy. We also note that whatever kind of energy is being reduced (we call it "chemical energy"), it is not simply heat energy.

For a given ball and valley (Figure 1.1), we need to know only one parameter to determine the potential energy of the ball (its height above the base level, or the bottom of the valley). In our "chemical energy" analogy, we know that there must be *at least* one other parameter, to take care of those processes that have no energy change. Determining the parameters of our "chemical energy" analogy is at the heart of chemical thermodynamics.

Exercises

E1.1 In Figure 1.4 assume that y is distance in meters. Calculate the change in potential energy of a ball having a mass of 100 g as it rolls from its position at $y = 4.0$ m down to the minimum of the valley at $y = 2.0\,\text{m}$. The acceleration due to gravity is 9.81 m/s^2.

E1.2 Apparently the ball-and-valley system has lost energy. But the first law of thermodynamics (Chapter 3) says that energy is conserved. Clausius said (see p. 26) "the energy of the universe is constant." What's going on?

Additional Problems

A1.1 The definition of the differential in Equation (1.4) is for a function having only one independent variable, x. We will see in Chapter 4 that all our important thermodynamic functions have either two or three independent variables. For these we just expand the definition from $u = f(x)$ to $u = f(x, y)$ or $u = f(x, y, z)$. For example, the ideal-gas equation $V = RT/P$ shows that the volume per mole of ideal gas is a function of two independent variables, temperature T and pressure P, R being a constant. We form the derivative of V with respect to each variable separately, keeping the other variable constant, giving what is called a *partial derivative*. It's partial because we operate on only one of the two variables. So we have $(\partial V/\partial T)_P$ and $(\partial V/\partial P)_T$, where we use ∂ instead of d to remind us that it is a partial derivative only, and the subscript tells us about the variable held constant. The *total differential* is then

$$dV = \left(\frac{\partial V}{\partial T}\right)_P dT + \left(\frac{\partial V}{\partial P}\right)_T dP \qquad (1.8)$$

which is basically just an example of Equation (1.4) with an extra term added. This has a very simple geometrical meaning we will see in Chapter 4.

(a) Calculate the derivatives $(\partial V/\partial T)_P$ and $(\partial V/\partial P)_T$ for the ideal-gas equation, $V = RT/P$.

(b) Show that dV is an exact differential by integrating Equation (1.8) from (P_1, T_1) to (P_2, T_2) by two different paths: (i) $(P_1, T_1) \to (P_1, T_2) \to (P_2, T_2)$ and (ii) $(P_1, T_1) \to (P_2, T_1) \to (P_2, T_2)$. The result for both paths should of course be $(V_2 - V_1)$.

A1.2 Find the total differentials of the following functions:

(a) $u = xyz/(x + y + z)$,
(b) $u = x/y + y/x + z/x$, and
(c) $u = e^{x+y^2}$.

2 Defining Our Terms

2.1 Something Is Missing

We mentioned in Chapter 1 that an early idea for understanding chemical reactions held that spontaneous reactions would always be accompanied by the loss of energy, because the reactants were at a higher energy level than the products, and they wanted to go "downhill." This energy was thought to be in the form of heat, but this idea received a setback when it was found that some spontaneous reactions in fact absorb heat. Also, there are some processes, such as the mixing of gases, where the energy change is virtually zero yet the processes proceed very strongly and are highly nonreversible. Obviously, something is missing. If the ball-in-valley analogy is right, that is, if reactions do proceed in the direction of decreasing chemical energy of some kind, something more than just heat is involved.

To learn more about chemical reactions, we have to become a bit more precise in our terminology and introduce some new concepts. In this chapter, we will define certain kinds of *systems*, because we need to be careful about what kinds of matter and energy transfers we are talking about; *equilibrium states*, the beginning and ending states for processes; *state variables*, the properties of systems that change during reactions; *processes*, the reactions themselves; and *phases*, the different types of matter within the systems. All these terms refer to real systems, but they also refer to the equivalent things in our models of these systems. To be quite clear about thermodynamics, it is a good idea to keep the distinction in mind.

2.2 Systems

2.2.1 Real-Life Systems

In real life, a *system* is any part of the universe that we wish to consider. If we are conducting an experiment in a beaker, then the contents of the beaker is our system. For a petrologist, a crystallizing magma could be the system. In considering geochemical, biological, or environmental problems here on Earth, the choice of system is usually fairly obvious, and depends on the kind of problem in which you are interested.

Figure 2.1 shows a seashore environment with three possible choices of natural system. At (a), we might be interested in the exchange of gases between the sea and the atmosphere (e.g., if the sea warms by one degree, how much CO_2 will be released to the atmosphere?). At (b), we might be interested in the dissolved material in the sea itself (e.g., the reactions between dissolved CO_2 and carbonate and bicarbonate ions). And at (c), we might be

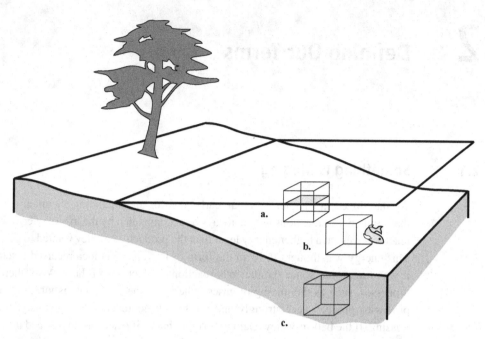

Figure 2.1 A seashore environment. The locations of three natural systems are shown.

interested in reactions between the sediment and the water between the sediment particles (e.g., dissolution or precipitation of minerals in the sediment). The chosen systems are shown as boxes, but in most cases we are not concerned with the dimensions or shape of the box; we normally define the system in terms of the masses or moles of components in the system, as well as the nature of its contacts with whatever is outside the system (see Section 2.2.2).

These are examples of *inorganic* systems. Thermodynamics can also be applied to organic systems, including living organisms. A single bacterium could be our system, or a dish full of bacteria, or a single organelle within a bacterium. The choice depends on your particular interests and is obviously very wide. However, they are all similar in one respect. Because natural systems exist in the real world, whatever system you choose is bounded by (in contact with) other parts of the world and may exchange energy and matter (liquids, solids, gases) with these other parts of the world. Systems of this type are said to be "open." All living organisms are thus open systems because they take in nutrients, and get rid of waste products. All three systems in Figure 2.1 are obviously open, because water can flow in and out of (a) and (b), and, even in (c), compaction of the sediments squeezes water out, and diffusion allows solutes to move in and out.

2.2.2   Thermodynamic Systems

Our goal is to understand energy changes in natural systems. We will do this by considering much simpler models of these systems, having variables that represent what we think are the essential elements of the natural systems. Using such models is a universal practice in science, as discussed in Section 1.5. Our models will be not material, but mathematical and conceptual. If we do it right, then the behavior of the model system will help us understand and predict the behavior of the real system.

Although most natural systems are open and quite complex, our models of these systems can be much simpler and yet still be valuable. The kinds of thermodynamic or model systems that have been found to be useful in analyzing and understanding natural (real-life) systems are as follows, and are illustrated in Figure 2.2. These thermodynamic systems are essentially defined by the types of walls they have. This is because we must be able to control (conceptually) the flow of matter and energy into and out of these systems.

- *Isolated systems* have walls or boundaries that are rigid (thus not permitting transfer of mechanical energy), perfectly insulating (thus preventing the flow of heat), and impermeable to matter. They therefore have a constant energy and mass content, since none can pass in or out. Perfectly insulating walls and the systems they enclose are called *adiabatic*. Isolated systems do not occur in nature, because there are no such impermeable and rigid boundaries. Nevertheless, this type of system has great significance because reactions that occur (or could occur) in isolated systems are ones that *cannot* liberate or absorb heat or any other kind of energy. Therefore, if we can figure out what causes *these* reactions to go, we may have an important clue to the overall puzzle.

- *Closed systems* have walls that allow transfer of energy into or out of the system but are impervious to matter. They therefore have a fixed mass and composition but variable energy levels.

- *Open systems* have walls that allow transfer of both energy and matter to and from the system. The system may be open to only one chemical species or to several.

As mentioned above, most natural systems are open. There are equations for open systems, those that can change composition, and we will introduce them (in Section 4.13), but we don't use them to any great extent. It is possible and more convenient to model them as closed systems; that is, to consider a fixed composition, and simply ignore any possible changes in total composition. If what happens because of changes in composition is important, it can often be handled by considering two or more closed systems of different compositions. Thus we will be dealing mostly with closed systems in our efforts to understand chemical reactions. Basically this means that we will be concerned mostly with individual chemical reactions, rather than with whole complex systems. In other words, even though a bacterium is an open system, it can be treated (modeled) as a closed system while considering many individual reactions within it. The reactants may need to be ingested and the products eliminated by the organism, but the reaction itself can be modeled independently of these processes. This greatly simplifies the task of understanding the biochemical reactions. The same is true of most geochemical and environmental systems.

The most common kind of open system in chemical thermodynamics is represented in Figure 2.2(b), that is, two open subsystems within an overall closed system. There can be any number of these "open subsystems" or phases, and finding out how many there are and what their compositions are, given some physical conditions, is a common problem in the application of thermodynamics.

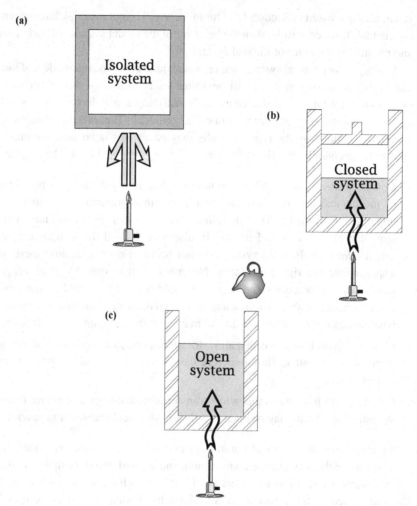

Figure 2.2 (a) Isolated system. Nothing can enter or leave the system (no energy, no matter). Whatever is inside the walls (which could be anything) will have a constant energy content and a constant composition. (b) Closed system. The closure is a piston to indicate that the pressure on the system is under our control. Energy can enter and leave the system, but matter cannot. The system here is shown as part liquid, part gas or vapor, but it could be anything. Both the liquid and the gas could also be considered as open systems, inside the closed system. Each may change composition, although the two together will have a constant composition. (c) Open system. Both matter and energy may enter and leave the system. The system may have a changing energy content and/or a changing composition. The pitcher shows one way of adding matter to the system.

The Isolated System Paradox

It is one of the paradoxes of thermodynamics that isolated systems, that have no counterpart in the real world, are possibly the most important of all in terms of our understanding of chemical reactions. You will have to wait until Chapter 4 to see why.

2.3 Equilibrium

2.3.1 Introduction

The concept of the equilibrium state is, along with entropy, arguably the most important of all thermodynamic concepts. But, in contrast to entropy, which has retained its somewhat mysterious nature ever since Clausius discovered its importance, equilibrium seems to be an intuitively simple idea. It's just a state having a balance of forces, a state which does not change, right? Well, yes, but thermodynamic equilibrium needs more discussion. First, because we have real and thermodynamic systems, we must distinguish between real and thermodynamic equilibrium states. Then we introduce the various varieties – stable, metastable, partial, and local. From the point of view of someone doing experiments, equilibrium is vital, because results must represent equilibrium states, otherwise they are not reproducible and are useless.

In natural systems it is the concept of *local equilibrium* which is vital. Natural systems may be overall far from equilibrium, and we have no control over them. Here the importance is not reproducibility, but the fact that systems which are not at equilibrium do not have fixed values of their thermodynamic properties, so we apparently cannot use our thermodynamic methods with them. The thermodynamic data we use, such as those in Appendix B, are derived from experiments in equilibrium states, so how can they be used in any other states? The answer is that we don't apply our thermodynamic methods to systems in complete disequilibrium, like a turbulent fluid or an exploding volcano. But systems in overall disequilibrium, such as acid mine drainage, crystallizing magmas, metallic ore deposition, or the carbon cycle, operate over long time periods and have small volumes throughout the system which *are* close to equilibrium, and samples from those places provide thermodynamic information. Even active geothermal systems, having large temperature gradients, boiling, alteration reactions, and so on (see Chapter 11 problem sets), have small volumes of local equilibrium, or so we believe. Other systems, like the ocean, have volumes of local equilibrium which are much larger.

We begin with thermodynamic equilibrium as an idealization. We have stressed the idea that there is a difference between real and thermodynamic systems, and the state of equilibrium is defined differently in the two cases.

2.3.2 Equilibrium in Thermodynamic Systems

The definitive definition of the equilibrium state is given by Gibbs (1961a). Based on the fundamental role of entropy discovered by Clausius (1867) which we get to in Chapter 4, he shows in his equations 14–21 (shown in Section C.9.2, Thermodynamic Equilibrium, of the Topics in Mathematics, in the online resources) that a system at equilibrium can have no gradients in temperature T, pressure P, or chemical composition. But all real systems have inhomogeneities and gradients, however small. Herzfeld (1962) says

But no real measurement is reversible, and no real system is completely in equilibrium.

Given the absence of *perfect* equilibrium in real systems, the definition of equilibrium as expressed by Gibbs and others is an idealization. In model systems which include a solution of variable composition, it is a state occurring at a maximum or a minimum of a function called a thermodynamic potential, an important topic introduced in Section 1.4. See Figure 13.3 for an example. In other systems, such as those having only pure solid phases, thermodynamic potentials do not exhibit a maximum or a minimum, but the system can nevertheless be at equilibrium. These situations are explored further in Chapter 9.

Some real systems approach this state more or less closely, but probably never attain it. When real systems or parts of real systems do approach thermodynamic equilibrium, thermodynamics can be applied to them. Obviously, we need to have some way of telling whether real systems are "at equilibrium" or have closely approached equilibrium.

2.3.3   Equilibrium in Real Systems

Equilibrium states in real systems have two attributes:

1. A real system at equilibrium has none of its properties changing with time, no matter how long it is observed.
2. A real system at equilibrium will return to that state after being disturbed, that is, after having one or more of its parameters slightly changed, then changed back to the original values. This is a common experimental technique, but is of limited use in the investigation of natural systems.

The hope is, of course, that a system which obeys these criteria will be close to the idealized equilibrium state of absolutely uniform T, P, and composition. Many real systems do satisfy these criteria. For example, a crystal of diamond sitting on a museum shelf obviously has exactly the same properties this year as last year, and, if we warm it slightly and then put it back on the shelf, it will gradually recover exactly the same temperature, dimensions, and so on that it had before we warmed it. The same remarks hold for a crystal of graphite on the same shelf, so that the criteria can apparently be satisfied for various forms of carbon. Many other natural systems just as obviously are not at equilibrium. Any system having macroscopic temperature, pressure, or compositional gradients will tend to change so as to eliminate these gradients, and is not at equilibrium until that happens. A cup of hot coffee, for example, is not at equilibrium with the air around it until it has cooled down to the temperature of its surroundings.

Despite the simplicity and importance of the equilibrium state, it is very often difficult to know whether any particular real system has achieved that state, or is acceptably close to that state. Experimenters constantly worry whether their experimental systems have achieved equilibrium, and, if so, is it a metastable or stable equilibrium? In the case of a geologist interpreting ancient igneous and metamorphic systems (rocks), the question becomes, did the system achieve equilibrium under some other, perhaps largely unknown, conditions? In such cases thermodynamics is applied, and if the results make sense the system is assumed to be in equilibrium, or to have been in equilibrium. Callen (1985, p. 15) says

In practice, the criterion for equilibrium is circular. Operationally, a system is in an equilibrium if its properties are consistently described by thermodynamic theory!

This might seem like a serious drawback, but the fact that thermodynamics has proved immensely useful and is a cornerstone of scientific training would indicate otherwise.

So, if diamond and graphite are both at equilibrium, do we have two kinds of equilibrium? In our ball-in-valley analogy, the ball in any valley would fit our definition. What distinction do we make between the lowest valley and the others?

2.3.4   Stable and Metastable Equilibrium

In this section we use the simple mechanical analogy in Section 1.3 and Figure 2.3 to distinguish between *stable* and *metastable* equilibrium. This explanation is satisfactory for an intuitive understanding, but we return to this subject for a better theoretical understanding in Section 4.8.

Stable and metastable are the terms used to describe the system in its lowest equilibrium energy state and any other equilibrium energy state, respectively. In Figure 2.3, we see a ball on a surface having two valleys, one higher than the other. At (a), the ball is in an equilibrium position, that fulfills both parts of our definition – it will stay there forever, and will return there if disturbed, as long as the disturbance is not too great. However, it has not achieved its lowest possible potential energy state, and therefore (a) is a *metastable equilibrium* position. If the ball is pushed past position (b), it will roll down to the lowest available energy state at (d), a *stable equilibrium* state. During the fall, for example, at position (c), the ball (system) is said to be *unstable*. In position (b), it is possible to imagine the ball balanced and unmoving, so that the first part of the definition would be fulfilled, and this is sometimes referred to as a third type of equilibrium, admittedly a trivial case, called *unstable equilibrium*. However, it does not survive the second part of the definition, so we are left with only two types of equilibrium, stable and metastable.

Of course, we find that the stable form of substances is different under different conditions. For example, the stable form of H_2O is water at $+5\,°C$, and ice at $-5\,°C$ (Figure 2.4). The freezing and melting of H_2O is normally fairly rapid, so, although it does happen, we don't often see metastable ice above its melting temperature, or metastable water below its freezing temperature. But many such phase changes are not so rapid, in fact they may not happen at all, even though energy would be released (lowered) if they did. These reactions, which get "stuck" in a high energy state, are usually not melting/freezing reactions, but solid-state reactions – for example a reaction in which a mineral having one

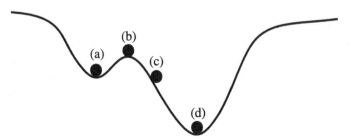

Figure 2.3 Four positions of a ball on a surface, to illustrate the concept of equilibrium. Position (a) – metastable equilibrium. Position (b) – unstable equilibrium. Position (c) – unstable. Position (d) – stable equilibrium.

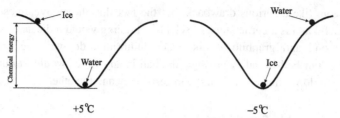

Figure 2.4 The mechanical analogy for H_2O at $-5\,^\circ C$ and $+5\,^\circ C$ and atmospheric pressure. At $-5\,^\circ C$, water is unstable and releases energy until it becomes ice at $-5\,^\circ C$. At $+5\,^\circ C$, ice is unstable and releases energy until it becomes water at $+5\,^\circ C$. The problem is, what kind of energy is being minimized?

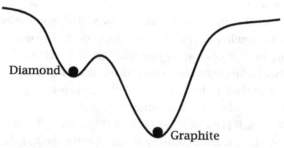

Figure 2.5 The mechanical analogy for carbon at Earth surface conditions. Graphite is the stable form of carbon because it has the lowest energy content of any form of carbon (under Earth surface conditions). Diamond has a higher energy content but is prevented from changing to graphite by an energy barrier.

crystallographic structure should change to a mineral having the same composition but with a different structure but does not.

A good example of this is the diamond/graphite reaction. We know that the stable form of pure carbon at Earth surface conditions is the mineral graphite, but that at high temperatures and pressures, such as are found deep in the Earth's mantle, graphite will spontaneously react to form diamond. However, when tectonic and igneous processes bring the diamond back to the surface, the diamond does not (fortunately) change back to graphite, so we say that diamond is a metastable form of carbon at Earth surface conditions (Figure 2.5). When we develop this subject further, we should be able to predict or calculate under what conditions it is the stable form of carbon.

The importance of metastable equilibrium states tends to be overlooked. This is because discussions of thermodynamic theory commonly assume stable equilibrium, or that there is only one kind of equilibrium. Metastable states are thought of as things like supersaturated solutions – temporary deviations or fluctuations away from the most stable state. In fact, real metastable states include almost everything we see around us in daily life (Lambert, 1998), and thermodynamic metastable equilibrium states have an important place in thermodynamic theory which we will explore in coming chapters. In petrology metastable minerals like diamond and aragonite are common, and metastable-to-stable transitions will be an important part of our thermodynamic theory.

2.3.5 Partial and Local Equilibrium

There are two other terms that are commonly used in connection with equilibrium states.

Partial Equilibrium

"Partial equilibrium" is intended to indicate that part or parts of the system have reached equilibrium, but those parts have not reached equilibrium with each other. Prigogine and Defay (1954, p. 38) give the example of a system in which equilibrium has been established for T and P (there are no measurable gradients in T and P), but not for the "redistribution of matter" by chemical reactions. The problem with this is that any such reaction, say a slowly dissolving solid phase, will be exothermic or endothermic and will inevitably disturb the temperature distribution. Because there will be some volume change, pressure is also disturbed. If such disturbance of T and P is very small, partial equilibrium may be said to apply, but, even if the reaction is very slow, the system may actually be very far from true stable equilibrium. Strictly speaking, there is no such thing as partial equilibrium in thermodynamics, or the systems that thermodynamics deals with. So-called partial equilibrium can often be usefully modeled as a succession of metastable equilibrium states, as will be discussed in Chapter 13. If the reaction in question is not proceeding at all, the system is not in partial equilibrium, but in a metastable equilibrium state.

2.3.6 Local Equilibrium

There are probably, in fact, very few applications of thermodynamic reasoning to natural phenomena where the concept of local equilibrium does not enter the analysis in some way, even though not formally acknowledged.

Thompson (1970)

Real-world systems are in constant flux, and never really achieve thermodynamic equilibrium, but we want to apply thermodynamics to them anyway, so if possible we choose parts of real systems which are reasonably close to thermodynamic equilibrium.

For example, you cannot apply thermodynamics to the ocean as a whole. Calcite is slightly supersaturated at the surface, but significantly undersaturated at 5 km depth, a fact discussed in Section 10.4.1 and illustrated in Figure 2.6. Thermodynamics cannot be applied to a system which is both supersaturated and undersaturated. You can apply thermodynamics to volumes in assumed metastable equilibrium states at the surface or at depth, not both together, so we say we apply thermodynamics to areas of "local equilibrium." It is obviously important to apply thermodynamics appropriately, and generally we do this, but the point is that local equilibrium is not part of thermodynamics, it is a concept we need, a property that real systems must have, in order to apply thermodynamics.

Understanding thermodynamics does not depend in any way on local equilibrium, but applying it to natural systems does. The question then naturally arises as to how one distinguishes between places that have local equilibrium and places that do not. This question does not have a good answer. Places having large gradients in temperature, pressure, or composition can be ruled out, but how large is "large"? Often the practice is to apply thermodynamics and see how it works out, as mentioned above. If it seems to

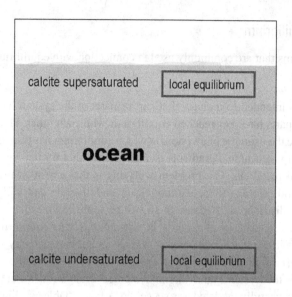

Figure 2.6 Calcite is both slightly supersaturated and slightly undersaturated in the ocean. Local equilibrium must be assumed in order to apply thermodynamics.

work well, then local equilibrium is assumed. Obviously some better approach would be desirable.

There have been several attempts at providing a quantitative criterion for local equilibrium. The most accessible for Earth scientists appears to be that of Knapp (1989), which is summarized in Zhu and Anderson (2002, Chapter 3), who also cite a number of other references on the subject. This analysis by Knapp is useful in defining and clarifying the local equilibrium problem in a quantitative way. Unfortunately, despite the rather drastic simplification, most of the parameters required to define the problem in real situations at the present time are poorly known. The quantitative results are then of questionable significance in any practical sense, but they are worth reflecting on. All applications of thermodynamics in natural systems assume local equilibrium, but defining exactly what that is has proven difficult.

2.4 State Variables

Systems at equilibrium have measurable properties. A property of a system is any quantity that has a fixed and invariable value in a system at equilibrium, such as temperature, density, or refractive index. Every system has dozens of properties. If the system changes from one equilibrium state to another, the properties therefore have changes that depend only on the two states chosen, and not on the manner in which the system changed from one to the other. This dependence of properties on equilibrium states and not on the processes that occur between equilibrium states is reflected in another name for them, *state variables*. State variables are all related by continuous functions, so they all have exact differentials. Several important state variables which we consider in later chapters are not measurable in an absolute sense in any particular equilibrium state, though they do have fixed, finite values in these states. However, the *changes* in these state variables between equilibrium states are measurable.

Reference in the above definition to "equilibrium states" rather than "stable equilibrium states" is deliberate, since as long as metastable equilibrium states are truly unchanging they will have fixed values of the state variables. Thus both diamond and graphite have state variables, but of course the numerical values of these variables are different in the two phases.

2.4.1 Total versus Molar Properties

Many physical properties, such as the volume and various energy terms, come in two forms – the total quantity in a system and the quantity per mole or per gram of pure substance considered. We use a different typeface for these total and molar properties. For example, water has a volume per mole (V) of about $18.0686\,cm^3\,mol^{-1}$, so if we have 30 moles of water in a beaker, its volume ($\mathbf{V}$) is $18.068 \times 30 = 542.06\,cm^3$. This relationship for a pure substance such as H_2O is $Z = \mathbf{Z}/n_i$, where $\mathbf{Z}$ is any total property, Z is the corresponding molar property, and n_i is the number of moles of the substance.

These two types of state variables have been given names.

- *Extensive* variables are proportional to the quantity of matter being considered, and will be written in bold-face type. For example, total volume ($\mathbf{V}$), total internal energy $\mathbf{U}$.

- *Intensive* variables are independent of the total size of the system and include not only all the *molar* properties just mentioned but also temperature, pressure, concentration, viscosity, and density. These will be written in italic type, for example temperature T, molar volume V, molar internal energy U.

In other words an extensive property such as $\mathbf{V}$ refers to a whole system, no matter what size and no matter how many components or phases. The system might be a partially crystallized magma, so $\mathbf{V}$ is the volume of the whole magma. Some intensive properties such as temperature and pressure also refer to the whole system (assuming equilibrium), but more often they refer to individual phases, such as the density of one of the phases. Thus the most common intensive properties, besides temperature and pressure, are the *molar* or *specific* properties of pure substances, that is, the property per mole or per gram. These are the properties that are listed in tables of data such as Appendix B, and are

Box 2.1 Scientific vs. Engineering Units

In science, *molar* properties, such as molar volumes and molar energies, are most commonly used. In engineering, on the other hand, *specific* properties are more common. Specific properties are mass-related rather than mole-related. Thus the specific volume of water at $25\,°C$ is $1.0029\,cm^3\,g^{-1}$. Molar and specific properties are related by the molar mass (or so-called gram formula weight, gfw) of the substance. That for water is $18.0153\,g\,mol^{-1}$, so $1.0029\,cm^3\,g^{-1} \times 18.0153\,g\,mol^{-1} = 18.068\,cm^3\,mol^{-1}$.

used to calculate the changes in such properties in individual chemical reactions. A more mathematical and hence more useful way of describing the difference between intensive and extensive variables is the concept of degree of homogeneity, discussed in Section C.7, Euler's Theorem for Homogeneous Functions, of the Topics in Mathematics, in the online resources.

Many equations in thermodynamics look much the same with total and molar properties because ratios are involved. That is, if $(\partial \mathbf{U}/\partial \mathbf{S})_\mathbf{V} = T$, then it is also true that $(\partial U/\partial S)_V = T$; or, if $(\partial \mathbf{G}/\partial P)_T = \mathbf{V}$, then $(\partial G/\partial P)_T = V$, so that the distinction may seem to be unimportant. However, in some important cases only the extensive form is true, so we will see in coming chapters that thermodynamic potentials apply only to the whole system (are extensive), and chemical potentials are defined using the extensive form of an energy term, not the intensive form. In other important cases, only the intensive form is true, such as "the most useful equation in thermodynamics" that we will see in Chapter 9. Generally speaking, experimentally measured quantities like amounts of heat and work are extensive and used to derive extensive thermodynamic variables. These are then converted to the intensive (molar) form for use in calculations and in collections of data such as Appendix B. In the pages to follow, where both intensive and extensive forms of equations are equally valid, we will often write only the molar form (intensive; italic type).

Partial Molar Properties

In addition to total and molar properties, we have *partial* molar properties, which are a little trickier to understand. It's relatively easy to see that the volume (extensive variable) of a system depends on how much stuff you have in the system, but that its temperature or density (intensive variables) do not. This is true no matter how many different phases there are in the system, as long as you are considering the *whole* system, not just parts of it.

A problem arises, though, when you consider the properties of solutions, which can have variable concentrations of solutes. The volume per gram of halite, crystalline NaCl, is the same whether you consider 10 or 20 grams of it. But what is the volume per gram of 10 grams of NaCl dissolved in a liter of water? This property depends on the concentration of NaCl – the volume per gram of 20 dissolved grams is different from that of 10 dissolved grams. And what *is* the volume of something dissolved in something else? How is it defined, or measured? These are important questions, and the answer is that we use *partial molar properties*. So we have partial molar volume, partial molar enthalpy, and so on. One of these, the partial molar Gibbs energy or chemical potential, is particularly important, and we get to it in Chapter 4.

2.5   Phases, Components, and Species

We must also have terms for the various types of matter to be found within our thermodynamic systems. A *phase* is defined as a homogeneous body of matter, having distinct boundaries with adjacent phases, and so is mechanically separable from the other phases. The shape, orientation, and position of the phase with respect to other phases are irrelevant, so that a single phase may occur in many places in a system. Thus the quartz in a granite is a single phase, regardless of how many grains of quartz there are. A salt solution is a single phase, as is a mixture of gases. There are only three very common types of phases – solid, liquid, and gas or vapor. A system having only a single phase is said to be *homogeneous*, and multiphase systems are *heterogeneous*.

The term generally used to describe the chemical composition of a system is *component*. The components of a system are defined by the smallest set of chemical formulae required to describe the composition of all the phases in the system. To take a simple example, consider a solution of salt (NaCl) in water (H_2O), in equilibrium with water vapor. This might look like Figure 2.2(b). There are two phases, liquid and vapor, and two components, NaCl and H_2O. A chemical analysis could report the amounts or concentrations of Na, Cl, H, and O in the system, and further analysis would find amounts of the ions H^+, Cl^-, Na^+, and several others, but only two chemical formulae are needed to describe the compositions of both phases.

We must be careful to distinguish between *components* and *species*, because they often have the same name. In the system NaCl–H_2O, H_2O is both a component and a chemical species. The ions just referred to are species, not components, because the composition of the system is described without reference to them. Quite often the term *component* is used to refer to the model of a system rather than the real system. Seawater, for example, has an incredibly complex composition, having numerous components and dozens of chemical species. But a thermodynamic model might consider seawater to have only two components, NaCl and H_2O, and this might be adequate for a particular application. Other applications might require the addition of, say, KCl or CO_2, giving three or more components.

As mentioned, the fact that real phases are more or less homogeneous, and that real systems achieve an approximate equilibrium, is what makes thermodynamics useful. The model is mathematically perfect, but real life comes close enough in many respects, so that the model is useful. The close similarity between reality and our models of reality, and the fact that we use the same terms to describe them, may lead to a certain degree of confusion as to what we are talking about. There is quite a bit more to say about phases, components, and the phase rule, which we get to in Chapter 12. There is also more to say about the way we use thermodynamics to investigate systems that are not even close to equilibrium.

2.6 Processes

Finally, we get to something that looks more interesting. *Processes* are what we are usually interested in – changes in the real world. In geology, these might be igneous, diagenetic, or metamorphic processes. In biology, they might be cellular processes. In the environmental world, they might be potentially harmful processes near waste-disposal sites – the possibilities are endless. However, most of the processes of interest to us have one thing in common – they are extremely complicated. The only hope we have of understanding them is to break complex processes down into their simpler component parts, and to construct simplified models of them. We have already begun to do this by defining several types of simple *systems* that we can use; we will now define a *process* in a way that will help us model real processes.

A thermodynamic process is what happens when a system changes from one stable or metastable equilibrium state to another stable or metastable equilibrium state. Any two equilibrium states of the system may be connected by any number of different processes because only the initial and final states are fixed; anything at all could happen during the act of changing from one to the other. A chemical reaction is one kind of process, but there are others. As noted in Chapter 1, simply warming or cooling a system is a process according to our definition. In spite of there seeming to be an endless number of kinds of processes in the world, we find that in thermodynamic models there are only two – reversible and irreversible.

2.6.1 Reversible and Irreversible Processes

There are several ways of discussing reversible and irreversible processes. Which one is preferable depends a lot on what your interests are. This book is focussed on chemical reactions and particularly those in natural systems, so an approach based on the decreasing energy in metastable to stable reactions and the constraints on such reactions is used. At various points we will point out the links to the other ways of looking at the same subjects.

The various ways of explaining reversible and irreversible are as follows.

1. Reversible processes as mathematical functions.
2. Cycles and the entropy of the universe.
3. Uncompensated heat and the affinity.
4. The traditional way.
5. Constraints and changes from metastable to stable equilibrium states.

Reversible Processes as Mathematical Functions

We mentioned in Section 1.5 that many aspects of thermodynamics are unrealistic, the reason being that we need to use mathematics, specifically calculus. Well, here is one of the most unrealistic yet most central aspects of thermodynamics – the reversible process. In thermodynamics reversible does not mean you can simply reverse a process, like warming a crystal and then cooling it back down to its original state. Any such process, and all natural processes such as the diagenetic and metamorphic processes we mentioned, may start and end in an equilibrium state, but during the process they are out of equilibrium. The essence of thermodynamics is the ability to calculate the changes in the properties of a system, its state variables, during a process and therefore to predict what they will be when the system achieves a new equilibrium state. To do this there must be a function, a mathematical expression, for the value of each property *during the process*, that can be integrated. If the system is out of equilibrium its properties vary from place to place and are continually changing. The properties are not single-valued, so no single-valued continuous function can possibly represent them during a process while it is in disequilibrium. Put it this way: what is the temperature of a system when every time you measure it you get a different answer? As Bridgman (1961, p. 133) says

... thermodynamics is concerned with reversible processes and equilibrium states. ... The reason for the importance of equilibrium states is obvious enough when one reflects that temperature itself is defined in terms of equilibrium states.

You could say systems in states of disequilibrium do not actually have state variables.

The solution is to represent the process as a continuous succession of equilibrium states which *can* be represented by a function. This succession of equilibrium states is what we call a reversible process.[1] It obviously does not represent what actually happens during a real process, but it does allow calculation of property changes, and an extremum in such a function indicates a stable or metastable equilibrium state. We used this approach in Section 1.4.1 when we used a continuous function (Equation (1.1)) to represent a real valley. Klotz (1964, p. 44) summarizes this approach nicely when he says

The great virtue of the concept of reversibility is that it introduces the idea of continuity into our analysis of actual processes and hence permits the use of calculus.

Single-Valued Continuous Functions

Thermodynamics could not exist without the concept of entropy; neither could it exist without single-valued continuous functions. The meaning of the terms single-valued and continuous in mathematics is reasonably self-explanatory, but their significance in thermodynamics is fundamental and is explored further in Section C.4, Single-Valued and Continuous Functions, of the Topics in Mathematics, in the online resources.

Cycles and the Entropy of the Universe

A different way of defining a reversible process, more useful to engineers, is the following (Jones and Dugan, 1996, p. 279):

A process is reversible if, after it has occurred, both the system *and its surroundings* can by any means whatsoever be returned to their original states. Any other process is irreversible.

The importance here is that the "original states" referred to involve not just the system but the system plus its surroundings. Some processes, like mixing cream into coffee or breaking an egg, are obviously irreversible, but even cases where the initial state seems to be recovered, like heating a crystal and then cooling it back down, are also ruled out when the surroundings are included. All kinds of natural effects lead to irreversibility; not just mixing and breaking but friction, heat transfer across a finite temperature gradient, free expansion, inelastic deformation, and a few others. Demonstrating that each such effect leads to irreversibility, i.e., changes in the system plus environment, is a major topic in many thermodynamic texts. A good example is Section 5.4 in Jones and Dugan (1996), and identifying and reducing these effects is important in engineering design problems.

---

[1] We restrict the term reversible to a succession of *stable* equilibrium states. In Chapter 4 we will see that we can also have a very similar succession of *metastable* equilibrium states, which, however, have an extra constraint variable.

The idea of using a cycle, a return to the original state, to consider processes was first proposed by Sadi Carnot in 1824 (Carnot, 1960), and the Carnot cycle (Section 4.10.3) has had a profound effect on all subsequent presentations of thermodynamics. The fact that reversibility involves not only the system but also its surroundings, given that the "surroundings" might be called the universe, led Clausius in 1865 to make the most famous statement in the history of thermodynamics:

> The energy of the universe is constant.
> The entropy of the universe tends to a maximum.
>
> Clausius (1867, 1879)

Cosmologists are still debating the entropy of the universe (see Carroll, 2010, for a nice discussion), but fortunately we are concerned only with processes here on Earth. Entropy is introduced in Chapter 4 but, to be consistent with our ball-in-valley idea of decreasing energy of some kind, we do it in a way that concentrates on the system itself rather than the system plus environment.

Uncompensated Heat and the Affinity

Uncompensated heat or uncompensated transformation is a term used by Clausius (and not many others). Prigogine and Defay (1954, pp. 34–35) observe that, although they adopt the term "uncompensated heat" for historical reasons, it is not a particularly happy choice. They show that it is better described as the energy generated or absorbed by an irreversible process *within* a system, as opposed to energy transferred *to or from* a system due to differences between the system and its environment. Prigogine and Defay relate this to the creation of entropy within a system and to a property called the *affinity*, discussed in Section 4.9.1. All the natural irreversible effects just mentioned such as friction can be interpreted as creating entropy, and the creation of entropy within a system is one way to explain the difference between reversible and irreversible processes; reversible processes, being always at equilibrium, create no entropy. That is, a reversible process will change the entropy of the system but will not increase the entropy of the surroundings. We take a different approach to discussing irreversibility, entropy, and affinity, but the distinction between energy generated or absorbed *within* a system and energy transferred *to or from* a system is important and will be discussed in different terms in Section 4.9. The affinity is a thermodynamic concept that will be fully explored in Chapter 13.

The Traditional Way

This point of view originated with Clausius. Badger (1967, p. 128), summarizing a lengthy passage in an article by Clausius, put it this way:

A *reversible process* is one in which the difference between any force of the system and the counterforce of the surroundings, ΔF, is zero. No real process can reach this limit. An *irreversible process* is one in which any $\Delta F > 0$ and for which ΔF may only *approach zero in the limit*. All real processes are therefore irreversible processes.

This idea is commonly made more concrete by considering a piston–cylinder arrangement in which a gas is compressed or expanded. When a weight is removed from the piston, the resulting expansion has to overcome not only the weight of the piston plus whatever weights are on it, but also the frictional force between the piston and the cylinder, so less work is produced than predicted by equations which do not include this frictional force, which of course depends on the particular circumstances. The system may start and end in an equilibrium state, but during the expansion the system is in a state of disequilibrium, and the process is irreversible. Then it is shown that, by performing the expansion more and more slowly, the friction which is the cause of the energy loss is gradually reduced, so that *in the limit* of an infinitely slow process the ideal is achieved, and the process is called a reversible process. It follows that all reversible processes are infinitely slow.

This kind of discussion has the somewhat dubious merit of seeming to relate the reversible process to real processes, but it comes with some problems. A process involving an infinite number of steps requiring an infinite amount of time is difficult to visualize. Furthermore, it makes it seem as if time is an important concept in thermodynamics, that time might be a thermodynamic variable. This is a common misconception that we discuss in Section 2.8. It is much simpler to say that the reversible process is simply a result of the fact that we use calculus, which requires continuous functions, which in turn require equilibrium states. Limit-taking is an integral part of calculus and need not be repeated in discussing thermodynamics.

Reversible processes are but one example of a host of concepts of a similarly idealized nature in chemistry and physics – for example, ideal gases and solutions, absolute zero temperature, infinitely dilute solutions, perfect black-body radiation, isolated systems, perfect insulators, friction-free surfaces, and so on. In every case, the adoption of the idealized case simplifies or makes possible the application of mathematics to physical reality.

Constraints and Changes from Metastable to Stable Equilibrium States

We have defined a metastable equilibrium state of a system as a state that has more than the minimum energy for the given conditions, but for some reason is prevented from releasing that energy and reacting or changing to the stable equilibrium state of minimum energy. An irreversible process is one that occurs when whatever constraint is holding the system in its high energy state is removed, and the system slides down the energy gradient to a lower energy state.

The only example we have given thus far of a metastable system is the mineral diamond, that could lower its energy content by changing into graphite, but there are many other similar examples of metastable minerals. The diamond is constrained from reacting because energy is required to break the strong carbon–carbon bonds in diamond before the atoms can rearrange themselves into the graphite structure. In Figure 1.2 this energy is represented by the fact that the ball must be pushed up a small hill before it can roll down to a lower valley. This energy, the barrier preventing change, is called an *activation energy*. We have also mentioned that most organic compounds, such as all the ones in living organisms, are metastable. When the life processes maintaining their existence cease, they quickly react (decompose) to form more stable compounds.

In most of the chemical reactions that we will be considering, particular minerals, or minerals plus liquids or gases, react to form different minerals under some given conditions. For example in reaction (2.5), the mineral corundum (Al_2O_3) is stable, considered by itself (i.e., there is no other form of Al_2O_3 that is more stable), but in the

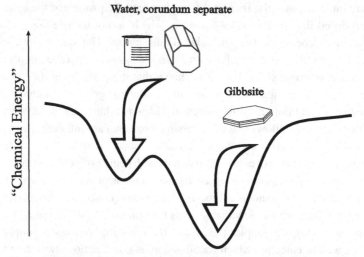

Figure 2.7 Water plus corundum can lower its energy content by reacting to form gibbsite.

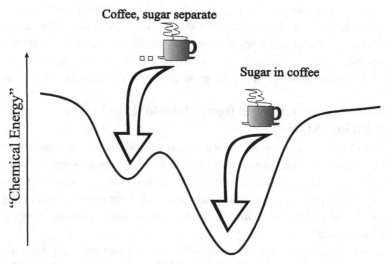

Figure 2.8 Sugar dissolves in coffee because the "chemical energy" of the dissolved state is less than that of the two coexisting separately.

presence of water it reacts to form gibbsite ($Al_2O_3 \cdot 3H_2O$), and the energy relationships are shown in Figure 2.7.

Do not confuse the metastability of diamond at Earth surface conditions with the metastability of corundum or water. Diamond is metastable because the same carbon atoms would have a lower energy in the crystal structure of graphite. But corundum by itself is not metastable, and neither is water, at 25 °C and atmospheric pressure. It is the *combination* of corundum and water that can be regarded as metastable, because their *combined* atoms would have a lower energy level in the form of gibbsite.

Another example, much like the cream in coffee example, is the dissolution of sugar in coffee (Figure 2.8). The assemblage of sugar lumps and a cup of coffee is a metastable assemblage in our usage. They are prevented from reacting (sugar dissolving) by the fact that they are separated, which constitutes a *constraint* on the system. When the constraint is removed by putting the sugar in the coffee, the reaction occurs, because the "chemical energy" is lowered.

The corundum plus water example and the sugar plus coffee example are different in an instructive way. If you actually put a crystal of corundum in a beaker of water, nothing at all happens, except that the corundum gets wet, whereas when the sugar is put in the coffee, it dissolves immediately. Both assemblages are metastable but the constraints are different. Like diamond, corundum is prevented from reacting by an activation energy barrier, meaning that the atoms in Al_2O_3 are too tightly bound to react, even though the system could lower its energy if they did. The sugar is prevented from dissolving in the coffee by a physical separation.

These are all examples of constraints in real systems. We will introduce the constraint used in thermodynamic systems in Chapter 4. You will not be surprised to learn that this constraint is not a physical situation but a mathematical variable.

Chemical Reaction Examples

Most people, especially Earth scientists, will readily agree that diamond and aragonite under ambient conditions are examples of metastable states which are constrained from reacting to a more stable state. In these cases the constraint is an activation energy. As mentioned, the sugar or cream into coffee examples have a different kind of constraint, the physical separation of reactants from products. In other words we consider sugar plus coffee to be in the same category as the corundum plus water, that is, a metastable equilibrium state.

If we carry this idea to its logical conclusion, we find that it includes *all* spontaneous chemical reactions such as

$$C\,(graphite) + O_2 = CO_2$$

$$CaSO_4\,(anhydrite) + 2H_2O = CaSO_4 \cdot 2H_2O\,(gypsum)$$

or reaction (2.5) on p. 31. If $A + B \rightarrow C$ is spontaneous at T, P, then $A + B$ is unstable unless constrained from reacting. If so, there is an extra (third) constraint such as separation of the reactants, and $A + B$ is then in a metastable state. The separation of A and B may be effected by various means, such as having a partition between them, or by having A and B in separate containers, but when this separation is ended, or *this constraint is released*, $A + B$ slides down the energy gradient, forming C. In the reverse case, where C spontaneously reacts to form $A + B$ at some different T, P conditions, then A and B are stable together, and no separation is implied.

Some metastable systems of course may have *more* than one extra constraint. Consider $A + B = C$ again, this time where A is H_2 gas, B is O_2 gas, and C is H_2O liquid, where the T is 25 °C and the P is 1 bar. If the H_2 and O_2 are in separate parts of a container separated by a partition, they are of course constrained from reacting, and the partition represents a first constraint. If the partition is then removed, the gases mix but they do not react to form water, because there is an activation energy barrier that must be overcome; this represents a second constraint. Finally, if a catalyst is introduced, removing this constraint, the gases react to form the stable phase, water.

So when we speak in coming chapters of metastable → stable reactions, we mean to include *any* spontaneous chemical reaction, not just those involving minerals. Whatever the nature of the constraint in reality, there is only one way of representing it in thermodynamics. In thermodynamics a constraint is a mathematical variable, and we see what this is in Chapter 4.

2.6.2 Reactions Involving Organic Compounds

Reactions involving organic compounds, whether in living organisms or not, are no different in principle from any other kind of reaction, such as those between minerals. The only difference is that, for organic compounds, the reaction usually proceeds from one metastable state to another metastable state of lower energy, rather than from a metastable state to a stable state. Consider for example the reaction

$$C_8H_{16}N_2O_3(aq) + H_2O(l) = C_6H_{13}NO_2(aq) + C_2H_5NO_2(aq) \qquad (2.1)$$

which represents the breaking of a peptide bond between two amino acids, one of the more fundamental processes in biochemistry. The (aq) here means that the compounds we are discussing are dissolved in water and, hence, the reaction takes place in water. If we use names rather than chemical formulae, this is

$$\text{leucylglycine} + \text{water} = \text{leucine} + \text{glycine} \qquad (2.2)$$

This reaction occurs spontaneously, and the energy relations can be depicted exactly as for simpler compounds. The only difference is that, rather than reacting to compounds in the lowest possible energy state, leucylglycine plus water reacts to form compounds in another metastable state (leucine plus glycine) of lower energy than the initial state, as

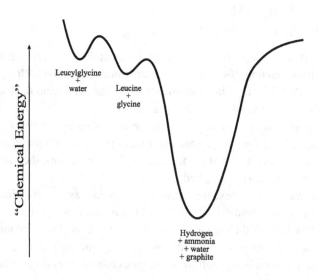

Figure 2.9 Energy relationships between organic compounds. Most organic compounds have much higher energy contents than do combinations of simple inorganic compounds of the same overall composition.

shown in Figure 2.9. Virtually all organic compounds are metastable with respect to simple inorganic compounds and elements such as water, nitrogen, hydrogen, and graphite. Thus the reaction

$$C_6H_{13}NO_2(aq) + C_2H_5NO_2(aq) = 2\,H_2(g) + 2\,NH_3(g) + 4\,H_2O(l) + 8\,C_{graphite} \qquad (2.3)$$

is also spontaneous, as shown in Figure 2.9.

Living organisms have developed (evolved) mechanisms (involving enzymes) for overcoming the energy barriers separating products and reactants of reactions required for the life processes of the organisms. Obviously no enzymes have been developed to enable the breakdown of the organisms to the simple inorganic compounds of which they are composed, as this would be fatal.

2.7 Notation

. . . .

2.7.1 Reaction Deltas

We have now set up the general framework within which thermodynamics is able to deal with processes. Any given process or chemical reaction within a chosen system will proceed from an initial equilibrium state (quite often a metastable equilibrium state) to another equilibrium state more stable than the initial one. During this process or reaction any real system is out of equilibrium. The system has a number of properties or state variables, such as volume and energy content, that have fixed values in equilibrium states and that therefore have fixed amounts of change between equilibrium states. These changes are always written using a delta notation, where the delta refers to the property in the final state minus the property in the initial state. For example, if the system undergoes a process during which its molar volume (V) changes from $V_{initial}$ to V_{final}, we write

$$\Delta V = V_{final} - V_{initial} \qquad (2.4)$$

If the process is a chemical reaction, a number of compounds may be involved. A generalized chemical reaction could be written as

$$a\,A + b\,B + \cdots = m\,M + n\,N + \cdots$$

Here A, B, M, and N are chemical formulae representing any compounds or elements we happen to be interested in, and each can be solid, liquid, gas, or a dissolved substance (a solute). An example is

$$Al_2O_3(s) + 3\,H_2O(l) = Al_2O_3 \cdot 3H_2O(s) \qquad (2.5)$$

where A is Al_2O_3 (corundum), B is H_2O (water), and M is $Al_2O_3 \cdot 3H_2O$ (gibbsite), there is no N; a and m are 1 and b is 3. We use (s), (l), (g), and (aq) after each formula to indicate whether they are in the solid, liquid, gas, or aqueous (dissolved in water) state.

One side of the reaction will usually be more stable than the other, and a reaction will tend to occur, unless the reactants and products are separated, or unless there is an energy barrier preventing the reaction, or unless the compounds are all at equilibrium together. In any case, the volume change between reactants and products is $\Delta_r V$ or $\Delta_r V^\circ$ and is equal to the sum of the volumes of the reaction products (the final state) minus the sum of the volumes of the reactants (the initial state). We insert a subscript r to indicate a chemical reaction and superscript ° to mean standard state conditions, which we will discuss later. These are all just numbers available in collections of data such as Appendix B.

Following this convention, the change in energy of the ball rolling down the hill in Figure 1.1 would be a negative quantity, as shown in Figure 1.3 (energy in state B minus energy in state A is negative). It follows, then, that the change in the "chemical energy" term we are looking for will always be a negative quantity in spontaneous reactions, as also shown in Figure 1.3 (energy of products minus energy of reactants).

Chemical Equations

For the most part, when we write reactions such as (2.5) we use the $=$ sign to indicate only that the reaction is "balanced," meaning that the same number and kinds of atoms appear on both sides, and that any electrical charges are also the same on both sides. If we want to emphasize that the reaction proceeds strongly or irreversibly we may use an arrow, as in $A \rightarrow B$, and if we want to emphasize that the two sides are in equilibrium, we might use $A \rightleftharpoons B$. However, the $=$ sign includes these possibilities, and all others.

2.7.2   Egg Reactions

We have not discussed all the examples we used in Chapter 1. To conclude our discussion of various common chemical reactions (Section 1.2.1), we should discuss the thermodynamics of frying eggs. We are tempted to say that the egg in the refrigerator is in a metastable state, and that frying it promotes an irreversible reaction to a more stable state, analogous to the leucylglycine + water → leucine + glycine reaction in Figure 2.9.

Strictly speaking, however, we know that eggs in the refrigerator won't last indefinitely; they will eventually "go bad." This means that they are not in a metastable equilibrium state in the refrigerator, but in an unstable, slowly changing state. This means that, because the raw egg occupies no "valley" for the egg components to roll into, it is very unlikely that we could restore the raw egg state, even if we had an appropriate energy source.

In studying natural systems such as eggs, it is often quite difficult to distinguish stable, metastable, and unstable states from each other without a considerable amount of work and ingenuity, but it can be done. When you get numbers from tables of data all this work has been done for you, although you have to realize that, because of the difficulties involved, some of the data may be inaccurate and may be revised at some future date. A compound believed to be stable under given conditions may later be found to be metastable after more careful work has been done.

Reactions in these complex systems are actually made up of a number of simpler reactions, and applying thermodynamics requires that the individual reactions be treated separately. The individual biochemical reactions in many organic systems still have not been figured out. Nevertheless, we are confident that any particular reaction, once defined, will follow the logic and the systematics described in this book.

Box 2.2 Volume Change

The volume data in Appendix B are listed under $V°$. In reaction (2.5) then,

$$\Delta_r V° = V°_{Al_2O_3 \cdot 3H_2O} - V°_{Al_2O_3} - 3\, V°_{H_2O}$$
$$= 63.912 - 25.575 - 3 \times 18.068$$
$$= -15.867 \, cm^3 \, mol^{-1} \tag{2.6}$$

There is therefore a net decrease in volume of $15.867 \, cm^3 \, mol^{-1}$ for the reaction as written. But you could equally well write

$$\Delta_r V° = \tfrac{1}{3}V°_{Al_2O_3 \cdot 3H_2O} - \tfrac{1}{3}V°_{Al_2O_3} - V°_{H_2O} \tag{2.7}$$
$$= -5.289 \, cm^3 \, mol^{-1}$$

Or you could write

$$\Delta_r V° = 2\, V°_{Al_2O_3 \cdot 3H_2O} - 2\, V°_{Al_2O_3} - 6\, V°_{H_2O} \tag{2.8}$$
$$= -31.734 \, cm^3 \, mol^{-1}$$

Note that each volume must be multiplied by its corresponding stoichiometric coefficient in the reaction. Despite some controversy on this subject (Craig, 1987), stoichiometric coefficients have no units. They are dimensionless. All these results are cm^3 *per mole*, so the question is, per mole of *what?*

The lesson here is that $\Delta_r V°$, or the reaction delta of any other state variable, refers to the reaction *as written*, which can be variable. Generally though, the delta referred to is the volume change per mole of whatever species have a stoichiometric coefficient of 1.0. The volume change is $-15.867 \, cm^3$ per mole of Al_2O_3 consumed or $Al_2O_3 \cdot 3H_2O$ formed (Equation (2.6)), and $-5.289 \, cm^3$ per mole of H_2O consumed (Equation (2.7)). For reactions where *no* species has a stoichiometric coefficient of 1.0, like reaction (2.8), we seem to have a problem. Obviously the volume change here is per *two* moles of $Al_2O_3 \cdot 3H_2O$, not one. It would be quite unusual to write a reaction this way, but, if it bothers you, divide by two and rewrite the reaction as reaction (2.6).

2.8 Time as a Thermodynamic Variable

McGlashan (1979, p. 102) says

There are those who say that time has no place in thermodynamics. They are wrong.

According to the presentation of thermodynamics in this text, McGlashan is wrong. Time is mentioned, but (significantly) not used mathematically, by many authors. These authors are all of course aware that there is no variable called "time" in equilibrium thermodynamic theory, so why is this idea so prevalent? One reason might be that entropy as "the arrow of time" seems to be firmly established.

A simpler example is the hot cup of coffee which cools down to equilibrium with its surroundings. But, it is said, evaporation continues until the water is gone, then the remaining coffee grounds slowly oxidize, and after a few eons the glass or ceramics in the cup may recrystallize to more stable forms, so time must be specified when defining equilibrium. All such cases are examples of the failure to distinguish between real systems and thermodynamic models of these systems.

As discussed in Section 2.3, real systems *never* achieve thermodynamic equilibrium as defined by Gibbs, though they often closely approach this state. Thermodynamic models of these systems, however, represent only thermodynamic equilibrium states, stable or metastable. Given appropriate data, we could calculate the properties of the cooling coffee such as its vapor pressure (or fugacity) at any point in its cooling history, but each calculation would necessarily assume the coffee to be in a metastable equilibrium state. We actually do this kind of calculation in Chapter 13. In thermodynamics integration and differentiation are never performed with respect to time, only with respect to state variables, with equilibrium assumed to be present throughout. There is no time variable in equilibrium thermodynamics, and that should be the end of it.

Well, not quite. In engineering thermodynamics, time is definitely a variable. However, it is used to denote the rate of mass and energy fluxes, and has no connection with equilibrium or the increase in entropy.

2.9 Thermodynamics and Natural Systems

We have emphasized that thermodynamic data and thermodynamic relationships are derived for and in theory can only be applied to idealized equilibrium states, but that real equilibrium states come sufficiently close to the ideal that our thermodynamic methods are useful. These real equilibrium states in natural systems occur in places having local equilibrium, which means that even a system in overall disequilibrium will have places where equilibrium is closely approached, and thermodynamics can be applied to samples from those places providing information about the system.

But there are other ways in which thermodynamics is useful, regardless of local equilibrium. One has to do with the development of thermodynamically consistent theories of geological processes. In Chapter 9 we consider a number of reactions involved in the formation of lead–zinc ore deposits. Field observations have provided an abundance of data on mineral compositions, fluid inclusions, structures, age relations, and so on. The problem is that of how to develop a theory of ore formation, a conceptual series of events that would result in the ore deposit as we see it; and these events must of course be thermodynamically possible. Thermodynamics provides the essential framework within which natural processes must work. We might propose that the ore solution contained A and B, but thermodynamics tells us that in that case we will have C, which doesn't fit with field observations. We could have A and D, but then we need E, but where does E come from? And so on. Note that in this work field observations are just as important as thermodynamic calculations.

Another way thermodynamics is used to better understand natural systems is geochemical modeling. A common practice in the geochemical literature is to follow, in a computer program, the irreversible reactions in some complex starting composition through any number of irreversible reactions and phase changes to the final stable equilibrium state in a series of small increments of reaction. The classic case, introduced by Helgeson (1968) is the dissolution of K-feldspar in water, where tiny amounts of the mineral, represented by a mathematical variable, are added to the solution until the final equilibrium state is achieved. After each increment all the species involved in all the reactions are adjusted to their equilibrium activities and any phase changes thus made necessary are performed at that point in the reaction. The bulk composition is unchanged. Each of these intermediate compositions represents a metastable equilibrium state within the overall irreversible reaction, and the series of such states represents an idealized reaction path. Such computed reaction paths do not necessarily represent what happens in reality, but they usually provide useful information nonetheless, and are routinely performed by petrologists, oceanographers, and others. A variety of these modeling programs, both free and commercial, is available. Steinmann *et al.* (1994) show how, with some assumptions, one particular reaction path can be calculated using a spreadsheet. In addition to the irreversible reactions and phase changes just described, recent developments have added kinetics, surface effects, advective–dispersive–diffusive transport, cooling, and other factors, but thermodynamics remains essential.

Just because thermodynamics is derived and is strictly speaking true only for idealized and unattainable equilibrium states does not make it useless. It is part of the learning process to find out how this can be.

2.10 Summary

If you look around the physical world today, you realize that there are incredibly many chemical and physical *processes* going on all around you, and as you look into these in more and more detail, as science has done, you find more and more complexity at all levels, right down to the atomic and subatomic levels. How can we systematize and understand these processes in such a way as to be able to control some of them for our own purposes?

Thermodynamics is one result of our attempts to do this, at the macroscopic level. It is not a description of any real process, but a rather abstract *model* that can be used for all real processes. Processes in the real world are incredibly complex, but our models of them are quite simple, containing a number of carefully defined concepts. *Processes* (reactions, changes) involve changes in energy and/or mass, and these must enter or leave the place where the process is occurring; so thermodynamics begins by defining several types of *systems*, depending on how the energy and/or mass is transferred. Processes must be defined by beginning and ending states, so thermodynamics defines *equilibrium* states, some having more energy (*metastable equilibrium* states) than others (*stable equilibrium* states), and processes or reactions that are able to go from higher energy states to lower energy states (*irreversible processes*), just like a ball rolling down a hill. Of course, a state of lower energy (stable) under one set of conditions may be a state of higher energy (metastable) under other conditions (diamond is metastable at the Earth's surface, but stable deep in the mantle). Corundum and water are, by themselves, perfectly stable and unreactive, but together they have a higher energy state than does gibbsite.

The only thermodynamic difference between organic reactions (including those in living organisms) and inorganic reactions is that both the reactants and the products of organic reactions are invariably metastable compounds; metastable, that is, with respect to simple inorganic compounds and elements. Inorganic reactions *may* involve metastable compounds, but more frequently they involve a metastable *assemblage* changing to a stable one (one having the lowest possible energy state).

Therefore, the determination of the energy states of substances and how they change under changing conditions is fundamental to understanding what processes are possible, and why they happen. The determination of the energy states of individual substances must be done for the most part by experiment and measurement, not by theoretical calculation, and the results are available in tables of data like those at the end of this book. Calculation of the change of these energy terms with changing conditions can be carried out only for hypothetical *reversible processes*, that are not possible in reality but are quite simple in the thermodynamic model.

The most important question now is what kind of energy is released during these reactions? If it is not heat energy, then what is it? We have called it "chemical energy," but that is just because we haven't said yet what it really is. This is the topic of the next two chapters.

Exercises

E2.1 Calculate the volume change for the reaction

$$Al_2O_3(s) + 3 H_2O(l) = 2 Al(OH)_3(s)$$

$Al(OH)_3(s)$ is another way of writing the formula for gibbsite.

E2.2 Calculate the volume change for the reaction

$$AlO_{1.5}(s) + 1.5 H_2O(l) = Al(OH)_3(s)$$

$AlO_{1.5}(s)$ is another way of writing the formula for corundum.

E2.3 Calculate $\Delta_r V°$ for reaction (2.1),

$$C_8H_{16}N_2O_3(aq) + H_2O(l) = C_6H_{13}NO_2(aq) + C_2H_5NO_2(aq)$$

E2.4 Calculate $\Delta_r V°$ for the reaction

$$NaCl(s) = Na^+ + Cl^-$$

Note that the standard volume of many ions is negative. How can any substance have a negative volume?

E2.5 (a) How many components are there in pure enstatite $MgSiO_3$?

(b) How many components are there in enstatite solid solution $(Mg, Fe)SiO_3$?

(c) How many components are there in olivine solid solution $(Mg, Fe)_2SiO_4$?

(d) How many components are there in coexisting enstatite and olivine solid solutions?

Additional Problems

A2.1 Why are the $V°$ values for all the gases the same in Appendix B? Calculate this $V°$ from data in Appendix A.

A2.2 Calculate $\Delta_r V°$ for the reaction

$$HCl(aq) = H^+ + Cl^-$$

You should get zero. Why?

A2.3 How many components are there in a beaker of water containing a piece of calcite, assuming equilibrium?

3 The First Law of Thermodynamics

3.1 Temperature and Pressure Scales

3.1.1 Temperature

One of the early milestones in the development of thermodynamics was the demonstration that there is an absolute zero of temperature. Nevertheless there are several different temperature scales, for historical reasons. All you need to know about this is that the Kelvin scale (named after William Thompson, Lord Kelvin) has an absolute zero of 0 K and a temperature of 273.16 K at the triple point where water, ice, and water vapor are at equilibrium together. The melting point of ice at one atmosphere pressure is 0.01 degrees less than this, at 273.15 K (Figure 3.1). The Celsius scale (named after Anders Celsius, a Swedish astronomer) has a temperature of 0 °C at the ice point (273.15 K) and absolute zero at −273.15 °C. This gives almost exactly 100 °C between the freezing and boiling points of water at one atmosphere, so water boils at 100 °C (373.15 K). Thus the numerical conversion between the two scales is

$$T \, \text{K} = T \, °\text{C} + 273.15$$

Remember that all equations in thermodynamics use the absolute or Kelvin temperature scale, so that, if you have temperatures in °C, you must convert them to the Kelvin scale before using them in equations. The "standard" temperature of 25 °C for example is 298.15 K. Standard IUPAC[1] usage is to refer to temperatures on the Kelvin scale as so many kelvins, not degrees kelvin or °K. One kelvin is defined as the fraction 1/273.16 of the temperature of water at the triple point. More detail on temperature scales can be found in Anderson and Crerar (1993, Section 4.3).

3.1.2 Pressure

Force is measured in newtons (N), where 1 newton will give a mass of 1 kg an acceleration of $1 \, \text{m s}^{-2}$. Pressure is defined as force per unit area, and a pressure of 1 newton per square meter ($1 \, \text{N m}^{-2}$) is called 1 pascal (1 Pa). This is a very small pressure, and older, larger pressure units are still in use. The bar, for example, is 10^5 Pa and is almost equal to the standard atmosphere (1 atm = 1.01325 bar). Weather reports in many countries give the atmospheric pressure in kilopascals (kPa), and it is usually close to 101 kPa, or 1 atm, or 1 bar. These units are summarized in Appendix A.

---

[1] International Union of Pure and Applied Chemistry.

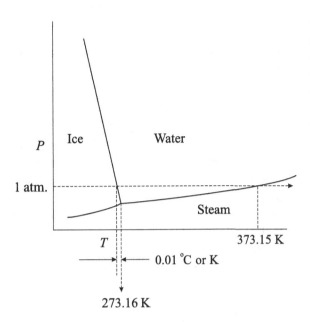

Figure 3.1 Schematic *P–T* phase diagram for the system H_2O. The temperature of the triple point is defined as 273.16 K. Illustrative, not to scale.

The standard temperature and pressure chosen for reporting values of thermodynamic variables are now 25 °C and 0.1 MPa. A pressure of 0.1 MPa is 100 kPa and 10^5 Pa, or 1 bar. It is convenient to use bars instead of pascals, because the bar is essentially the same as atmospheric pressure, and the notation is slightly simpler.

3.2 Internal Energy

In everyday conversation we use words like heat, work, and energy quite frequently, and everyone has a sufficiently good idea of their meaning for our ideas to be communicated. Unfortunately, this type of understanding is not sufficient for the construction of a quantitative model of energy relationships like thermodynamics. To get quantitative about anything, or, in other words, to devise equations relating measurements of real quantities, you must first be quite sure what it is you are measuring. This is not too difficult if you are measuring the weight of potatoes and carrots; it is a more subtle problem when you are measuring heat, work, and energy. Historically, it took several decades of effort by many investigators in the nineteenth century to sort out the difficulties that you are expected to understand by reading this chapter!

3.2.1 Energy

Everyone knows what energy is, but it is an elusive topic if you are looking for a deep understanding. In fact, a Nobel Prize-winning physicist has affirmed that

It is important to realize that in physics today, we have no knowledge of what energy *is*.

Feynman *et al.* (1963), p. 4–2

An eminent French scientist said

> As we cannot give a general definition of energy, the principle of the conservation of energy simply signifies that there is *something* which remains constant.
>
> Poincaré (1952), p. 166

If you consult a dictionary as to the meaning of energy, you find that the scientific meaning is *"the ability to do work, i.e., move a body."* In physics, work is not what you do from 9 to 5 every day, but the action of a force moving through a distance. So, if you lift a book from the floor and put it on the table, you are performing work (the mass of the book, multiplied by the acceleration due to gravity, times the distance from the floor to the table), and we say that we expended energy to lift the book. It has proved tremendously useful to take the view that the energy we expended has not disappeared, but has been transferred to the book. In other words, the book on the table has more energy (potential energy) than it had on the floor, and the increase is exactly equal to the work we did in lifting it. Thus we can use energy to do work, and we can do work on a system to increase the energy of that system. Work and energy are thus very closely related concepts (note that they have the same dimensions in Appendix A).

If only things were that simple! However, we know that they are not, because the energy in a stick of dynamite on the table is not equal to the work expended in lifting it from the floor. Similarly, the energy in water is not the same as that in ice, irrespective of whether it is on the floor or on the table. These complications are actually of two types.

1. There are many ways of doing work, because there are many kinds of forces. We are particularly concerned with the work involved in chemical reactions.
2. Although work and energy are indeed closely related, doing work is not the only way of changing the energy of something, and changing the energy of something does not always produce work. For example, we could change the energy in our book by warming or cooling it.

We have to consider both work (in all its forms) and heat to get a consistent picture of energy changes.

3.2.2   Absolute Energy

In discussing energy, we always seem to be talking about *changes* in energy. The book has more energy on the table than on the floor, and presumably more energy on the roof than on the table. And we can add energy by warming the book too. But how much energy does the book have in any particular state – say, on the table at 25 °C? What is the absolute energy content of the book? This was a difficult question until 1905, when Einstein proposed the essential equivalence of mass and energy in his famous equation

$$\mathbf{E}_r = mc^2 \tag{3.1}$$

where $\mathbf{E}_r$ is the rest energy of a system, m is the mass, and c is the speed of light. Therefore, the energy contained in any macroscopic system is extremely large, and adding energy to a system (for instance by heating it) will in fact increase its mass. However, ordinary (i.e., non-nuclear) energy changes result in extremely small and unmeasurable changes in mass, so that relativity theory is not very useful to us, except in the sense that it gives energy an absolute kind of meaning, which is sometimes helpful in trying to visualize what energy *is*.

So, in considering ordinary everyday kinds of changes and chemical reactions, we will continue to deal with energy *changes* only, never with how much energy is in any particular equilibrium state. This is entirely sufficient for our needs, but it does introduce some complications that would be avoided if we had a useful absolute energy scale.

3.2.3   The Internal Energy

Thermodynamics does not use the term E_r, but it does have a term for the energy in a system. All that is required to develop our model of energy relationships is that every equilibrium state of a system has a fixed but unknown energy content, which is called the *internal energy*, U. The numerical value of this energy content is not known, and not needed. It could be thought of as identical to the rest energy E_r, if that helps, or as some small subset of E_r; it doesn't really matter. All that matters is that, when the system is at equilibrium, its energy content or energy level is constant. Since we do not use absolute values of U or U, we cannot use absolute values of any of the several quantities having the internal energy in their equations of definition.

Notice that, although we have been illustrating energy and work by using the ball-in-valley idea (Chapter 2) and the book-and-table idea (this chapter), which emphasizes *potential* energy, this particular kind of energy/work is actually irrelevant in thermodynamics, except as an analogy. We will define the energy content of systems of importance to us to be the same whether they are on the floor or the table. Thermodynamics is concerned only with the internal energy U or its molar form U, no matter where the system is. This gets a little more complicated if we are considering a magma rising rapidly through the Earth's gravitational field, but for now we will stick to U and its changes due to work and heat. For the role of gravity as well as electric and magnetic fields in thermodynamics see Guggenheim (1959) and Pippard (1966).

The Meaning of ΔU

Although the principles of thermodynamics are independent of any assumptions regarding the atomic and molecular structure of matter, it is helpful to think of internal energy as the summation of the kinetic and potential energies of the molecules of a substance. The kinetic energy of the molecules is associated with the translational, rotational, and vibratory motions of molecules, which increase as heat is added. Potential energy is related to the forces between the atoms, ions, and molecules, being large in solids, smaller in liquids and gases. A chemical reaction causes the atoms, molecules, and ions in a system to rearrange themselves, breaking some bonds and forming others, resulting in changes in their potential and kinetic energies, and therefore changes in U.

If the system is *isolated* (Section 2.2.2) this will result in a rise or fall in pressure and temperature, which are simply manifestations of the average kinetic energy of the particles (McQuarrie and Simon, 1997, Chapter 27), but the total energy U of the system does not change. In a *closed* system, energy is permitted to enter or leave the system, so changes in the particle energies might result in energy leaving or entering the system as heat (as when the heat flow is measured in a calorimeter, Section 5.3) or as work (as when a system expands against an external pressure, Section 3.5.1).

The Internal Energy Paradox

Somewhat paradoxically, in spite of being one of the most fundamental of thermodynamic quantities, changes in $\mathbf{U}$ or U are little used in geochemical applications. It is never listed in tables of thermodynamic values such as those in Appendix B, for example, and one rarely needs to calculate ΔU. The reason for this will become apparent as we proceed. It has to do with the fact that we, the users of thermodynamics, have a great predilection for using temperature, pressure, and volume as our principal constraints or measured system parameters. It turns out that this requires that we use ΔU in slightly modified forms, that is, ΔU modified by what are often relatively small correction factors (such as $P\,\Delta V$), and these modified forms are given different names and symbols. It is then quite possible to rarely think about ΔU, since it seems only to arise in the development of the first law. For a better understanding of the subject, however, it is best to realize that, in most energy transfers in the real problems that we will be considering, ΔU is by far the largest term involved. The fact that we do not usually calculate its value does not mean it is not important.

3.3   Energy Transfers

In the discussions in the previous chapters, we proposed the idea that changes or reactions occur because systems can lower some kind of energy by such changes. However, we mentioned that the most obvious kind of energy, heat energy, was not the right kind of energy. There is another very common kind – energy expended as *work*, as when dynamite is used to break rock. However, work energy is not the answer to our questions either, nor is the combination of heat and work. Nevertheless, they are extremely important, and together form the basis of the first law.

- *Heat* (q) is energy that crosses a system boundary in response to a temperature gradient.
- *Work* (w) is energy that crosses a system boundary in response to a force moving through a distance (such as happens when a system changes volume).

These statements describe heat and work as energy transfers, but are not good definitions. In fact it has proved quite difficult to provide a rigorous definition of heat in this respect. According to the experts (e.g., Canagaratna, 1969), there is really only one. Heat is that part of any energy transfer which is not accounted for by mechanical work (which has a satisfactory definition: force × distance), and assuming that other forms of energy transfer are negligible. That is,

$$q = \Delta \mathbf{U} - w$$

This turns out to be not very different from the way we actually do think about heat in thermodynamics, as when we subtract a $P\,\Delta \mathbf{V}$ term from some calorimetry results, and describe the remaining energy term as heat.

We use q and w for increments or amounts of heat and work in joules. If this heat or work is related to a chemical reaction we could use q and w ($J\,mol^{-1}$), but this is rarely necessary. Normally total heat and work (q and w) are measured quantities which are used to calculate the change in some state variable such as ΔU or ΔH. These state variables are then converted to their molar forms, ΔU and ΔH, for use in calculations. An example of the use of q is given on page 117, where q is measured in a calorimeter for small amounts of solid Al and Al(OH)$_3$, then converted into q_P, which is enthalpy (Equation (3.32)). See the boxes on pages 46 and 49 for (total) work calculations.

The Pond Analogy

Heat and work are forms of energy that is transferred in different ways. An enlightening analogy has been offered by Callen (1985). In Figure 3.2 we consider the water in a very deep pond. The amount of water in the pond is quite large, but finite, and could in principle be measured, but is in fact unknown. It corresponds to the internal energy **U** of a system.

Water may be added and subtracted from the pond either in the form of stream water (heat) or precipitation/evaporation (work). Both the inlet and the outlet stream water can be monitored by flow gauges, and the precipitation can be measured by a rain gauge. Evaporation would be trickier to measure, but we may assume that we have a suitable measure for it. Now, if the volume of stream inlet water over some period of time is q_i, with volumes for the stream outlet water of q_o, for the rain of w_r, and for the evaporation of w_e, then, if there are no other ways of adding or subtracting water, clearly

$$\Delta U = (q_i - q_o) + (w_r - w_e)$$

where ΔU is the change in the amount of water in the pond, which could be monitored by a level indicator as shown. Thus

$$\Delta U = q + w \tag{3.2}$$

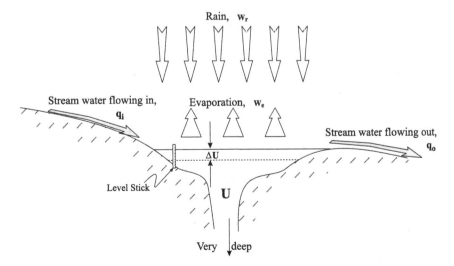

Figure 3.2 The pond analogy for the first law.

where

$$q = q_i - q_o$$

and

$$w = w_r - w_e$$

Once water has entered the pond, it loses its identity as stream or rain water. The pond does not contain any identifiable stream water or rain water, simply water. Similarly systems do not contain so much heat or work, just energy. Just as the water level in the pond can be raised *either* by stream water alone *or* by rain water alone, Joule showed in the nineteenth century that a temperature rise in a water bath of so many degrees can be caused *either* by heating (transferring energy due to a temperature difference) *or* by thrashing a paddle wheel about in it (transferring energy by doing work on the system).

Another implication or assumption in our pond analogy is that water is conserved; that is, it cannot simply appear or disappear as if by magic. The same proposition regarding energy is known as the first law of thermodynamics. We invoked this principle when we said that the energy expended in lifting a book from the floor to the table was not lost, but transferred to the book.

Internally Generated Energy

Note finally that in our pond model water enters and leaves the pond from the environment; there is no provision for water (energy) to be generated *within* the pond. However, in real systems energy *can* be generated or absorbed within a system due to chemical reactions. We will have to develop some way to take care of this. We noted in Section 2.6.1 that Prigogine and Defay (1954) refer to this energy as uncompensated heat, but that's because they specifically exclude work energy, other than whatever work is done by a change in the system volume. It is actually just energy, which does not become work or heat until it leaves the system.

3.4   The First Law of Thermodynamics

The first law of thermodynamics is the law of conservation of energy. If U is the energy content of a system, and it may gain or lose energy only by the flow of heat (q) or work (w), then clearly, as in the pond analogy, ΔU must be the algebraic sum of q and w. In order to express this algebraically, we must have some convention as to what direction of energy flow $+q$, $-q$, $+w$, and $-w$ refer to. In the pond analogy we assumed implicitly that addition of water to the pond was positive, whether as stream water or rain water. Thus heat added *to* a system is positive, and work done *on* a system is positive. This convention may be represented as in Figure 3.3(a) and is what we call the "scientific" convention – scientists like it because it is internally consistent. It results in the equation previously found,

$$\boxed{\Delta U = q + w}$$

(3.3)

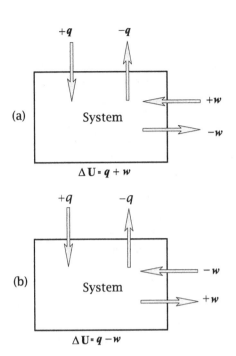

Figure 3.3 The two commonly used conventions for the sign of *q* and *w*, leading to two formulations of the first law.

Another convention (Figure 3.3(b)) is to say that heat added to a system is positive, but that work done *by* a system is also positive, or that work done *on* a system is negative. This we call the "engineering" convention, because engineers prefer to think in terms of heat engines, and an engine doing work is something positive. This results in the relation

$$\Delta U = q - w$$

and also results in slightly simpler equations expressing pressure–volume work (the minus signs in Equations (3.5) to (3.8) on pages 51 and 52 would be missing). In this text we will use the scientific sign convention. Any additions of matter and energy to the system are positive in sign and all losses are negative.

Note that we have not "proved" the first law. It is a principle that has been deduced from the way things work in our experience, but the fact that it has never been known to fail does not constitute a proof. Neither does the fact that the sun has never failed to rise in the east constitute a proof that it will rise in the east tomorrow, but I wouldn't bet against it.

3.5 Work

3.5.1 Types of Work

There are many different ways of doing work on a system depending on what kinds of forces are available, and many different ways of having a system do work. For example, consider the following examples.

- The force of gravity means that we have to do work to lift objects, as mentioned above. If the mass is m, the acceleration due to gravity is g, and the distance is dh, the work w is $w = mg \cdot dh$. As mentioned in Section 3.2.3, if this work is just used to change the potential energy of the system, it is not usually of interest in thermodynamics, but in rare cases it can be used to transfer energy to a system, as when Joule used a falling weight to drive a paddle wheel in a tank of water.

- Tensile force can be used to stretch a wire. If the tensile force is f and the increase in length is dl, $w = f \cdot dl$, assuming the wire deforms elastically.

- Tensile force can be used to increase the area of a soap film. If the surface tension is γ and the increase in surface area is dA, $w = \gamma \cdot dA$.

- Expansion due to the heating of a gas, or indeed of anything at all, produces a force. This case is of special interest to us, because the work done by expansion or contraction of systems cannot be avoided. We can choose to eliminate other forms of work, but not this one (unless we consider only constant-volume systems, which is useful in theory but not very practical). It is treated in more detail below.

- Chemical work. For example, a battery can be used to do work, because a chemical reaction occurs in it, which produces a voltage. The work done by chemical reactions is of course a principal focus of chemical thermodynamics, and the equations for it will be developed in later chapters.

There are others, such as work done by centripetal and frictional forces, that you can review in a physics text. Thermodynamics can accommodate all kinds of forces and types of work, but, because they are in principle all the same, and are treated in the same way, it is simpler to develop the subject by considering only those forms of work that we cannot

---

Box 3.1   Other Forms of Work

A block weighing 10.0 kg is lifted 4 m at a place where $g = 9.80\,\mathrm{m\,s^{-2}}$. The work done is

$$w = mg \cdot dh$$

$$= 10.0 \times 9.80 \times 4$$

$$= 98.0\,\text{newtons} \times 4\,\text{meters}$$

$$= 392\,\text{joules}$$

A film of water has a surface tension of $\gamma = 72 \times 10^{-3}\,\mathrm{N\,m^{-1}}$. The work done in expanding its area by $1\,\mathrm{cm^2}\ (= 10^{-4}\,\mathrm{m^2})$ is

$$w = \gamma \cdot dA$$

$$= 72\,\mathrm{N\,m^{-1}} \times 10^{-4}\,\mathrm{m^2}$$

$$= 72 \times 10^{-4}\,\text{joules}$$

avoid. Therefore the basic structure of thermodynamics is always developed using heat and pressure–volume work, and other forms of work considered afterward. In our case, the only other form of work of any importance is chemical reaction work.

3.5.2   Pressure–Volume Work

Work in natural environments is for the most part only of one kind – the work of expansion, or pressure–volume work. Pressure–volume work is always discussed using a piston–cylinder arrangement as shown in Figure 3.4. This seems natural to engineers, but may seem rather artificial or even useless to someone interested in processes that happen in nature or in the environment. You have to realize that virtually *all* processes in *all* natural systems involve some change in volume, and therefore work is done against the pressure on the system, whatever that is (it is very often atmospheric pressure). We use a piston–cylinder arrangement for convenience – any system that changes volume could be used. Once we have found the appropriate equations for pressure–volume work, we can use them in our models of any system, irrespective of whether or not the system has pistons and cylinders.

The piston–cylinder arrangement as shown in Figure 3.4 is not a real piston in a real cylinder, of course, but a conceptual one, so we can give it whatever properties we like. We must be careful about this, however, otherwise the results will be useless. The cylinder is fitted with some devices that can hold the piston in position at various levels. When the piston is held stationary, the forces tending to move the piston are balanced (force pushing up equals force pushing down). If this were not the case, the piston would move. The two forces are acting on opposite sides of the same piston, having the same area (and force/area = pressure), so the pressure of the gas, P_{int}, is exactly balanced by the external pressure, P_{ext}. The external pressure is provided partly by the stops that are holding the piston in place and partly by the weight of the piston itself, plus any weights on the piston. If the stops are removed, then all of a sudden P_{ext} is reduced to that produced by the piston

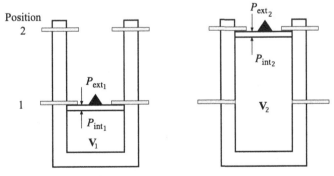

T constant

Figure 3.4 Irreversible expansion of a gas from external pressure P_{ext_1} to P_{ext_2}. Expansion occurs when the stops holding the piston at position 1 are released. During expansion, the external pressure is fixed by the weight of the piston plus the weights on the piston.

and weights only, $P_{int} \gg P_{ext}$, and the piston moves up until it encounters more stops – WHAP! – and all of a sudden $P_{int} = P_{ext}$ once more, though at a different (lower) pressure (the experiment has been arranged such that the gas pressure is 10 pressure units at the upper stops, which is position 2, and 20 pressure units at the lower stops, position 1). Real gases tend to cool during expansion, so, if we want the initial and final states to be at the same temperature, some heat must flow into the cylinder from the surroundings.

At this stage, one normally says something like "If the piston is well-lubricated and well-constructed, we can ignore friction effects, ...," but we know we are conducting a model experiment, so we just say there is no friction in our model. The pressure–volume history of the change can be illustrated as in Figure 3.5. The external pressure during expansion is constant, since it is fixed by the mass of the piston plus whatever weights are on it. The work done during the expansion is

$$-w = \text{force} \times \text{distance}$$
$$= (\text{total mass} \cdot g) \cdot \Delta h$$
$$= (P_{ext} \cdot A) \cdot \Delta h$$
$$= P_{ext}(A \cdot \Delta h)$$
$$= P_{ext} \cdot \Delta \mathbf{V} \tag{3.4}$$

where A is the area of the piston and Δh the distance it travels, so $-w$ is the area under the path of expansion or expansion curve in Figure 3.5. The minus sign is because the system is doing work. If we repeat the process, but this time we place a larger weight on the piston, exactly the same thing will happen, but more work is done because a greater mass was lifted through the same volume.

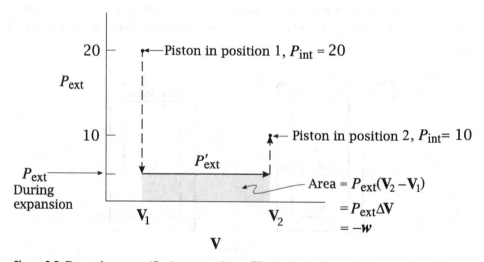

Figure 3.5 External pressure (P_{ext}) versus volume (**V**) plot for the irreversible expansion of the gas in Figure 3.4. The vertical dashed lines indicate an instantaneous change in pressure. The solid horizontal lines indicate change in volume at constant pressure.

Box 3.2 $P \Delta V$ **Work**

In Figure 3.5, suppose the pressure units are bars, $\mathbf{V}_1 = 1000\,\text{cm}^3$ of ideal gas, and during expansion P_{ext} is 5 bars. Assume T is constant. How much work is done?

The pressure on the gas is halved (from 20 to 10 bars), so the (ideal) gas will expand to twice its volume ($PV = $ constant), so $\mathbf{V}_2 = 2000\,\text{cm}^3$. Then, from Equation (3.4),

$$-w = P_{ext}(\mathbf{V}_2 - \mathbf{V}_1)$$
$$= 5 \times (2000 - 1000)$$
$$= 5000\,\text{bar cm}^3$$

To convert this to joules, Appendix A gives the conversion $1\,\text{bar} = 0.10\,\text{J cm}^{-3}$, so

$$-w = 500\,\text{J or } w = -500\,\text{J}$$

Note the minus sign, which indicates the system is doing work. If $\mathbf{V}_2$ were less than $\mathbf{V}_1$, $\Delta \mathbf{V}$ would be negative and w would be positive, meaning work is done on the system.

If another weight is added for the next expansion, we may have a total weight that is too great to allow the piston to reach the upper stops (position 2) and it will come to rest (equilibrium) somewhere in between. Then, if the second weight is removed, the piston will proceed upward again as before, giving an expansion path as shown in Figure 3.6. If we use a lot of weights and remove them one at a time, letting the piston come to rest after each step, we will get a path such as the one shown in Figure 3.7.

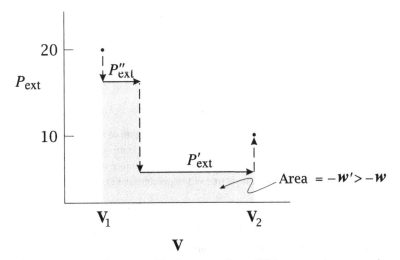

Figure 3.6 External pressure (P_{ext}) versus volume ($\mathbf{V}$) for a two-stage expansion of gas. After an initial expansion at P''_{ext}, some weight was removed from the piston and the expansion continued at P'_{ext}.

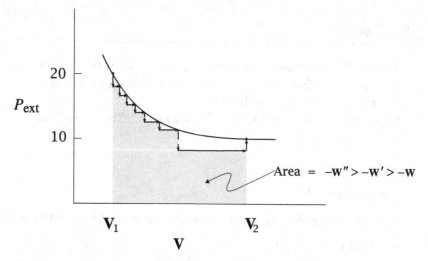

Figure 3.7 External pressure (P_{ext}) versus volume ($\mathbf{V}$) for a multistage expansion of gas. After each constant-P_{ext} expansion, some weight was removed, allowing a further expansion. If in each stage the weight was added rather than removed, each little arrow would be reversed, work would be done *on* the system, and the work variable would be w instead of $-w$.

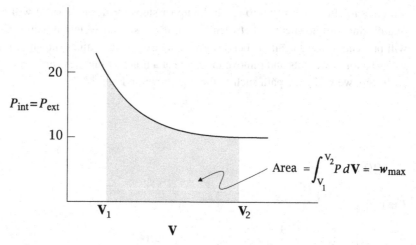

Figure 3.8 Pressure versus volume for the reversible expansion of a gas. The limiting case in which infinitely many constant-P_{ext} steps are taken gives the maximum area under the curve. During the expansion, internal pressure and external pressure are never more than infinitesimally different, or $P_{int} = P_{ext}$ at all times. If the same is true of temperature the expansion is isothermal, and the curve in the figure is an *isotherm*.

Clearly we are approaching a limit of maximum work obtainable from the expansion of our gas, and clearly too, the maximum will be when we take an infinite number of infinitesimally small incremental steps from $\mathbf{V}_1$ to $\mathbf{V}_2$, resulting in the curve in Figure 3.8.

Since we have been letting the piston come to rest or equilibrium after every weight removal, in the limit we will have an infinite number or continuous succession of equilibrium states, giving us an example of a reversible process. The name "reversible" is appropriate since at any stage in the expansion the direction of movement can be reversed by changing the external pressure infinitesimally – the piston is in a balanced condition. After each increment of movement of the piston we have been allowing heat to flow into the cylinder so that the initial and final temperatures are the same. In the limit we have also achieved a perfect *isothermal* expansion, where the temperature difference is never more than infinitesimal. We could perform the same reversible expansion or compression using a cylinder having *adiabatic* walls which prevent any heat flow to or from the gas. The resulting curve (Figure 3.8) is then an *adiabat* rather than an isotherm, and the work done is different because the functional relationship between P and V in Equation (3.5) is different.

In the limit when infinitesimal increments of $\mathbf{V}$ are taken, the work of expansion is (see Figure 3.8)

$$-w_{rev} = -w_{max} = \int_{\mathbf{V}_1}^{\mathbf{V}_2} P \, d\mathbf{V} \tag{3.5}$$

Here we need make no distinction between P_{ext} and P_{int} because they are never more than infinitesimally different in our continuous succession of equilibrium states. Again, note the negative sign required to comply with the scientific sign convention. In general, considering both reversible and irreversible expansions,

$$-w \le \int_{\mathbf{V}_1}^{\mathbf{V}_2} P \, d\mathbf{V} \tag{3.6}$$

Since the end positions 1 and 2 of our expansion in every case consisted of our gas at stable equilibrium at a fixed P and T, according to the first law there is a fixed energy difference $\Delta \mathbf{U}$ between the two states. We have gone to some length to show that there is no fixed "difference in work," or work available from the change from one state to the other. The work done depends on what values of P_{ext} are in effect at all stages of the expansion. In the example we used only two different values, P'_{ext} and P''_{ext}, but there could be any number, giving any value of the total work done. Thus we are led to believe that the amount of heat flowing into our thermostatted cylinder must at all times, once equilibrium had been established, have compensated for the variations in work performed, giving the same total $q + w$ in every case. We could verify this, of course, by making calorimetric measurements, but this is basically what Joule and many other workers have already done.

Our intent here is to illustrate not so much the constant energy change between equilibrium states, but that this energy change, while accomplished by heat and work, can be made up of an infinite variety of combinations of heat and work. When the process is made reversible, we get the maximum work of expansion, and this will be given by Equation (3.5), but even so, we are unable to calculate this amount of work (evaluate the integral) without more information (we need to know P as a function of $\mathbf{V}$ so that we can integrate Equation (3.5)).

The integration of (3.5) at a constant external pressure results in

$$-w = P_{ext}(V_2 - V_1) \tag{3.7}$$

$$= P_{ext}\,\Delta V \tag{3.8}$$

as in Equation (3.4). The *internal* pressure necessarily varies during this (irreversible) expansion, as discussed above.

A point worth emphasizing is that in any real expansion, which is necessarily irreversible, the work obtained is always less than the maximum obtainable (from a reversible expansion). This can also be expressed as

$$-w \le -w_{max} \tag{3.9}$$

or

$$-w \le -w_{rev} \tag{3.10}$$

For the opposite case of compressing the gas from position 2 back to position 1, the inverse series of steps can be employed. Thus, if at position 2 a heavy weight is placed on the piston, it will WHAP down to the stops at position 1, describing a path such as in Figure 3.9. Obviously, much more work has had to be done in compressing the gas than we obtained, even in the reversible case, from expansion. However, by adding a larger number of smaller weights one at a time we can reduce the amount of work required for the compression, gradually approaching the stable equilibrium curve from above, rather than from below as before. In the limit, of course, we find that for a reversible compression the work required is exactly the same as the work available from a reversible expansion.

Considering this work stuff in such detail may make it look complex, but it really is not. Just remember that, if you need to calculate work (which happens surprisingly little

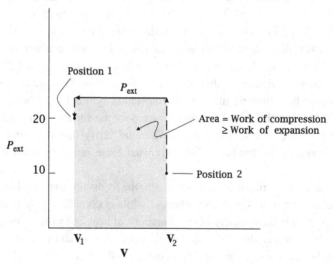

Figure 3.9 External pressure (P_{ext}) versus volume for the irreversible compression of gas at constant P_{ext}.

in geochemistry), you need either a constant pressure process, or a reversible one (so you can integrate). More important is the fact that any real work done is always less than the theoretical maximum (Equation (3.9)), usually much less, and usually of more interest to engineers than to geochemists.

3.5.3   An Inexact Differential

The limitless sequence of small increments of movement of the piston (and hence of work done) in Figure 3.7 resulting in Figure 3.8 might suggest an integration. Why not integrate dw? We can consider w to be made up of small or infinitesimal increments δw, then sum up all the increments between states 1 and 2, giving (McQuarrie and Simon, 1997, p. 773)

$$\int_1^2 \delta w = w \underline{\; not \;} \Delta w \;\; or \;\; w_2 - w_1 \tag{3.11}$$

This introduces the subject of inexact differentials. Integration of δw does not result in Δw because the differential of w is not the exact differential dw but the inexact differential δw. w does not have an exact differential because there is no function from which we could derive dw as we did for dy in the example in Section 1.4.1. As shown in Figures 3.4–3.8 all numerical values of w depend on the path followed by the piston during the expansion. This path is infinitely variable, and even if we had a function for one such path it would only be valid for that one path. w is a *path variable*, not a state variable. Change in a path variable depends on the series of states occupied between its initial and final states (its path), and its differential is therefore inexact. Change in a state variable depends only on the initial and final states (Section 2.4) and its differential is exact. We use δ rather than d for inexact differentials.

After every increment of change in Figures 3.4–3.8 some heat was exchanged, so this amount of heat will also depend on the path followed. q is also a path variable, and

$$\int_1^2 \delta q = q \underline{\; not \;} \Delta q \;\; or \;\; q_2 - q_1 \tag{3.12}$$

At each point on the isotherm in Figure 3.8 the small (or infinitesimal) change in work done (δw_{rev}) at pressure P (where $P_{int} = P_{ext}$) is

$$-\delta w_{rev} = P\, d\mathbf{V} \tag{3.13}$$

If we integrate this expression keeping P constant we have a constant *internal* pressure expansion

$$-\int_1^2 \delta w_{rev} = \int_1^2 P\, d\mathbf{V}$$
$$-w_{rev} = P\, \Delta \mathbf{V} \tag{3.14}$$

where in this case, because w is reversible, P is the reversible pressure, i.e., $P_{int} = P_{ext}$ at all times.

Having now discovered that work and heat have inexact differentials, we can write the first law (Equation (3.3)) in differential form as

$$d\mathbf{U} = \delta q + \delta w$$

(3.15)

3.5.4   The Meaning of δ

How can we integrate δw (Equation (3.11)) if w does not represent a function? δw represents a small increment of energy expended as work and is not really a differential. In Equation (3.11) $\int \delta w$ represents a summing up of all the δw increments, not a mathematical integration. We should perhaps say instead $\sum \delta w = w$. Alternatively, we can imagine that for one partcular path from $\mathbf{V}_1$ to $\mathbf{V}_2$ we have determined a function describing w as a function of P and $\mathbf{V}$ for that one path, and theat δw is the exact differential of that function. But for any others of the infinite number of possible paths we would have a different function and a different δw. So δw may be the exact differential of some imaginary function, but clearly this is unsatisfactory, so we call δw inexact.

There are many other expressions containing differential terms which are not exact differentials. For example, $(x\,dy + y\,dx)$ is exact, being equal to $d(xy)$, but $(x\,dy - y\,dx)$ is not equal to the differential of any function of x and y, and is inexact. Some such inexact differential expressions can be transformed into exact differentials by multiplying or dividing the expression by some variable or combination of variables. It turns out that if δw and δq refer to a reversible process, they are still just variables representing increments of energy expressed as work or heat, but dividing each δw by its pressure P and each δq by its temperature T to get $\delta w_{rev}/P$ and $\delta q_{rev}/T$ we get the exact differentials $-d\mathbf{V}$ (Equation (3.13)) and $d\mathbf{S}$ (Equation (4.2)). See the Online Resources for more discussion.

In any case, its not that important. The important thing to know is that Equations (3.3) and (3.15) mean that a fixed difference in system energy can be achieved by any combination of heat and work, reversible or irreversible. We never actually use w or q except in those cases where they become state variables. This occurs when a process is reversible (Equation (3.5)) or when it is a constant pressure change in volume (Equation (3.4)) which results in q_P becoming enthalpy (Equations (3.22) and (3.31)). And, as Dickerson (1969, p. 387) says,

It is possible to know thermodynamics without understanding it.

Box 3.3   Deltas and Differentials

On page 33 we looked at the meaning of $\Delta_r V^\circ$, the value of which changes depending on how you write the reaction, and in Section 3.5.1, page 45, we used differential notation where we could easily have used delta notation. For example, we wrote $w = mg \cdot dh$ where we could just as easily have written $w = mg \cdot \Delta h$. Because there is a persistent tendency to see differentials as infinitesimals, we made the point that dh is not *necessarily* infinitesimal, so in a sense we have not changed anything in switching from Δ to d. What we said there is true enough, but it is not the whole story.

In mathematics Δ refers to some finite change in a quantity, such as the change in height as above, or your speed before and after accelerating. A differential dx, on the other hand, although it need not be infinitesimal, is not *just* a change in some quantity called x. It means that there is a mathematical function for which it represents some increment in the x variable. Therefore $dG_{T,P} = 0$ has a significantly different meaning from $\Delta G_{T,P} = 0$. $dG_{T,P} = 0$ signifies the extremum of a mathematical function, while $\Delta G_{T,P} = 0$ does not.

3.6 Heat

It might be expected that since

$$\Delta U = q + w \tag{3.3}$$

and

$$-w_{\text{rev}} = \int_{V_1}^{V_2} P \, dV \tag{3.5}$$

$$-w \leq \int_{V_1}^{V_2} P \, dV \tag{3.6}$$

perhaps there is a very similar story for the heat transfers in the gas expansion cases we have been considering. That is, perhaps

$$-q_{\text{rev}} = \int_{Z_1}^{Z_2} T \, dZ \tag{3.16}$$

$$-q \leq \int_{Z_1}^{Z_2} T \, dZ \tag{3.17}$$

where Z is some property of the gas. And, if this is true, then just as

$$-\frac{\delta w_{\text{rev}}}{P} = dV \tag{3.13}$$

$$-\frac{w_{\text{rev}}}{P} = \Delta V \tag{3.14}$$

we should have

$$-\frac{\delta q_{\text{rev}}}{T} = dZ \tag{3.18}$$

$$-\frac{q_{\text{rev}}}{T} = \Delta Z \tag{3.19}$$

and

$$-\frac{\delta q}{T} \leq dZ \tag{3.20}$$

$$-\frac{q}{T} \leq \Delta Z \tag{3.21}$$

where, again, the $<$ refers to any real or irreversible process. This is indeed the case (except for a sign change), but we must await the development of the second law, which will introduce us to entropy $(-\mathbf{Z})$.

3.6.1 Enthalpy, the Heat of Reaction

In processes at constant external pressure, the work done, as we have seen (Equation (3.8)), is $-P_{ext}\,\Delta \mathbf{V}$. In what follows it doesn't matter whether we use P_{ext} or P, as long as it is constant. Therefore the first law can be written

$$\Delta \mathbf{U} = \boldsymbol{q}_P - P\,\Delta \mathbf{V} \tag{3.22}$$

where $\boldsymbol{q}_P$ is the heat transferred in a constant-pressure process. Thus

$$\boldsymbol{q}_P = \Delta \mathbf{U} + P\,\Delta \mathbf{V}$$

from which we can see that the heat transferred in constant-pressure processes is equal to a function involving only state variables, and so it is itself a state variable. Don't forget that we have gone to some trouble to show that in general neither $\boldsymbol{q}$ nor $\boldsymbol{w}$ is a state variable; it is only in the special case of constant-pressure processes that they both become state variables. Because of this, it is useful to define a new term, enthalpy,

$$\mathbf{H} = \mathbf{U} + PV \tag{3.23}$$

and the molar form

$$H = U + PV \tag{3.24}$$

Equation (3.23) has the differential form

$$d\mathbf{H} = d\mathbf{U} + P\,d\mathbf{V} + \mathbf{V}\,dP \tag{3.25}$$

$$dH = dU + P\,dV + V\,dP \tag{3.26}$$

which at constant pressure becomes

$$d\mathbf{H}_P = d\mathbf{U} + P\,d\mathbf{V} \tag{3.27}$$

$$dH_P = dU + P\,dV \tag{3.28}$$

or

$$\Delta \mathbf{H}_P = \Delta \mathbf{U} + P\,\Delta \mathbf{V} \tag{3.29}$$

$$\Delta H_P = \Delta U + P\,\Delta V \tag{3.30}$$

Upon inserting Equation (3.22) we get

$$\Delta \mathbf{H}_P = \boldsymbol{q}_P \tag{3.31}$$

or in molar form

$$\Delta H_P = q_P \tag{3.32}$$

As mentioned earlier, it is generally q and ΔH that are measured in calorimeters, but it is the molar quantity ΔH or $\Delta H°$ that is used for pure substances and is found in tables of data.

All we have done here is notice that, because work becomes a fixed quantity in constant-pressure processes, heat does too, by the first law. Because constant-pressure processes are so common (including all reactions carried out at atmospheric pressure, such as most biochemical reactions), it is convenient to have a state variable defined to equal this heat term. Defining enthalpy as in (3.23) and (3.24) accomplishes this, and we now have a "heat of reaction" term, which will be useful in all constant-pressure processes.

Box 3.4   Heat of Reaction

The standard heat of reaction for reaction (2.5) is

$$\Delta_r H° = \Delta_f H°_{Al_2O_3 \cdot 3H_2O(s)} - \Delta_f H°_{Al_2O_3(s)} - 3\,\Delta_f H°_{H_2O(l)}$$
$$= -2586.67 - (-1675.7) - 3(-285.83)$$
$$= -53.48\,\text{kJ mol}^{-1}$$
$$= -53,480\,\text{J mol}^{-1}$$

Again, the minus sign means heat is evolved (exothermic reaction). This amount of heat would raise the temperature of a liter of water by about $12\,°C$.

Note that, because $\mathbf{H}$ is a state variable, $\Delta \mathbf{H}$ is perfectly well defined between any two equilibrium states. But, when the two states are at the same pressure, $\Delta \mathbf{H}$ and ΔH become equal to the total and molar heat flow during the process from one to the other, and in practice enthalpy is little used except in this context. Chemical reactions having a negative $\Delta_r H$ ($\Delta_r H < 0$) are termed *exothermic*, and those having a positive $\Delta_r H$ are termed *endothermic*.

3.6.2   Additivity of State Variables

At several points in our discussions so far, we have mentioned or assumed that we can add and subtract state variables such as ΔH, for example as shown in Figure 5.3 on page 116. This is perhaps obvious, but it is so fundamental that we will emphasize it here.

We said in Chapter 2 (Section 2.4) that a state variable is a property of a system that has a fixed value when the system is at equilibrium, whether we know the value of that property or not. For example, a mole of water at $25\,°C$, 1 atm has a fixed but unknown enthalpy H, and fixed values of all other state variables. We also said that this means that the changes in these properties between equilibrium states depend only on the equilibrium states, and not on what happens between the time the system leaves one equilibrium state and the time it settles down in its new equilibrium state. Therefore, if two different reactions produce the same compound, we can subtract the $\Delta_r H°$, for example, of these reactions to get the

$\Delta_r H°$ of the combined reaction, and the properties of that compound will cancel out. For example, carbon dioxide, CO_2, might be produced from the oxidation of either graphite or carbon monoxide, CO:

$$C + O_2 = CO_2 \qquad\qquad \Delta_r H° = -393.509 \, kJ \, mol^{-1}$$

$$CO + \tfrac{1}{2}O_2 = CO_2 \qquad\qquad \Delta_r H° = -282.984 \, kJ \, mol^{-1}$$

On subtracting the reactions and the $\Delta_r H°$ values (reverse the second reaction, change the sign of $\Delta_r H°$, and add), we get

$$C + \tfrac{1}{2}O_2 = CO \qquad\qquad \Delta_r H° = -110.525 \, kJ \, mol^{-1}$$

Thus we get the properties of a reaction that is impossible to carry out experimentally from two reactions that are relatively easy to do experimentally.

3.6.3   Enthalpy of Formation from the Elements

A major problem arises from the definition of enthalpy, Equation (3.23). The problem is that we cannot measure it. This arises from the nature of energy itself, because we can only measure energy changes, not absolute energies. Therefore we can only measure enthalpy changes, and changes in any other property which includes the energy **U** (or U).

The problem this creates is that we do not want to have to tabulate an enthalpy change for every process or chemical reaction which might become of interest to us – there are too many. We would like to be able to associate an enthalpy with every substance – solids, liquids, gases, and solutes – for some standard conditions, so that, having tabulated these, we could then easily calculate an enthalpy change between any such substances under those standard conditions, as illustrated in Box 3.4 on page 57. After that, we could deal with the changes introduced by impurities and other non-standard conditions. The method developed to allow this is to determine, for every pure compound, the difference between the enthalpy of the compound and the sum of the enthalpies of its constituent elements, each in its most stable state, which make up the compound. This quantity is called $\Delta_f H°$, the standard molar enthalpy of formation from the elements. For aqueous ions, the quantity determined is a little more complicated, but the principle is the same. It is this enthalpy quantity which is invariably tabulated in compilations of data.

Properties of Aqueous Ions

Because positive and negative ions cannot be separated from each other to any significant extent, their individual properties cannot be measured. To obtain such data, we assume the properties of the hydrogen ion H^+ to be zero. You can then measure some property of aqueous HCl which is completely ionized to H^+ and Cl^-, and assign the result to Cl^- alone. More discussion of this can be found in Chapter 15. Electrolyte Solutions, in the online resources.

For example, the standard enthalpy of formation of anhydrite is

$$\Delta_f H^\circ{}_{CaSO_4(s)} = H^\circ_{CaSO_4(s)} - H^\circ_{Ca(s)} - H^\circ_{S(s)} - 2\,H^\circ_{O_2(g)} \tag{3.33}$$

where the superscript $^\circ$ refers to the standard conditions (see below). None of the individual H° quantities is determinable, but the difference is determinable by calorimetry. Now, if we want to know the heat liberated or absorbed in a chemical reaction, we need only look up these $\Delta_f H^\circ$ values for each reactant and product. For example, for the formation of gypsum from anhydrite, we write

$$CaSO_4(s) + 2\,H_2O(l) = CaSO_4 \cdot 2\,H_2O(s) \tag{3.34}$$

for which the "standard molar heat of reaction," $\Delta_r H^\circ$, is

$$\Delta_r H^\circ = \Delta_f H^\circ_{CaSO_4 \cdot 2H_2O} - \Delta_f H^\circ_{CaSO_4} - 2\,\Delta_f H^\circ_{H_2O(l)} \tag{3.35}$$

$$= H^\circ_{CaSO_4 \cdot 2H_2O} - H^\circ_{CaSO_4} - 2\,H^\circ_{H_2O(l)} \tag{3.36}$$

Note that in balanced reactions the H° terms for all the elements cancel out, and we are left with the "real" enthalpy difference, Equation (3.36), between products and reactants, with no contribution from arbitrary conventions or assumptions. It is, however, a heat of reaction for standard conditions only.

3.6.4   Standard States

It is often assumed that the "standard conditions" are 25 °C, 1 bar. Actually it is a bit more complicated in two respects.

1. Knowing the T and P of the state is not sufficient – we must also specify the physical state of the substance. For solids and liquids, it is simply the pure substance (as in our anhydrite–gypsum–water example), but for gases it is the gas acting ideally at one bar (or 10^5 Pa), and for solutes it is the solute acting ideally at a concentration of one molal.
2. While the temperature and pressure of the standard conditions are indeed 25 °C and one bar for purposes of tabulating data, we can and often do have standard conditions at any T and P.

These more complete definitions of our "standard conditions" define our *standard states*, which will be seen to become particularly useful when we introduce the concept of *activity* in Chapter 8.

3.6.5   The Heat Capacity

An older name for the enthalpy is the "heat content." This name is somewhat discredited for good reasons, but nevertheless it helps a little in conveying the essential idea behind the next concept, the heat capacity. The molar heat capacity can be defined as the amount of heat required to raise the temperature of one mole of a substance by one degree. Of course, some substances require much more heat to do this than do others. As usual, there are the molar and total (intensive and extensive) versions of the heat capacity.

The formal definition is

$$\left(\frac{dH}{dT}\right)_P = C_P \tag{3.37}$$

or, in the extensive form,

$$\left(\frac{d\mathbf{H}}{dT}\right)_P = \mathbf{C}_P, \tag{3.38}$$

and of course we have also

$$\left(\frac{d\Delta H}{dT}\right)_P = \Delta C_P \tag{3.39}$$

and

$$\left(\frac{d\Delta H^\circ}{dT}\right)_P = \Delta C_P^\circ \tag{3.40}$$

Thus heat capacity is the rate of change in H or $\mathbf{H}$ with T. A large C_P means that H changes a lot for a given change in T, i.e., it takes a lot of heat to raise the temperature.

It takes a different amount of heat to raise the temperature of a system depending on whether the volume or the pressure is kept constant, giving two different quantities, C_P and C_V. C_V is rarely used in geochemistry, but the heat capacity at constant pressure, C_P, is a surprisingly important quantity. It can be used to calculate not only $\Delta_r H$ at high temperatures, as in the next section, but also the high-temperature values of several other thermodynamic quantities. The units of the molar heat capacity are $J\,mol^{-1}\,K^{-1}$, and for the specific heat capacity, or specific heat, are $J\,g^{-1}\,K^{-1}$. The extensive or total form is $\mathbf{C}_P$ or $\mathbf{C}_V$ in $J\,K^{-1}$.

3.6.6   Temperature Dependence of the Heat Capacity

If you are interested in a chemical reaction at any temperature beyond 25 °C, you need to know values of the heat capacity for all the species in your reaction. Many different equations have been suggested to represent the variation of C_P with temperature, and several are in current use. No differences in principle are involved, so we will consider only one. The equation suggested by Maier and Kelly (1932) is

$$C_P = a + bT - cT^{-2} \tag{3.41}$$

To know how C_P for a substance varies with T we need only look up the values of a, b, and c for that substance. However, it is important to note that these coefficients are available only for pure solids, liquids, and gases, because C_P for pure substances increases in a fairly simple way with T. Aqueous solutes, however, have a much more complex behavior.

For chemical reactions in which solutes are not involved, the change in each coefficient between products and reactants is evaluated in the usual way. For example, for reaction of anhydrite plus water to form gypsum, Equation (3.34),

$$\Delta_r a = a_{CaSO_4 \cdot 2H_2O} - a_{CaSO_4} - 2\, a_{H_2O(l)}$$

$$\Delta_r b = b_{CaSO_4 \cdot 2H_2O} - b_{CaSO_4} - 2\, b_{H_2O(l)}$$

$$\Delta_r c = c_{CaSO_4 \cdot 2H_2O} - c_{CaSO_4} - 2\, c_{H_2O(l)}$$

so the change in C_P between products and reactants is

$$\Delta_r C_P^\circ = \Delta_r a + \Delta_r b\, T - \Delta_r c\, T^{-2} \tag{3.42}$$

3.7   How Far Have We Got?

We have defined internal energy as some unspecified subset $\mathbf{U}$ of the rest energy $\mathbf{E_r}$ in a system and considered the two common ways of changing this energy. Along the way, we have noted that energy never disappears, and this is called the first law of thermodynamics. How far have we got toward finding the "chemical energy," that always decreases in spontaneous changes?

Well, we've made the first vital step, but if you think about the previous chapters you'll realize that we cannot have the answer yet. Why? Because we noted that some processes occur spontaneously with *no* energy change (ink spreading in water). Obviously, then, clarifying our thoughts about energy changes will not help in explaining processes that happen with no change in energy of any kind.[2] We still have some way to go toward defining a useful "chemical energy."

3.8   The Model Again

In this chapter we have discussed some very practical operations. There is nothing particularly theoretical about gases expanding in cylinders and performing work. It happens countless times every day all over the world. Equations such as (3.4) belong to the real world. However, the result of the limit-taking, when the increments of expansions or compressions between two equilibrium states are increased without limit, is a reversible process that belongs not to the real world but to the thermodynamic model. This is another illustration of the point made in Section 2.6.2, that energy differences between states can be calculated only for *reversible* processes.

The equation

$$-w_{rev} = \int_{\mathbf{V}_1}^{\mathbf{V}_2} P\, d\mathbf{V} \tag{3.5}$$

is a very simple one, considered mathematically. If P can be expressed as an integrable function of $\mathbf{V}$, then the integration is carried out and w_{rev} is determined for a given change from $\mathbf{V}_1$ to $\mathbf{V}_2$. This presents absolutely no conceptual difficulties (beyond those

---

[2] Actually, we will note in the next chapter that the internal energy U is in fact the energy we need to predict which way reactions will go under certain unusual conditions, but it is rarely used in this sense.

in understanding calculus) if P and V are mathematical variables. However, if P and V represent measured pressures and volumes from a real system in the real world, then, even if P has been determined as an integrable function of V for a number of individual measurements of P and V, the integration represents a variation of P with V that is impossible to carry out in the real system. It is, however, simple to carry it out in the thermodynamic model, that is essentially mathematical and in which P as a function of V is simply a (continuous) line in P–V space. This line represents a reversible process, a perfectly simple and understandable facet of the thermodynamic model.

3.8.1   Applicability of the Equations

Don't forget – this conclusion about the work done due to a change in volume is applicable not only to piston–cylinder arrangements. Virtually all chemical reactions involve some change in volume between reactants and products, and the equations are applicable no matter what the physical form of the reactants and products. In other words, when corundum and water react to form gibbsite (Figure 2.5), the volume occupied by the gibbsite is different from the sum of the volumes of the water and the corundum; therefore, some work is done during the reaction, and this work can be calculated using Equations (3.5) and (3.8). Even in reactions in living cells there will generally be a difference in volume between products and reactants, and a constant-pressure environment, so some work is done during each and every biochemical reaction. This work energy may be relatively small compared with the heat evolved or absorbed during the same reactions, but it must always be considered. In reactions at higher pressures, it of course becomes even more important.

3.8.2   Clarifying Notation

We have introduced quite a few subscripts and superscripts all at once here, which can be confusing. The logical relationships among these terms are shown in Figure 3.10, using H as an example; we could also use $\mathbf{H}$. The same relationships will hold for other parameters we will introduce later.

The most general term for a change in H is simply ΔH. This refers to any change in the enthalpy of any system between two equilibrium states (stable or metastable), not necessarily associated with a chemical reaction. A special case is the ΔH between the products and reactants of a chemical reaction, called $\Delta_r H$, so this represents a subset of the more general term ΔH. A special kind of chemical reaction involves only pure compounds, whose thermodynamic parameters can be found in tables, so a subset of all $\Delta_r H$ values can be called $\Delta_r H°$, to indicate that all products and reactants are in their pure reference states.[3] A special case of $\Delta_r H°$ is the reaction in which a compound is formed from its elements, all in their pure reference states, and this is called $\Delta_f H°$.

---

[3] Later on (Chapter 8) we will find that, strictly speaking, superscript ° refers to a more general "standard state," and that "pure reference states" are just one kind of standard state.

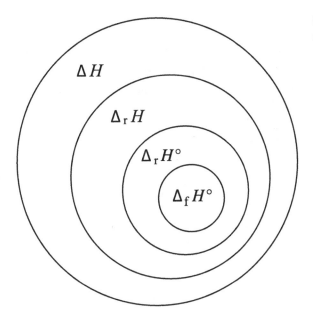

Figure 3.10 The hierarchy of ΔH terms (from Anderson (2015)).

3.9   Summary

This chapter attempts to make precise our use of the terms *energy*, *heat*, and *work*. The line of thought we are pursuing has to do with systems that spontaneously decrease their energy content, so we have started to get quite clear about what kinds of energy we mean. Relativity theory tells us that the absolute energy of any system at rest (i.e., excluding potential and kinetic energy) is given by multiplying the mass of the system by the square of the speed of light, but this approach is not very useful except in the study of nuclear processes. None of the chemical reactions we are interested in are of this type. However, apart from relativity theory there is no way of knowing the energy content of a system, so thermodynamics defines the absolute energy of a system as $\mathbf{U}$, an unknown quantity, and we are concerned only with *changes* in this energy.

When we consider by what means the energy content of systems can change, we find that there are only two – we can heat/cool the system, or we can do work on the system/have the system do work. There are several ways of doing work on systems, depending on the forces we choose to consider (magnetic, electrostatic, surface tension, etc.), so we start out by choosing the most common, pressure–volume work. The others are all handled in the same way and can be brought in when the situation calls for them.

Then, by appealing to long experience with energy transfers, we propose the first law of thermodynamics, the law of conservation of energy. Systems (that is, *any* system) can change their energy content by having energy subtracted or added in the two forms – heat and work. Any combination of the two can result in the same total energy change; there is no specific "difference in heat" or "difference in work" between two different states of the same system.

Finally, we went into some detail on one special kind of energy transfer – the heat absorbed or released during a chemical reaction, where the initial and final states have the same pressure, generally referred to as a constant-pressure reaction. This quantity of heat is the enthalpy, and it is one of the fundamental building blocks of our model. Because it includes U there are no absolute values for H, which is a decided nuisance, but a very simple way around this is by using the "formation from the elements" convention. This means that, for every compound, we measure ΔH for the reaction in which a compound is formed from its elements, each in its most stable form, and these quantities are given the symbol $\Delta_f H°$, where the subscript f stands for "formation from the elements," and the superscript ° means all substances are in their pure (except for dissolved substances) reference states. The ΔH for any other reaction can then be found by combining these $\Delta_f H°$ terms for compounds.

Defining what we mean by energy and energy transfers is, of course, important, but it does not by itself answer our questions about why reactions go one way and not the other.

Exercises

E3.1 (a) Calculate the work done (in joules) by the expansion in Figure 3.6 of an ideal gas if $V_1 = 1000\,cm^3$, $P'_{ext} = 5\,bar$, and $P''_{ext} = 18\,bar$. After the expansion at P''_{ext}, the piston is at equilibrium with no stops. However, during the expansion at P'_{ext}, the piston hits the upper stops, at which point P'_{ext} suddenly increases to 10 bars.

(b) Calculate the work done if there are four expansions having P_{ext} values of 18, 16, 14, and 10 bars.

(c) Calculate the maximum amount of work available from the expansion of this gas.

(d) Calculate the work done if the gas is compressed from V_2 to V_1 by placing a large weight equivalent to $P_{ext} = 22$ bars on the piston.

(e) Why don't we use real gases instead of ideal gases in this type of problem?

E3.2 How much work is done when one mole of corundum combines with three moles of water to form one mole of gibbsite (reaction (2.5)), at atmospheric pressure?

E3.3 Calculate the change in internal energy ($\Delta_r U°$) for reaction (2.5),

$$Al_2O_3(s) + 3\,H_2O(l) = Al_2O_3 \cdot 3H_2O(s) \tag{2.5}$$

E3.4 Calculate the standard heat of reaction for reaction (2.1),

$$C_8H_{16}N_2O_3(aq) + H_2O(l) = C_6H_{13}NO_2(aq) + C_2H_5NO_2(aq) \tag{2.1}$$

E3.5 Calculate the maximum amount of useful work available from reactions (2.1) and (2.3) under standard conditions.

E3.6 Calculate the work of expansion against atmospheric pressure for reaction (2.3), and compare your answer with the maximum useful work available.

Additional Problems

A3.1 The unit of energy used by nutritionists is the Calorie (with a capital C), which is 1000 calories, or 1 kcal. In Appendix A, we see that 1 calorie = 4.184 joules, so 1 Cal = 4184 J. The basal metabolic rate for humans varies from about 1300 to 1800 kcal/day. Any physical activity adds to this basal energy requirement. In other words, a person at rest would use the energy in a glass of orange juice (100 Cal) in about 100/1500 days, or 1.6 hours. How long would the energy ($\Delta_r H°$) from reaction (2.5) last an average human, if it were available?

A3.2 The C_P for (supercooled, metastable) water between 0 and $-10°C$ is 76.1 $J\,mol^{-1}\,K^{-1}$. The C_P for ice between 0 and $-10°C$ is 36.5 $J\,mol^{-1}\,K^{-1}$. If the heat of fusion of ice at 0 °C is 6008 $J\,mol^{-1}$, what is it at $-10°C$? (see Equation (3.40)).

A3.3 During a process 1 → 2, 4000 kJ of heat is added to a closed system and the system's internal energy increases by 3000 kJ. During the return process 2 → 1 that restores the system to state 1, 1000 kJ of work is done on the system. Determine the heat transferred during the process 2 → 1.

A3.4 Show that Equation (3.5) makes sense both for expansions ($V_2 > V_1$) and for compressions ($V_2 < V_1$).

A3.5 Show that Equation (3.6) also makes sense for irreversible expansions and irreversible compressions.

A3.6 In many experimental petrology labs, argon is used to pressurize reaction vessels. The argon comes in cylinders 1.55 m long, 11 cm in diameter (internal dimensions). The argon pressure in the cylinders is 2200 psi (pounds per square inch) at 25 °C.

(a) What is the mass of the gas in the cylinder, assuming it is an ideal gas?
(b) What is the ideal-gas molar volume?
(c) If the argon is released into a reaction vessel until it reaches a pressure of 1000 psi, what is the maximum amount of work it can do during that process if it acts as an ideal gas? Under what circumstances would this maximum amount of work be done?
(d) What is the minimum amount of work that could be done? Under what circumstances?
(e) How high could the maximum amount of work lift a 100 kg weight, assuming an ideal gas?

A3.7 The differential expression $(x\,dy - y\,dx)$ is not the differential of any function of x and y, and is inexact. Show that it becomes the differential of a function, an exact differential, on dividing both terms by xy. What is that function?

4 The Second Law of Thermodynamics

4.1 Introduction

The first sentence of Gibbs' (1961a) classic memoir "On the Equilibrium of Heterogeneous Substances" is

The comprehension of the laws which govern any material system is greatly facilitated by considering the energy and entropy of the system in the various states of which it is capable.

Given the fact that virtually all of equilibrium thermodynamics can be derived by doing exactly that, as Gibbs did, this must rank as one of the world's great understatements. In this chapter, we begin to explore what Gibbs was referring to. By considering "the laws which govern any material system," we should be able to find out the answers to the questions we posed in Chapter 1.

4.1.1 The Entropy Paradox

In this chapter we introduce a thermodynamic parameter called entropy. Entropy is often described in the following way:

The entropy of an isolated system continuously increases during a spontaneous, irreversible process until it reaches a maximum value at equilibrium.

However, if at the same time we consider the fact that in classical thermodynamics entropy is defined only in equilibrium states we have a paradox, stated for example by Tisza (1966, pp. 120–121):

The nature of this problem is illustrated by the following paradox: How are we to give a precise meaning to the statement that entropy tends toward a maximum, whereas entropy is defined only for systems in equilibrium? Thus in an isolated system, the entropy is constant, if it is defined at all.

The discussion of constraints and metastable equilibrium states in this chapter will provide an answer to this paradox.

4.2 The Problem Restated

Having taken a couple of chapters to get our terminology settled and to get used to discussing energy changes in systems, we must now get back to our main problem – what determines whether chemical processes will go or not go? Our method of determining this might be considered a bit simple-minded – we will simply determine the "chemical energy" differences between equilibrium states. Processes can take place spontaneously if they are in the direction of lowering the chemical energy. They cannot take place spontaneously in the opposite direction.

We have seen that the first great principle of energy transfers is that energy never disappears; it simply takes on different forms. It is the second principle or law that more directly addresses our main problem. It is observed that, once the conditions of the beginning and ending states have been determined, processes can proceed spontaneously in only one direction between these states and are never observed to proceed in the other direction unless they are "pushed" with an external energy source. Thus, for beginning and ending conditions of $P = 1$ bar and $T = 5\,^{\circ}\text{C}$, ice will melt, but water will never spontaneously change to ice. We are looking for a "chemical energy" term that will always decrease in such spontaneous reactions and will enable us to systematize and predict which way reactions will proceed under given conditions. This may seem like a simple problem, but it is not.

The greatest single step forward in the development of thermodynamics was the recognition and definition by Clausius of a parameter, entropy, that enables such predictions and systematizations to be made. And yet, entropy still is not the energy term we have been looking for; the energy that always decreases in spontaneous reactions. In fact, it is not even an energy term. Nevertheless, it is the secret to an understanding of spontaneous reactions.

4.2.1   What's Ahead

In this chapter, we will explain why this is so. As with the first law, there is no way of proving the second law. It is a principle that is distilled from our experience of how things happen. It can be stated in many different ways, usually having something to do with the impossibility of perpetual motion or the availability of energy or the entropy of the universe, topics that seem to have little to do with the problem we have set for ourselves – that of finding an energy term that always decreases in spontaneous reactions. We will choose to state it in a way that emphasizes its role as a directionality parameter. This leads to the shortest possible path to the practical applications we wish to consider.

So here's the plan. We will define entropy as a parameter in our model systems, having certain properties. This definition is based fundamentally on the work of Carnot (1960) and especially Clausius (1879), but stated in a way that avoids referring to the entropy of the universe, and which is consistent with all the other ways of explaining entropy. It should be accepted at first on faith, as simply a useful parameter, because it will not have any intuitive meaning as do other terms such as energy, work, and heat. Then we will show how the "chemical energy" term we have been looking for is related to entropy. Finally, we will discuss what entropy is (and what it is not), and, in Chapter 5, we will discuss how to measure it.

4.3 Thermodynamic Potentials

At the start we should note that "directionality parameters" have a technical name. They are called *thermodynamic potentials*, as introduced in Section 1.4. A potential in this sense is a quantity (a state variable) which is minimized (or maximized) at equilibrium, *subject to certain constraints*. This means that if you want to compare two states of the same system to see which is the more stable, i.e., in which direction the spontaneous change will "go", two state variables must be the same in both states, and we call these two variables the constraints on the system. But not just any two. Which two depends on which state variable you have chosen to be maximized or minimized, i.e., on your choice of thermodynamic potential. Why two, and not one or three? Fundamentally it is because we need two state variables to define an equilibrium state of a closed system, and that's because we chose to limit the ways our model systems can exchange energy to two – heat and only one kind of work. We need one constraint variable for each.

For example, an appropriate thermodynamic potential would have a lower value for calcite than for aragonite at 25 °C, 1 bar (note that in saying 25 °C, 1 bar we have chosen two constraints; one related to heat and one to work). Generally, however, we have problems similar in principle but more complex. We may have several phases, including a solution with many compositional variables, and we will want to know the equilibrium compositions of all the phases at T and P. In these cases there is a range of values for the thermodynamic potential, one for every possible composition of the phases involved, and we need to find the *minimum* value of the potential. For any other value, greater than this minimum, some change in compositions (phases will dissolve, precipitate, etc.) will take place until the minimum value is achieved. We then speak of *minimizing* the potential at a given T and P, i.e., subject to the given constraints. This is not a hypothetical problem, but a real problem in applied mathematics. We will see one way of doing this (*speciation*) in Chapter 10.

Box 4.1 Potentials in Science

In science a potential is a measure of the available energy, where energy is defined as the capacity to do work. Potentials exist because of gravitational, elastic, electromagnetic, and other forces. The term "thermodynamic potential" is so named because it is analogous to potential quantities in other branches of science, and several state variables having appropriate constraints can be interpreted as potentials for work or heat, although only two (the Gibbs and Helmholtz energies) are commonly shown to be such potentials. Badger (1967) gives a complete discussion of this subject.

In mechanics a body has a potential for doing work because a force is acting on it that is capable of causing the body to move. For a given mass and force, the potential energy is a function only of the position of the body. Force is a vector quantity, but, if we restrict ourselves to one dimension, the functional relationship among the potential energy of a body $E_{potential}$, its position r, and the force $\mathcal{F}$ is

$$dE_{\text{potential}}/dr = -\mathcal{F} \tag{4.1}$$

Since the potential $E_{\text{potential}}$ is a result of the force $\mathcal{F}$ and will decrease if the body is allowed to move, the terms are given opposite signs. The functional form in Equation (4.1) is common to all potential quantities. Any change in the potential will appear as work, either done on the body to increase the potential, or by the body in lowering its potential. Thus

$$w = \int_{r_1}^{r_2} dE_{\text{potential}}$$

$$= E_{\text{potential}_{r_2}} - E_{\text{potential}_{r_1}}$$

where r_1 and r_2 are two positions of the body, and work is considered negative when done by the body.

We will see that in thermodynamics there are quite a few equations in the form of (4.1), such as

$$(\partial U/\partial S))_V = T \tag{4.20}$$

and

$$(\partial U/\partial V))_S = -P \tag{4.21}$$

and, in all these cases, the numerator (in this case U) is a thermodynamic potential, the denominator (in this case S or V) is a configuration term analogous to distance, and the right-hand side (in this case T or $-P$) is analogous to a force.

In this chapter we will identify four thermodynamic potentials, those associated with internal energy, entropy, Gibbs energy, and Helmholtz energy. The number could in fact be extended – several other state variables with certain unusual constraints are also thermodynamic potentials. However, they are rarely used in this sense, so we won't bother with them. As a matter of fact, only one thermodynamic potential is ever used in geochemistry, but understanding is increased by learning about the others.

Constraint Variables and Natural Variables

As just mentioned, each thermodynamic potential to be defined is a state variable which will give the direction of change, provided that two other state variables are defined as having fixed values. These two other variables, the variables which are included in the definition of a thermodynamic potential, we call constraint variables, but in most texts they are called natural variables. For example, in the following section we define a state variable called entropy, S. Entropy itself is not a thermodynamic potential, though we commonly refer to it as such for convenience, nor is $S_{T,P}$ or $S_{X,Y}$, where X and Y are any two randomly chosen state variables. Only entropy constrained to constant energy U and volume V, or $S_{U,V}$, is a thermodynamic potential. U and V are the natural variables of entropy. The term constraint variable, which in this text means the same thing, becomes more meaningful when we consider, later in this chapter, the fact that quite often there are more than two constraints on the system.

4.4 Entropy

4.4.1 Analogy

The first and most important thermodynamic potential we need is entropy, constrained to a constant internal energy and volume. One way to define entropy would be to simply say that the $\mathbf{Z}$-term in Equation (3.16) does indeed exist, where entropy is called $\mathbf{S}$, and $\mathbf{S} = -\mathbf{Z}$. This provides a useful analogy between pressure–volume and temperature–entropy, and we will see these terms linked together in many equations. They represent work and heat energy in many processes we will be considering.

This way of defining entropy is also useful in explaining a somewhat puzzling feature of thermodynamics. In the next section, we will see that, although entropy is a state variable of the kind we are looking for (one that can be used to tell which way reactions will go), it is unfortunately one that *increases* in spontaneous reactions, rather than decreasing, as we had supposed. This turns out to be simply because entropy was historically defined as $-\mathbf{Z}$ in Equation (3.16), rather than as $\mathbf{Z}$. In other words, Equation (3.18)

$$\frac{\delta q_{\text{rev}}}{T} = -d\mathbf{Z} \tag{3.18}$$

is actually written

$$\frac{\delta q_{\text{rev}}}{T} = d\mathbf{S} \tag{4.2}$$

that is, without the minus sign. This is a sort of "historical accident." Integrating this expression then gives (as suggested by Equation (3.19))

$$q_{\text{rev}} = T \, \Delta\mathbf{S} \tag{4.3}$$

which is analogous to our work equation from Chapter 3

$$w_{\text{rev}} = -P \, \Delta\mathbf{V} \tag{3.14}$$

and similarly, including irreversible processes

$$\frac{\delta q}{T} \leq d\mathbf{S} \tag{4.4}$$

$$q \leq \int_1^2 T \, d\mathbf{S} \tag{4.5}$$

Equation (4.2) is the relationship discovered by Clausius by analyzing Carnot cycles. Atkins (2010) says

The key equation of classical thermodynamics is the Clausius expression $d\mathbf{S} = \delta q_{\text{rev}}/T$.

He adds that students commonly have a problem seeing that it is intuitively plausible. In this book more attention is given to how entropy is measured and used, rather than to how one can derive Equation (4.2) by considering Carnot cycles and heat engines. Nevertheless, learning about the historical development of our understanding of entropy will certainly give one a deeper appreciation of its significance (see, e.g., Carnot, 1960; Purrington, 1997).

4.4.2   Definition

For our purposes a better way to define entropy is as follows. If there is indeed "something missing," that is, only one thing missing from the energy-decreasing analogy, it is something that causes reactions to "go," *even when no energy change whatsoever occurs*. Now, we have defined a type of system (the *isolated* system, Section 2.2.2) that does not permit energy to enter or leave the system. Therefore, all we have to do is define a parameter with which we can predict reaction directions in this kind of system, then combine it with the energy-decreasing idea, and we should have our answer. This is what we do with the following definition, paraphrased after Callen (1985, pp. 27–28). It can also serve as a statement of the second law of thermodynamics. The statement reflects Callen's "postulational" approach to thermodynamics; that is, let's postulate that there is a parameter that includes everything we need, and see how it works out:

There exists an extensive property of systems, entropy (S), which for isolated systems achieves a maximum when the system is at stable equilibrium. Entropy is a smoothly varying function of the other state variables and is an increasing function of the internal energy U.

As mentioned (in Section 4.3) two state variables are required to define the equilibrium state. Using an isolated system means that these two are U and V. This ensures that we will be comparing states having the same energy and volume, and that no energy change can occur in the system. They are also our two constraint or natural variables, because by our definition entropy with these constraints will give the direction of change. We insert the postulate that entropy increases with U to ensure that the other directionality parameters to be derived decrease (have minima) rather than increase.[1] This can be shown by considering the isolated (model) system in Figure 4.1.

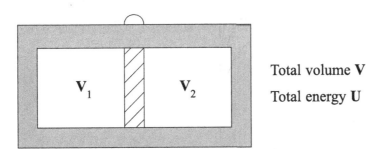

Figure 4.1 An isolated system having a movable partition. The partition is impermeable to matter but conducts heat. The volume V of the system is the sum of the volumes of the two subsystems, that is, $V = V_1 + V_2$.

---

[1] We seem to have a contradiction here. We say no energy change in the system can occur, then we say entropy is an increasing function of the energy. This just means we can have many systems like Figure 4.1 with various energy contents, as shown in Figure 4.2.

4.4.3   The U–S–$V_1/(V_1 + V_2)$ Surface

The exterior wall in Figure 4.1 is impermeable to energy and rigid, so the system has constant U and V. The piston is movable and can be locked in any position. It is impermeable, but it conducts heat so that the two sides are at the same temperature. If there are equal amounts of the same gas in the two compartments, the equilibrium position of the piston when it is free to move is where $V_1 = V_2$. Also, according to our definition of S, the equilibrium position of the piston is one of maximum entropy for the system, and any other position has lower entropy.

In the equilibrium position, the piston is not locked in place, i.e., there is no constraint other than U and V, no third constraint. In any other position the piston must be locked in place, because the pressure on one side is greater than the pressure on the other side.[2] Nevertheless, such locked positions are unchanging, and are equilibrium states. We want to distinguish between these states having an extra constraint and those equilibrium states that have only two constraints, so we call the three-constraint states *metastable equilibrium states*.

Then, if we consider the same situation but with successively greater energy contents U', U'', and U''' (which would be the case with higher gas temperatures), we will have entropy–volume curves as in Figure 4.2, where S is plotted against $V_1/(V_1 + V_2)$, which varies between 0 and 1. The maximum value of S is at $V_1/(V_1 + V_2) = 0.5$, where $V_1 = V_2$, and the curves for U', U'', and U''' are arranged with increasing entropies because we defined S to be an increasing function of U. In Figure 4.3, the curves of Figure 4.2 are drawn in three dimensions, and in Figure 4.4 the complete surface is shaded with a number

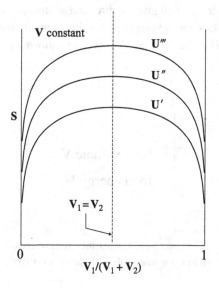

Figure 4.2 Entropy (S) versus volume fraction $V_1/(V_1 + V_2)$ for the system in Figure 4.1 at three different energy levels, where $U''' > U'' > U'$. Volume V is constant.

---

[2] If you find yourself wondering how one could lock and unlock the piston if the system is truly isolated, you have not yet fully grasped the fact that thermodynamics deals with mathematical models, not real things. In the model, the position of the piston is a mathematical variable.

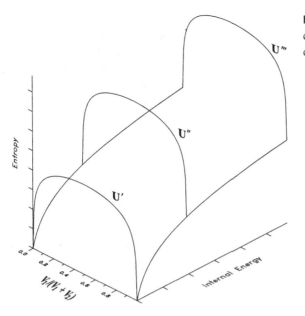

Figure 4.3 The U–S–$V/(V_1 + V_2)$ curves of Figure 4.2 in three dimensions.

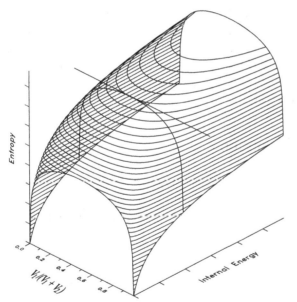

Figure 4.4 The U–S–$V/(V_1 + V_2)$ surface, calculated for an ideal gas. All points on the surface other than those at maximum S, minimum U represent metastable equilibrium states of the system.

of contours – the horizontal ones being contours of constant S and V, the vertical ones contours of constant U and V (recall that the whole diagram is for conditions of constant V).[3] In Figure 4.5 two of these contours are abstracted to show more clearly that the two contours, which meet at a point, have a common tangent.

---

[3] Don't confuse this surface, U–S–$V/(V_1 + V_2)$, which includes both stable and metastable equilibrium states, with the U–S–V (or U–S–V) surface (Section 4.6), which has only stable equilibrium states. This is important.

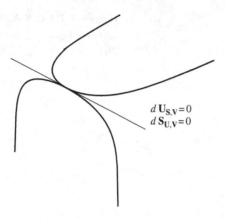

Figure 4.5 Constant **S**, **V**, and constant **U**, **V** sections from the **U–S–V/(V$_1$ + V$_2$)** surface shown in Figure 4.4, with their common tangent, which is simultaneously d**S**$_{U,V}$ and d**U**$_{S,V}$. The tangent point represents a position of stable equilibrium for the system.

d **U**$_{S,V} = 0$
d **S**$_{U,V} = 0$

This tangent is located at the extremum in both curves, and so is in mathematical terms both d**U**$_{S,V} = 0$ and d**S**$_{U,V} = 0$. In other words, the condition that at equilibrium **S** is a maximum for given **U**, **V** *implies* the condition that **U** is a minimum for given **S**, **V** at a given equilibrium point.

Not only is **S** a parameter that always *increases* when a metastable state changes to a stable state at given values of **U** and **V**, but also **U** is a parameter (a *state variable*) that always *decreases* when a metastable state changes to a stable state (i.e., when a third constraint is released) at given values of **S** and **V**. Thus, as long as we consider only systems at constant **U** and **V** or constant **S** and **V**, both **S**$_{U,V}$ and **U**$_{S,V}$ are parameters of the type we have been looking for – they are thermodynamic potentials. Systems of this kind are very rare; still, we're making progress.

4.5   The Fundamental Equation

The first and second laws can be combined into a single equation, which lies fairly close to the very heart of thermodynamics, called the fundamental equation.

On combining Equation (3.26) ($\Delta U = q + w$) with Equations (3.14) ($w_{rev} = -P \Delta V$) and (4.3) ($q_{rev} = T \Delta S$) we obtain

$$\Delta \mathbf{U} = T \Delta \mathbf{S} - P \Delta \mathbf{V} \tag{4.6}$$

and

$$\Delta U = T \Delta S - P \Delta V \tag{4.7}$$

These can also be written in differential notation as

$$\boxed{d\mathbf{U} = T\, d\mathbf{S} - P\, d\mathbf{V}} \tag{4.8}$$

and

$$\boxed{dU = T\, dS - P\, dV} \tag{4.9}$$

It is worth mentioning that the "fundamental" nature of this equation does not mean that we often use it directly. We usually use it after a little manipulation, so that we can use constraints or integration limits T and P, rather than S and V. In other words, we usually use equations involving the Gibbs energy $\mathbf{G}$ (Section 4.11.1), rather than the internal energy $\mathbf{U}$. Equation (4.8) is fundamental because it is directly linked to the first and second laws and because many other equations are derived from it.

From Equation (4.8) we can write

$$d\mathbf{U}_{S,V} = 0 \quad \text{at equilibrium} \tag{4.10}$$

which is the equation for an extremum in a continuous function (remember Figure 1.4), and we know from Figures 4.4 and 4.5 that it is a minimum rather than a maximum. Because $\mathbf{U}$ is a thermodynamic potential that will always decrease in a spontaneous (irreversible) process, we can also write

$$d\mathbf{U}_{S,V} < 0 \quad \text{for an irreversible process} \tag{4.11}$$

so that in general

$$d\mathbf{U}_{S,V} \le 0 \tag{4.12}$$

Equations (4.10)–(4.12) actually imply the existence of a third constraint. They say there is a function $\mathbf{U}$ with independent variables $\mathbf{S}$ and $\mathbf{V}$ and a third independent variable, because, if $\mathbf{S}$ and $\mathbf{V}$ are constant, the system cannot change its energy by heat or P–V work, and cannot exhibit an extremum with respect to either variable. $\mathbf{U}$ can change only because of increments in a third constraint variable, and that change is always negative for irreversible processes (Equation (4.11)). In Figure 4.1 the locked piston is the third constraint. We must derive equations which include a constraint variable more general than $\mathbf{V}_1/(\mathbf{V}_1 + \mathbf{V}_2)$ and for other more useful thermodynamic potentials (Section 4.9).

Starting from Equation (4.8) written instead as

$$d\mathbf{S} = (1/T)d\mathbf{U} + (P/T)d\mathbf{V} \tag{4.13}$$

similar reasoning results in

$$d\mathbf{S}_{U,V} \ge 0 \tag{4.14}$$

4.6   The U–S–V Surface

4.6.1   Geometrical Meaning of the Fundamental Equation

The easiest way to get a clear geometric picture of the fundamental equation is to realize that, by virtue of the first law, every system at equilibrium, whether as simple as an ideal gas or as complex as a bacterium, has a single fixed energy content for given values of two independent variables, and by virtue of the second law we are able to use $\mathbf{S}$ (or S) as one of these variables. Every system can therefore be represented by a surface in $\mathbf{U}$–$\mathbf{S}$–$\mathbf{V}$ space, such as that shown in Figure 4.6. Note that, in order to make the surface easier to draw, $\mathbf{U}$ increases downward in Figure 4.6. At every point on the surface, such as point

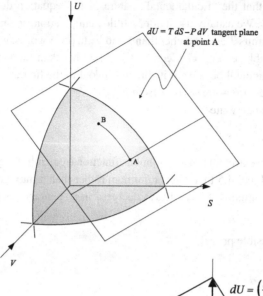

Figure 4.6 Every system has a unique **U–S–V** surface. The fundamental equation (4.9) represents a tangent to this surface when $d\mathbf{S}$ and $d\mathbf{V}$ are of arbitrary magnitude, and it can be integrated to give the change in $\mathbf{U}$, $\Delta\mathbf{U}$, between any two points on the surface, such as A and B. The tangent plane is illustrated further in Figure 4.7.

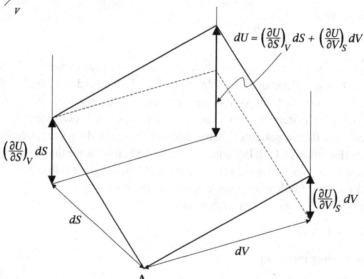

Figure 4.7 The tangent surface at point A in Figure 4.6, showing how $d\mathbf{U}$ is geometrically related to $d\mathbf{S}$ and $d\mathbf{V}$.

A, there will be a tangent surface, the equation for which is the fundamental equation, $d\mathbf{U} = T\,d\mathbf{S} - P\,d\mathbf{V}$. Figure 4.7 shows how, starting at point A, increments of $d\mathbf{S}$ and $d\mathbf{V}$ are combined with the slopes $\partial\mathbf{U}/\partial\mathbf{S}$ and $\partial\mathbf{U}/\partial\mathbf{V}$ to produce the total change in $\mathbf{U}$ at any other point on the tangent plane. In this case, each of $d\mathbf{U}$, $d\mathbf{S}$, and $d\mathbf{V}$ has any magnitude, however large. Because of this, you may visualize Equation (4.9) as having the form $Z = aX + bY$, which is a combination of the straight-line equations $Z = aX$ and $Z = bY$, and is the equation of a plane in X–Y–Z space.[4]

---

[4] Z here has no connection with the Z in Equation (3.16).

On the other hand, the fundamental equation can also be used to calculate values of $\Delta \mathbf{U}$ between any two points on the $\mathbf{U}$–$\mathbf{S}$–$\mathbf{V}$ surface itself, such as between points A and B. Because $\mathbf{U}$ follows a continuous function of $\mathbf{S}$ and $\mathbf{V}$ between A and B, the fundamental equation must be *integrated* between A and B, and this means allowing $d\mathbf{S}$ and $d\mathbf{V}$ to take on infinitesimal values and performing a summation, symbolized by the $\int$ symbol. This is written

$$\Delta \mathbf{U} = \mathbf{U}_B - \mathbf{U}_A$$
$$= \int_A^B T\, d\mathbf{S} - \int_A^B P\, d\mathbf{V}$$

The *calculation* of this difference follows the *reversible* path shown on the $\mathbf{U}$–$\mathbf{S}$–$\mathbf{V}$ surface. We can integrate this path only if it represents a continuous function, and it lies on an equilibrium surface, so that it is a continuous succession of equilibrium states. However, the calculated ΔU is the same no matter how the change from A to B is actually carried out.

The difference in $\mathbf{U}$, or in fact any thermodynamic property, between a quartz crystal at $25\,°C$ and the same crystal at $50\,°C$ has nothing to do with how that difference is achieved, but the calculation of that difference follows a reversible path. That just means we calculate the difference by integrating a continuous function. If you have difficulty understanding differential equations as used here, study Figures 4.6 and 4.7.

4.7   Applicability of the Fundamental Equation

For such a simple relationship, Equation (4.8) traditionally generates quite a bit of confusion. This is of two types, or perhaps two aspects of the same problem. That problem is reversibility versus irreversibility.

Equations (4.6) and (4.8),

$$\Delta \mathbf{U} = T\, \Delta \mathbf{S} \quad P\, \Delta \mathbf{V} \tag{4.6}$$
$$d\mathbf{U} = T\, d\mathbf{S} - P\, d\mathbf{V} \tag{4.8}$$

appear to be limited to reversible processes, because they are derived from Equation (3.3), in which both (3.14) and (4.3),

$$w_{\mathrm{rev}} = -P\, \Delta \mathbf{V} \tag{3.14}$$
$$q_{\mathrm{rev}} = T\, \Delta \mathbf{S} \tag{4.3}$$

refer to reversible processes. But this is not the case, which can be explained in various ways.

1. For one thing, Equations (4.6) and (4.8) contain only state variables ($\mathbf{U}$, $\mathbf{S}$, and $\mathbf{V}$, in addition to T and P). Therefore, because the changes in state variables do not depend on the nature of the change, Equation (4.6) (or the integration of Equation (4.8)) is true

for any change between two equilibrium states which have the same composition, with one important exception (see Section 4.7.1).

2. Although in an irreversible process the inequalities in Equations (3.6) and (4.5) apply, the magnitude of the inequalities will cancel out when they are added together. We could give examples of this, but it must be true because of item 1.

3. In Figure 4.6, points A and B represent two stable equilibrium states of a system. Integration of the fundamental equation from A → B takes place on the U–S–V surface, as shown, which is necessarily a reversible process. An irreversible process A → B would cause the system to leave the surface at A, be not representable in U–S–V space between A and B, and then reappear at B. In either case, ΔU, ΔS, and ΔV are related by Equation (4.6).

4.7.1   Two Kinds of Irreversibility

So Equations (4.8) or (4.9) apply to reversible and irreversible processes. There remains the "important exception" just mentioned. The problem here is that there are two very different kinds of irreversible processes, though both conform to the Carnot/Clausius criterion of irreversibility, i.e., an entropy increase in system plus environment (page 24). Both kinds of irreversibility involve a system changing from one equilibrium state to another equilibrium state.

1. In one case both of the two equilibrium states can be stable or both can be metastable, as long as there is no change from one to the other, usually metastable → stable. For example, Equation (4.8) can be applied to increments of change such as the irreversible gas expansion in Figure 3.4, or perhaps the irreversible heating of a crystal from 25 to 50 °C by simply putting it in an oven at 50 °C. The crystal could be (stable) calcite or (metastable) aragonite. Integrating Equation (4.8) applies to all these processes. They can be thought of as processes in response to a change in the first and/or second constraint. No chemical reaction is involved.

2. But Equation (4.8) does *not* apply to irreversible processes in systems in which a chemical reaction occurs, i.e., a metastable state changing to a stable state, for example a crystal of aragonite at some T and P within the calcite stability field recrystallizing to calcite, or indeed to any chemical reaction. We need to develop equations to deal with this kind of process, which evidently involves an energy change *within* the system.

The most general way of expressing the applicability of the fundamental equation is that it applies to any process between equilibrium states which does not involve a third constraint. Such a change is usually the release or removal of the constraint, which allows a reaction to proceed. The constraint can be removed all at once (like putting sugar in coffee, Figure 2.8) or incrementally (as discussed in Chapter 13).

4.8   Constraints and Metastable States

We introduced the idea of constraints and metastable states in Section 4.4.3. Distinguishing between real systems and our model systems becomes even more important here.

4.8.1   Constraints

Real systems, such as aragonite, are said to be constrained from reacting to form a more stable state (calcite) by an activation-energy barrier. The usage is rather imprecise, because we don't actually know whether the aragonite is changing on some very long time scale or not. In model systems, the meaning is much more precise.

Box 4.2   Constraints in Mechanics and Economics

Constraints, along with potentials, work, and energy, constitute another topic common to thermodynamics and mechanics. The motion of bodies in mechanics is subject not only to the force(s) applied, but also to whatever constraints are present. A marble rolling on the surface of a bowl is constrained to remain on the surface. Beads on a rod are constrained to move in a straight line, and so on.

Constraints in mechanics can be classified into various types, for example as to whether the equation of constraint contains time as a variable or not. In thermodynamics, which has only scalar variables, and which has no time variable, constraints are simpler, and are identified with ways in which systems can change their energy content.

Constrained extremal problems are common in economics, where profits are maximized subject to constraints such as the cost of production. If x and y represent the quantities of two products your company makes, the profit P realized is some function $P = f(x, y)$. Quantities produced are limited by the cost of production c, so a constraint is another function $g(x, y) = c$, and the problem is to maximize $f(x, y)$ among those x and y which satisfy $g(x, y)$ (Marsden and Troma, 1988, p. 265).

A constraint in mathematics (and in thermodynamics) is a condition that must be observed. For example, you might want to find the largest volume a rectangular box can have subject to the constraint that the surface area is fixed at $10\,m^2$ (Marsden and Troma, 1988, p. 270). In thermodynamics we minimize a function of several variables while constraining two state variables to constant (known) values. We constrain two, because in the first law we define only two ways of changing the energy of a system, heat and P–V work. If there is a third way of changing energy, a third state variable must be specified – a third constraint. The system must be in some equilibrium state for that variable, or any variable, to be defined. The only thermodynamic equilibrium state that is not a stable equilibrium state is a metastable equilibrium state. A thermodynamic constraint, then, is defined as *a state variable associated with some method of changing the energy of a system* (an exception is of course any third constraint in an isolated system). This state variable may be one of the two used to define a stable equilibrium state or it may be a third constraint variable which results in a metastable equilibrium state. Changing one of the first two constraint variables results in a change between two equilibrium states, not accompanied by a chemical reaction, as when a gas expands due to a change in P or T, or when aragonite is heated by putting it in an oven. Change in a third constraint variable results in a transition from a metastable to a stable state or vice versa, such as between aragonite and calcite or, as discussed in Section 2.6.1, any spontaneous irreversible reaction from reactants to products.

4.8.2   Real Metastable States

The distinction between stable and metastable equilibrium according to most sources is that the stable equilibrium state is "truly unchanging," or unchanging given indefinite time, whereas the metastable state may be changing, but perhaps too slowly for the change to be observed. This distinction clearly refers to real systems, and is often very difficult to make. We know that at 25 °C, 1 bar, calcite is the most stable form of $CaCO_3$. Aragonite is another form, and, although it never changes to calcite on museum shelves, it does change in sedimentary basins under some conditions (probably those conditions that allow dissolution/precipitation) given very long time periods, so is it metastable or unstable? Is volcanic glass an unstable or a metastable phase? These questions can start arguments among geochemists. There are many reactions for which kinetic rate constants are known at high temperatures (the reacting assemblage is therefore unstable), but not at 25 °C, where the assemblage is considered metastable. At what temperature does metastable change to unstable? If anyone were interested, the answer would of course be completely arbitrary. The kinetics of very slowly changing systems is a problem for real systems, but it is not a problem in thermodynamics.

It was mentioned earlier (page 18) that in most thermodynamic discussions metastable states are not considered very important. In this text they are called metastable equilibrium states, emphasizing the term equilibrium, and are a central feature of thermodynamic theory.

4.8.3   Thermodynamic Metastable States

As mentioned in Section 4.8.1, a metastable equilibrium state has at least three constraints. The first two of these constraints are independent state variables such as $\mathbf{S}$ and $\mathbf{V}$ associated with two ways of changing the energy of a system, such as $T\,d\mathbf{S}$ and $P\,d\mathbf{V}$. A metastable equilibrium state has an additional or third independent state variable associated with a third way of changing the energy of the system. We have seen several examples of metastable equilibrium states so far in Chapters 1 and 2, but only as ball-in-valley analogies or qualitative references to metastable minerals. In Figures 4.1 and 4.2 we have a quantitative example where the thermodynamic potential is $\mathbf{S_{U,V}}$ and the third constraint variable is the variable position of the piston. This example follows from our definition of entropy, and shows entropy acting as a thermodynamic potential.

Real systems may be truly unchanging and metastable, or they may be unstable and changing very, very slowly. Often we don't know which. But our models of systems have no such uncertainty. Metastable systems are modeled as in complete thermodynamic equilibrium, with (at least) three constraints. If the real system is really unstable on some very long time scale, our model is in this respect incorrect, but useful nonetheless. We should also note that, because equilibrium states have single-valued properties, states not

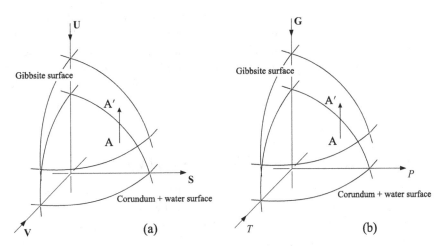

Figure 4.8 (a) Two equilibrium states of the system Al_2O_3–H_2O in **U**–**S**–**V** space: metastable corundum plus water, and stable gibbsite. The process A → A′ is reaction (2.5). Changes in **U** along either surface can be described by Equation (4.8). The change in **U** between the surfaces (A → A′) is an irreversible process. (b) The same system in **G**–T–P space. Note that **U** and **G** increase downwards.

in stable or metastable equilibrium cannot be represented on a diagram. Gibbs (1961b, p. 39) says, referring to the **U**–**S**–**V** surface,

> When the body is not in a state of thermodynamic equilibrium, its state is not one of those which can be represented by our surface.

This also applies to lines, which are, or can be, part of surfaces such as those in Figure 4.8.

The isolated system is useful for developing the relationship between **S** and **U**, but such systems are not very useful for modeling real processes. The task now is to develop equations for other thermodynamic potentials more useful than $U_{S,V}$. We need to deal with the usual type of metastable → stable process, such as any chemical reaction where the first two constraint variables are T and P, which invariably involves a decrease in the "chemical energy" of the system.[5]

4.9 The Energy Inequality Expression

Stable equilibrium states are the target destination for all metastable states, once their constraints have been released. What equations refer to this process? We try here to make these important relationships fairly intuitive.

Recall that on page 44 we noted that we had made no provision for energy generated *within* a system; the $T\,d\mathbf{S}$ term in Equation (4.8) accounts only for energy crossing the

[5] Some would prefer to say "... which invariably involves an increase in the entropy of the universe."

system boundary due to changes in the temperature difference between the system and its surroundings, independently of any chemical reaction, and the $P\,d\mathbf{V}$ term accounts only for work done due to a change in system volume. In a chemical reaction there is an energy change which is not accounted for in Equation (4.8). There must be a third term on the right-hand side of this Equation, or any of the fundamental equations, for this new source of energy. When we define this third energy source we will see what the third constraint variable is.

4.9.1 The Third Energy Source and Third Constraint Variable

The energy *produced* within the system due to chemical reactions might leave the system as heat or as work. Or energy could be *consumed* within a system (an endothermic reaction, say), causing heat or work energy to enter the system rather than leaving it. We won't bother to keep repeating both possibilities, but just discuss the common case, energy production within the system.

The Third Energy Source as Heat

Prigogine and Defay (1954, Chapter 3) use the term "uncompensated heat" called q' (using our notation), which they attribute to De Donder (1920). This is a positive quantity of energy generated within a closed system due to an irreversible process such as a chemical reaction. If work other than P–V work is ruled out, this energy can leave the system as heat (calorimetry measures such quantities of heat), and changes the inequality in Equation (4.4) into an equality. Thus

$$T\,dS > \delta q \tag{4.4}$$

$$= \delta q + \delta q' \tag{4.15}$$

or

$$\delta q = T\,dS - \delta q' \tag{4.16}$$

where, you will recall, δq is the heat term in the first-law equation (3.15), $d\mathbf{U} = \delta q + \delta w$. So with $\delta q = T\,dS - q'$ and $\delta w = -P\,dV$, we can write

$$d\mathbf{U} = (T\,dS - \delta q') - P\,d\mathbf{V}$$

or

$$d\mathbf{U} = T\,dS - P\,d\mathbf{V} - \delta q' \tag{4.17}$$

which is a version of the fundamental equation (4.8) valid for systems which include chemical reactions, or metastable $\rightarrow$ stable transitions. Prigogine and Defay (1954) explicitly exclude all non-P–V work (their Equations (2.2) and (2.3)), so from Equation (4.17) we can write

$$d\mathbf{U}_{S,V} = -\delta q' \tag{4.18}$$

which is their Equation (3.11). This shows again that an irreversible process in a system having constant values of $\mathbf{S}$ and $\mathbf{V}$ is accompanied by a decrease in the internal energy $\mathbf{U}$.

Logically one could ask, as both δq and δw have inequality expressions, why choose δq (in Equation (4.15)) rather than δw for the extra term? Why not have uncompensated work, and write

$$P\,d\mathbf{V} = -\delta w - \delta w'$$

or

$$\delta w = -P\,d\mathbf{V} - \delta w'$$

where $\delta w'$ is a positive quantity. This would result in

$$d\mathbf{U} = T\,dS - P\,d\mathbf{V} - \delta w'$$

instead of (4.17), but with the same effect. Prigogine and Defay (1954, Section 3.8) discuss this possibility, which was apparently proposed by Schottky *et al.* (1929) and is essentially the approach taken in Anderson (2005).

A More General Third Energy Source

So we have equations incorporating a source of energy within a system for cases where the energy appears as work or as heat. But there is a more general form of this extra energy which includes both cases. We give a more complete discussion of this subject in Chapter 13.

Equation (4.8) can also be written in the form of a total differential as

$$d\mathbf{U} = \left(\frac{\partial \mathbf{U}}{\partial \mathbf{S}}\right)_{\mathbf{V}} d\mathbf{S} + \left(\frac{\partial \mathbf{U}}{\partial \mathbf{V}}\right)_{\mathbf{S}} d\mathbf{V} \tag{4.19}$$

showing that, on comparing with (4.8),

$$\left(\frac{\partial \mathbf{U}}{\partial \mathbf{S}}\right)_{\mathbf{V}} = T \tag{4.20}$$

and

$$\left(\frac{\partial \mathbf{U}}{\partial \mathbf{V}}\right)_{\mathbf{S}} = -P \tag{4.21}$$

which are the slopes of the $\mathbf{U}$–$\mathbf{S}$–$\mathbf{V}$ surface (Figure 4.6) in the $\mathbf{V}$ and $\mathbf{S}$ directions.

De Donder (1920) suggested that any energy-generating process within a system proceeds from a starting state to a finished state, and defined the *extent-of-reaction variable* ξ which has values from 0 at the start of the process to 1 at the completion of the process. The thermodynamic potential, in this case $\mathbf{U}_{\mathbf{S},\mathbf{V}}$, will vary by $d\mathbf{U}_{\mathbf{S},\mathbf{V}}$ for each increment $d\xi$, so the change in $\mathbf{U}$ because of this process will be the rate of change times the amount of change, or

$$d\mathbf{U}_{\mathbf{S},\mathbf{V}} = \left(\frac{\partial \mathbf{U}}{\partial \xi}\right)_{\mathbf{S},\mathbf{V}} d\xi \tag{4.22}$$

and Equation (4.8) becomes, in the most general case,

$$dU = \left(\frac{\partial U}{\partial S}\right)_{V,\xi} dS + \left(\frac{\partial U}{\partial V}\right)_{S,\xi} dV + \left(\frac{\partial U}{\partial \xi}\right)_{S,V} d\xi \tag{4.23}$$

We then define $(\partial U/\partial \xi)_{S,V}$, and the corresponding term for any other thermodynamic potential, as the affinity $-\mathcal{A}$, so Equation (4.23) becomes

$$dU = T\,dS - P\,dV - \mathcal{A}\,d\xi \tag{4.24}$$

which is the equation for the changes in U for any system having an internal generation of energy. This is always due to the occurrence of an irreversible process in the system, usually a chemical reaction. The $\mathcal{A}\,d\xi$ term appears as the extra or third term in equations for any of the thermodynamic potentials, and ξ is the most general third constraint variable. Finally, on comparing Equations (4.24) and (4.17) we see that

$$\delta q' = \mathcal{A}\,d\xi \tag{4.25}$$

which is Equation (3.21) of Prigogine and Defay (1954).

In summary, in any irreversible process in a system having fixed values of S and V in which the third constraint is released, U decreases. The third term on the right-hand side is negative and if we don't include this negative term Equation (4.8) becomes

$$dU < T\,dS - P\,dV \tag{4.26}$$

Combining (4.26) and (4.8), a general fundamental equation for all states of the system is

$$dU \le T\,dS - P\,dV \tag{4.27}$$

If S and V are constant, $dS = 0$ and $dV = 0$, so it follows from (4.27) that

$$dU_{S,V} \le 0 \tag{4.28}$$

which shows once again that U is minimized at equilibrium in systems having constant values of S and V, and implies the existence of a third constraint variable. We have already seen this in Equation (4.12) and in graphical form in Figure 4.5.

A Common Misconception

A common misconception is that the inequality in Equations (4.26)–(4.28) and similar ones we will see is due to the fact that for an irreversible process $\delta q < T\,dS$. As explained on page 77, although $\delta q < T\,dS$ in any irreversible process, it is compensated for by the fact that $\delta w < -P\,dV$ so that if no chemical reactions are involved $dU = \delta q + \delta w$ remains true. The inequality is caused rather by the internally generated energy, which is uncompensated.

4.10   Entropy and Heat Capacity

So far, all we know about entropy is that it increases in spontaneous reactions in isolated systems, and that it appears in equations such as (4.8) and (4.9). Hidden in the equations we have derived so far is an important relationship between entropy and heat capacity, which we will see in Chapter 5 serves as a basis for the measurement of entropy.

On combining Equations (3.26) and (4.9),

$$dH = dU + P\,dV + V\,dP \tag{3.26}$$

and

$$dU = T\,dS - P\,dV \tag{4.9}$$

we find

$$dH = T\,dS + V\,dP \tag{4.29}$$

or its extensive form

$$d\mathbf{H} = T\,d\mathbf{S} + \mathbf{V}\,dP \tag{4.30}$$

which, incidentally, defines enthalpy as another thermodynamic potential which is minimized for processes at constant $(\mathbf{S}, P)$. It is rarely used as such, but Stolper and Asimow (2007) provide one example. If we choose constant-pressure conditions, dP becomes zero, so

$$dS = \frac{dH}{T} \tag{4.31}$$

and, on substituting $C_P\,dT$ for dH (from Equation (3.37)), we have

$$C_P\,dT = T\,dS \tag{4.32}$$

Here we have the entropy defined in terms of something measurable, the heat capacity. Upon integrating (4.32), we have

$$S_{T_2} - S_{T_1} = \int_{T_1}^{T_2} \frac{C_P}{T}\,dT \tag{4.33}$$

or

$$\Delta S = C_P \ln\left(\frac{T_2}{T_1}\right) \tag{4.34}$$

which allows calculation of ΔS between equilibrium states at two different temperatures.

4.10.1 Illustrating Entropy Changes

Our statement of the second law (page 71) implicitly includes the following relationships (4.35)–(4.38):

$$\Delta \mathbf{S_{U,V}} > 0 \quad \text{for spontaneous processes} \tag{4.35}$$

or, switching to molar units and differential notation,

$$dS_{U,V} > 0 \quad \text{for spontaneous processes} \tag{4.36}$$

and, at the maximum value of S,

$$dS_{U,V} = 0 \quad \text{at equilibrium} \tag{4.37}$$

or, combining (4.36) and (4.37),

$$dS_{U,V} \geq 0 \quad \text{for any constant-}U,V \text{ process} \tag{4.38}$$

The normal reaction for students reaching this point is to have not much idea what all these equations really mean, if anything. Before going on to discuss entropy in other terms, we can illustrate what they mean in terms of some simple measurements.

EXAMPLE 4.1

Take Equation (4.31), which can also be written $\Delta S = \Delta H/T$. It says that, for example, if you melt ice reversibly at $0\,°C$,[6] then ΔS will equal the heat of fusion, ΔH, divided by the temperature, 273.15 K. The heat of fusion of ice is $+6008\,\mathrm{J\,mol^{-1}}$ (positive, because we must *add* heat to the system), so

$$\Delta S_{ice \to water} = S_{water} - S_{ice}$$
$$= \frac{6008}{273.15}$$
$$= 21.995\,\mathrm{J\,mol^{-1}}$$

Now the enthalpy change for the melting of ice changes very little from 273.15 to 274 K, in fact hardly at all, so we can assume it is still $+6008\,\mathrm{J\,mol^{-1}}$. So, if you melt ice irreversibly, say at 274 K, then ΔS will be *greater* than the enthalpy change divided by the temperature, so

$$\Delta S_{ice \to water} = S_{water} - S_{ice}$$
$$> \frac{6008}{274}$$
$$> 21.927\,\mathrm{J\,mol^{-1}}$$

Similarly, if you freeze water at 272 K, the enthalpy change is now $-6008\,\mathrm{kJ\,mol^{-1}}$, and

$$\Delta S_{water \to ice} = S_{ice} - S_{water}$$
$$> \frac{-6008}{272}$$
$$> -22.088\,\mathrm{J\,mol^{-1}}$$

which means that

$$\Delta S_{ice \to water} < +22.088\,\mathrm{J\,mol^{-1}}$$

In other words,

$$21.927 < \Delta S_{ice \to water} < 22.088\,\mathrm{J\,mol^{-1}}$$

so we have determined the entropy change between ice and water at equilibrium at 273.15 K to within 1% with two irreversible measurements.

EXAMPLE 4.2

As another example, consider the problem of determining the change in entropy of a substance X between 300 K and 350 K at one bar. We will suppose that the heat capacity of X is constant at exactly $10 \, \mathrm{J \, mol^{-1}}$ so that 500 J are required to heat one mole of X from 300 to 350 K, 250 J to heat it from 300 to 325 K, and so on. It follows that, on heating one mole of X from 300 to 350 K in a thermostat at 350 K,

$$S_{350} - S_{300} > \frac{500}{350}$$

and on cooling back to 300 K in a similar thermostat at 300 K

$$S_{300} - S_{350} > \frac{-500}{300}$$

from which we conclude that

$$\frac{500}{300} > S_{350} - S_{300} > \frac{500}{350}$$

or

$$1.67 > S_{350} - S_{300} > 1.43 \, \mathrm{J \, mol^{-1}}$$

which means we have determined the ΔS to about one part in six with two measurements. To improve our precision, we could double the number of measurements, and heat and cool in two stages each. That is, because

$$S_{350} - S_{300} = (S_{350} - S_{325}) + (S_{325} - S_{300})$$

we have

$$\frac{250}{325} + \frac{250}{300} > S_{350} - S_{300} > \frac{250}{350} + \frac{250}{325}$$

or

$$1.60 > S_{350} - S_{300} > 1.48 \, \mathrm{J \, mol^{-1}}$$

Given enough patience, we could make 50 measurements at one-degree intervals, in which case

$$\frac{10}{349} + \frac{10}{348} + \cdots + \frac{10}{301} + \frac{10}{300} > S_{350} - S_{300} > \frac{10}{301} + \frac{10}{302} + \cdots + \frac{10}{349} + \frac{10}{350}$$

or

$$1.543890 > S_{350} - S_{300} > 1.539128 \, \mathrm{J \, mol^{-1}}$$

Clearly we are approximating an integral, which is of course Equation (4.34), which in this case becomes

$$\Delta S = S_{350} - S_{300}$$

$$= \int_{300}^{350} \frac{C_P}{T}\, dT$$

$$= \int_{300}^{350} C_P\, d\ln T$$

$$= 10 \times \ln(350/300)$$

$$= 1.541507 \text{ J mol}^{-1}$$

This type of calculation is similar to the one we did for work using the piston–cylinder arrangement in Chapter 3, in the sense that we approach the reversible process by taking more and more steps. We reiterate that the physical operation implied by the integral $\int (C_P/T)dT$ is a reversible process, which is impossible. This doesn't bother us, however, because the integration involves surfaces in the thermodynamic model, not physical reality. It is helpful though to see, as above, what sequence of physically real measurements could lead to the same result.

4.10.2   T–S Diagrams

We mentioned a couple of times the complete analogy between w–P–V and q–T–S relationships. Thus isobars on a T–S diagram (Figure 4.9) are quite analogous to isotherms on a P–V diagram (Figures 3.4–3.9). The isobars in Figure 4.9 have a positive slope because $(\partial S/\partial T)_P$, or C_P/T, is always positive.

We won't go through all the details as we did with work and P–V diagrams in Chapter 3, but you can see from Equation (4.3) that the area under an isotherm in Figure 4.9 between two points on the isotherm will equal the reversible heat between those states, just as the area under an isotherm in Figure 3.8 is the reversible work.

4.10.3   Carnot Cycles

The only purpose in showing this T–S diagram is to complete the analogy with P–V diagrams. However, it is worth noting that this particular type of diagram is very useful in illustrating Carnot cycles, which we will not discuss in any detail. A Carnot cycle consists of an isothermal expansion (shown at 600 °C), then an adiabatic (S constant, $q = 0$) expansion, then an isothermal compression (shown at 500 °C), then an adiabatic compression back to the starting conditions. Each expansion and compression is reversible (continuous succession of equilibrium states), and the fact that it is a cycle, ending up at the starting point, means that the total energy change is zero. By the first law, if $\Delta U = 0$, then q and w must be equal and opposite, or $q = -w$.

However, the fact that $q = -w$ does not mean that you have 100% heat-to-work conversion in a Carnot cycle. Gases tend to cool when expanding, so during the high-temperature isothermal expansion heat must be supplied to the system (the gas) to maintain

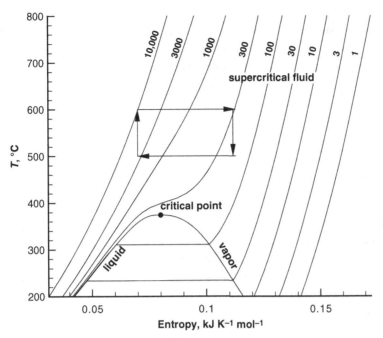

Figure 4.9 The entropy of water. The contours are labeled in bars. The arrows outline a Carnot Cycle. Data from Program STEAM (Harvey *et al.*, 2000).

its constant temperature, and conversely heat will be lost from the system during its low-temperature isothermal compression. The heat transferred from the cycle under the lower isotherm (500 °C in this case) is "lost," and is unavailable for doing work. Therefore the work accomplished by the Carnot heat engine (the area of the cycle) must always be less than the heat input (the area under the upper isotherm on a *T–S* diagram), and the "efficiency," or the fraction of heat transformed into work, is governed entirely by how far apart the upper and lower isotherms are.

An analysis of this combination of reversible adiabatic and isothermal expansions and compressions of a gas or any other "working substance" arranged in cycles and producing work, found in many texts, leads to the result that

$$\frac{q_1}{q_2} = -\frac{T_1}{T_2}$$

for any heat engine operating reversibly, absorbing an amount of heat q_1 at temperature T_1, losing an amount of heat q_2 at a lower temperature T_2, and doing some work in between. Two fundamental conclusions can be deduced from this equation. Neither is immediately obvious, unless you are already steeped in the subject. The first conclusion is that there exists a *thermodynamic temperature scale* with fixed ratios of temperature between any two equilibrium states. Fixing the temperature of any one equilibrium state then fixes the temperature of all others. This follows from the fact that q_1/q_2 is a fixed number for any two equilibrium states, being independent of the size, shape, and working substance of the (hypothetical) Carnot engine used. All that need be done is to somehow determine q_1 and q_2 for a Carnot cycle operating between any two isotherms, then arbitrarily choose a number for the temperature of one of these states. The temperature of the other state will then be established, and these two temperatures establish a linear scale according to which all other equilibrium states can be measured with a suitable thermometer.

The second fundamental conclusion following from

$$\frac{q_1}{q_2} = -\frac{T_1}{T_2}$$

is that

$$\frac{q_1}{T_1} + \frac{q_2}{T_2} = 0$$

for a single reversible cycle, or

$$\sum_i \frac{q_i}{T_i} = 0 \tag{4.39}$$

for a number (i) of linked reversible cycles.

This is sufficient to suggest the existence of a state variable equal to q_{rev}/T, since this function is conserved in heat engine cycles carried out reversibly. This new state variable is of course the entropy, S, where

$$\Delta S = \frac{q_{rev}}{T} \tag{4.3}$$

for a single reversible process represented by the Δ. In differential notation

$$dS = \frac{\delta q_{rev}}{T} \tag{4.40}$$

and

$$S_B - S_A = \int_A^B \frac{\delta q_{rev}}{T} \tag{4.41}$$

We wrote these equations previously simply by analogy. They were in fact discovered by Clausius (1879) by analysis of Carnot cycles. Because steam engines and internal combustion engines operate in cycles of expansion and compression of some working substance, you can readily imagine that this idealized cycle is of great interest to engineers. It represents the ideal, or maximum, work that can be attained by any heat engine.

4.11 A More Useful Thermodynamic Potential

Now we have two parameters, $S_{U,V}$ and $U_{S,V}$ (or $S_{U,V}$ and $U_{S,V}$), that will tell us which way processes will go, but they refer to processes which virtually never occur, except perhaps in classroom exercises – that is, processes which occur at constant values of U and V, or of S and V, our two pairs of constraints. We need a parameter which will refer to processes at constant T and P, our most common case.[7]

---

[7] It is usual to speak of processes occurring at constant U and V, or constant T and P. It would be more accurate to speak of processes having the same values of U and V, or of T and P, before and after the process. It doesn't really matter what the system does between the two states; that is, the system need not be at constant T and P during the process.

4.11.1   Gibbs Energy

We now proceed to introduce the Gibbs energy, the energy that always decreases in spontaneous processes at given values of T and P. Before doing that, pause to consider this. In a sense we don't have to develop this concept by definition (Equation (4.42)) or by manipulating other functions. Having postulated in Section 4.4.2 the existence of the entropy function, which we called a thermodynamic potential, it turns out that all other thermodynamic potentials (that is, potentials for all valid combinations of constraint variables other than $\mathbf{U}$ and $\mathbf{V}$) are automatically implied. The Gibbs energy in this sense has already been defined. This is because of a bit of mathematical magic called the Legendre transform, discussed in the online resources (Topics in Mathematics, C.8 Legendre Transforms). For mathematically minded people this is elegant and entirely sufficient, and the following development is unnecessary. The same idea is shown in graphical terms in Figure 4.13.

From Equation (4.27)

$$d\mathbf{U} - T\,d\mathbf{S} + P\,d\mathbf{V} \leq 0 \tag{4.27}$$

we can see that if we define a function

$$\mathbf{G} = \mathbf{U} - TS + PV \tag{4.42}$$

called the Gibbs energy, the differential of which is

$$d\mathbf{G} = d\mathbf{U} - T\,d\mathbf{S} - \mathbf{S}\,dT + P\,d\mathbf{V} + \mathbf{V}\,dP \tag{4.43}$$

or

$$d\mathbf{G}_{T,P} = d\mathbf{U} - T\,d\mathbf{S} + P\,d\mathbf{V} \tag{4.44}$$

which, combined with (3.27), is

$$d\mathbf{G}_{T,P} = d\mathbf{H} - T\,d\mathbf{S} \tag{4.45}$$

then we find by comparing (4.44) and (4.27) that

$$d\mathbf{G}_{T,P} \leq 0 \tag{4.46}$$

for a whole system or

$$dG_{T,P} \leq 0 \tag{4.47}$$

for an individual reaction.

Thus our definition of the second law has led to a function which will always decrease to a minimum in spontaneous processes in systems having specified values of T and P. It is an extremely useful thermodynamic potential. All we have to do is find a way to get measurable values of this function for all pure compounds and solutes and how they change with T, P, and concentration, and we will then be able to predict the equilibrium configuration of any system by minimizing $\mathbf{G}$.

To see how $\mathbf{G}$ changes with T and P is fairly simple. To see how it changes with composition is a little more difficult. We will get to that in Chapters 7 and 8. On combining (4.8) and (4.43), we find another fundamental equation,

$$d\mathbf{G} = -\mathbf{S}\,dT + \mathbf{V}\,dP \tag{4.48}$$

Because (4.48) could also be written as a total differential,

$$d\mathbf{G} = \left(\frac{\partial \mathbf{G}}{\partial T}\right)_P dT + \left(\frac{\partial \mathbf{G}}{\partial P}\right)_T dP \tag{4.49}$$

we see from (4.48) that

$$(\partial \mathbf{G}/\partial T)_P = -\mathbf{S} \tag{4.50}$$

and

$$(\partial \mathbf{G}/\partial P)_T = \mathbf{V} \tag{4.51}$$

which are the slopes of the G–T–P surface in the T and P directions (Figure 4.8). We will see how to integrate these expressions in Chapter 5.

In addition,

$$d\mathbf{G}_{T,P} = 0 \tag{4.52}$$

Equation (4.52) is simply the condition for a minimum in $\mathbf{G}$, that is, the tangent is horizontal (see Figure 4.10). It follows too, that $\Delta \mathbf{G}_{T,P} < 0$ for spontaneous processes in systems having the same T and P before and after the process, because any spontaneous process must head toward this minimum from some higher point (some point of greater $\mathbf{G}$ value). Note too that the mathematical conditions for a function minimum, such as $d\mathbf{U}_{\mathbf{S},\mathbf{V}} = 0$ and $d\mathbf{G}_{T,P} = 0$, are all true simultaneously at the minimum, as illustrated in Figure 4.5, and they imply nothing whatsoever about how the function (i.e., the system) reached or achieved that minimum.

By combining Equation (4.43) with Equation (4.24), we get

$$d\mathbf{G} = -\mathbf{S}\,dT + \mathbf{V}\,dP - \mathcal{A}\,d\xi \tag{4.53}$$

We will see that this $\mathcal{A}\,d\xi$ term can appear in all the thermodynamic potential equations.

4.11.2   Gibbs Energy in Chemical Reactions

Let's review what we know about Equation (4.53). The $d\mathbf{G} = -\mathbf{S}\,dT + \mathbf{V}\,dP$ part describes an equilibrium surface in G–T–P space for a particular system. Recalling the discussion in Section 4.7.1, $d\mathbf{G} = -\mathbf{S}\,dT + \mathbf{V}\,dP$ describes *either* the stable equilibrium surface *or* a metastable equilibrium surface. Its just that the functional relationship between $\mathbf{S}$ and T

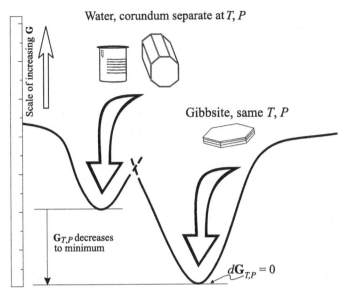

Water, corundum separate at T, P

Gibbsite, same T, P

$\mathbf{G}_{T,P}$ decreases to minimum

$d\mathbf{G}_{T,P} = 0$

Scale of increasing G

Figure 4.10 The decrease in Gibbs energy for the reaction of water and corundum to form gibbsite. A modification of Figure 2.7.

and that between V and P will be different in the two cases. In other words, you can apply $d\mathbf{G} = -\mathbf{S}\,dT + \mathbf{V}\,dP$ to either calcite or aragonite (e.g., path A → B in Figure 4.11).

The (integrated) $\mathcal{A}\,d\xi$ term represents the energy difference between reactants and products, e.g., between the aragonite and calcite surfaces, path B → C in Figure 4.11. From Equation (4.53),

$$d\mathbf{G}_{T,P} = -\mathcal{A}\,d\xi$$

$$\Delta\mathbf{G}_{T,P} = -\int_{calc}^{arag} \mathcal{A}\,d\xi$$

The equation thus opens the way to considering chemical reactions trying to achieve stable equilibrium, not just the equilibrium states themselves, and is therefore quite important. This is explored further in Chapter 13.

4.11.3   Helmholtz Energy

Similarly, if we define a function

$$\mathbf{A} = \mathbf{U} - T\mathbf{S} \tag{4.54}$$

called the Helmholtz energy, or the Helmholtz work function, the differential of which is

$$d\mathbf{A} = d\mathbf{U} - T\,d\mathbf{S} - \mathbf{S}\,dT \tag{4.55}$$

or

$$d\mathbf{A}_T = d\mathbf{U} - T\,d\mathbf{S} \tag{4.56}$$

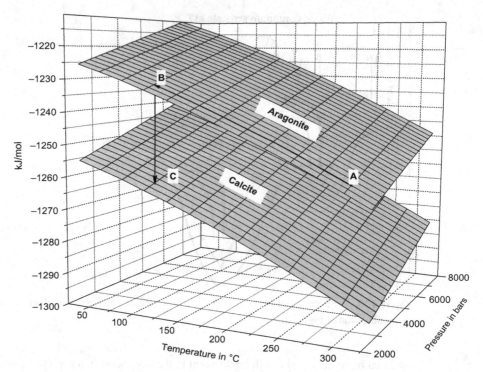

Figure 4.11 Calcite and aragonite surfaces in G–T–P space. The path A $\rightarrow$ B represents a reversible process on a metastable equilibrium surface. The path B $\rightarrow$ C represents an irreversible process corresponding to A $\rightarrow$ A$'$ in Figure 4.8. Under the conditions of the diagram, aragonite is a metastable form of the stable mineral calcite (CaCO$_3$) and becomes the stable phase at high pressures. The calcite surface has been moved down 30 kJ mol^{-1} for better visibility, so the intersection of these two surfaces is not shown, but a schematic diagram showing the intersection of two Gibbs energy surfaces for a different kind of phase change is shown in Figure 12.3. Data from Holland and Powell (2011).

we find, on comparing (4.27) and (4.56),

$$d\mathbf{A}_T + P\, d\mathbf{V} \leq 0 \tag{4.57}$$

or

$$d\mathbf{A}_{T,\mathbf{V}} \leq 0 \tag{4.58}$$

or

$$dA_{T,V} \leq 0 \tag{4.59}$$

Thus we have another function, **A**, which will always decrease to a minimum in spontaneous processes in systems having specified values of T and **V**, and is another thermodynamic potential. The usefulness of this function will be discussed below.

4.12   Gibbs and Helmholtz Functions as Work

. . . .

4.12.1   Gibbs Energy as Useful Work

A ball in a metastable equilibrium valley (e.g., Figure 1.2) is capable of doing work as it rolls down to lower elevations (once it has been pushed over the barrier). The maximum work it can do is exactly equal to the (minimum) work required to push the ball back up to its metastable elevation. One way of understanding the Gibbs energy is that it is equal to the maximum amount of useful work that chemical systems can do as they change from metastable states to stable states, underlining the usefulness of the ball-in-valley analogy.

However, we must first distinguish between *total* work and *useful* work. Chemical systems undergoing change (i.e., systems in which reactions occur) can do various kinds of work. For instance, batteries can do electrical work. While undergoing these reactions, the chemical system invariably undergoes some change in volume, because it is most unlikely that the reaction products would have exactly the same volume as the reactants. This change in volume ΔV takes place under some ambient pressure P, so $P \Delta V$ work is done during the reaction regardless of whether any other kind of work is done or not – if the reaction is to take place, it cannot be avoided. This "work against the atmosphere" (or against the confining pressure, whatever it is) usually is not *useful*; it simply takes place whether we like it or not, and at atmospheric pressure it is often a rather small part of the total energy change. Although we can decide to eliminate electrical work or other kinds of mechanical work from our systems, we cannot eliminate this $P \Delta V$ work (unless we consider only constant-volume systems, which is not usually very practical).

We can talk about the total work $\mathbf{w}$, or the total work per mole, w. We usually use the Gibbs energy for individual reactions, with data for each species in $J \, mol^{-1}$, so here we consider the molar form. The net work (per mole) other than $P \Delta V$ work, which we could also call (page 83) uncompensated work, can be written

$$w_{net} = w_{total} - w_{P \Delta V}$$
$$= w + P \Delta V$$

and, because $q \leq T \Delta S$ (Equation (4.5)), it follows from the first law ($\Delta U = q + w$) that

$$w \geq \Delta U - T \Delta S \tag{4.60}$$

By adding $P \Delta V$ to both sides, we obtain

$$w + P \Delta V \geq \Delta U - T \Delta S + P \Delta V$$

giving

$$w_{net} \geq \Delta U - T \Delta S + P \Delta V \tag{4.61}$$

and, because

$$\Delta G_{T,P} = \Delta U - T \Delta S + P \Delta V \tag{4.62}$$

on combining (4.61) and (4.62), we obtain

$$w_{net} \geq \Delta G_{T,P} \tag{4.63}$$

or, the other way around,

$$\Delta G_{T,P} \leq w_{\text{net}} \tag{4.64}$$

However, don't forget that, according to our sign convention, $-w$ is the work done by a system, or available from a system, so we should perhaps write

$$-\Delta G_{T,P} \geq -w_{\text{net}}$$

for a reaction, or

$$-\Delta \mathbf{G}_{T,P} \geq -\mathbf{w}_{\text{net}}$$

for a whole system.

In other words, the net useful work available from a chemical reaction cannot be greater than the decrease in G that the reaction undergoes. For example, if a battery is doing work by lighting the bulb in a flashlight, the maximum amount of useful work it can do is given by its decrease in G toward stable equilibrium, when the battery is dead. If the system does *no* work other than expanding or contracting against its confining pressure (no work other than $P\,\Delta V$ work), then $w_{\text{net}} = 0$, and

$$\Delta G_{T,P} \leq 0 \tag{4.65}$$

This result is not surprising, as it agrees with our conclusion in Section 4.11.1, but it does serve to link the Gibbs energy with an intuitive concept, the available work.

The van 't Hoff Equilibrium Box

This mathematical demonstration that the change in Gibbs energy at a given T and P is the maximum amount of non-$P\,\Delta V$ work available from a chemical reaction is of more interest to engineers than to geochemists. It does, however, give a meaning to the older term for the Gibbs energy, the "Gibbs *free* energy." It is the energy that is free to do useful work. A better demonstration of this fact using an idealized physical system is the van 't Hoff equilibrium box. See the online material under this title.

Box 4.3   Gibbs Energy as Maximum Work

The maximum amount of work, other than the work done by the change of volume against a confining pressure, available from reaction (2.5) is

$$\Delta_r G^\circ = \Delta_f G^\circ_{Al_2O_3 \cdot 3H_2O(s)} - \Delta_f G^\circ_{Al_2O_3(s)} - 3\,\Delta_f G^\circ_{H_2O(l)}$$

$$= -2310.21 - (-1582.3) - 3(-237.129)$$

$$= -16.523 \, \text{kJ mol}^{-1}$$

$$= -16{,}523 \, \text{J mol}^{-1}$$

By comparison, the $P \Delta V$ work done by atmospheric pressure during this reaction is $1.61\,\mathrm{J\,mol^{-1}}$.

This result, while technically true, is unrealistic because it is hard to imagine how this reaction might be arranged in such a way as to produce useful work. Only some reactions can actually be made to do non-P–V work. A better example would be the burning of octane (C_8H_{18}) in oxygen. If this is done in a calorimeter it does no useful work, but if done in an internal combustion engine, it does. This is related to Equation (4.63), which says that $\Delta G_{T,P}$ gives the maximum work that can be achieved from any reaction, such as burning octane, and Equation (4.64), which says that, even if you don't arrange things so that this work is actually done, $\Delta G_{T,P}$ will still be a negative quantity.

An even better example can be found in Chapter 11, where we see a link between how much work an electrochemical cell (a battery) can do and the change in its Gibbs energy.

4.12.2   Helmholtz Energy as Total Work

On comparing equations (4.60) and (4.56), we see that

$$w \geq \Delta A_T \tag{4.66}$$

which is analogous to (4.63), only in this case we say that the *total* work per mole (rather than the *available* work) cannot be greater than the decrease in A, or that ΔA_T is an upper limit to the total work done in isothermal processes.

If no work at all is done, then $w = 0$, which implies that $\Delta V = 0$, and

$$\Delta A_{T,V} \leq 0 \tag{4.67}$$

so that A always decreases in spontaneous processes under constant-T, V conditions. It is another thermodynamic potential.

It might seem useful to have a function related to the total work available from a system, but in fact A is little used in this sense. It (that is, $A_{T,V}$) is also not much used as a thermodynamic potential, despite the fact that replacement processes in weathering, metamorphism, and metasomatism are commonly interpreted as occurring at constant volume (Nahon and Merino, 1987; Carmichael, 1986). However, replacement processes do not, by definition, take place in a closed system, so the Helmholtz energy is not the appropriate potential. So what is the appropriate potential quantity in open systems? We consider this in Section 4.13. Where the Helmholtz energy frequently *is* used is in constructing *equations of state*. See Chapter 13, Equations of State in the online resources for more on this topic.

4.12.3   Notation Again

As a reminder, Figure 4.12 shows the logical relationship between all our various ΔG terms, just as Figure 3.10 did for ΔH terms. Refer to Section 3.8.2 for a discussion.

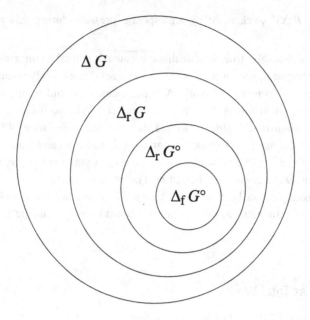

Figure 4.12 The hierarchy of ΔG terms.

4.13   Open Systems

4.13.1   The Open System Equation

If you look at Figure 3.3 and Figure 4.1, or if you think about the fundamental equation (4.9), you realize that we have expended all this time and effort in defining thermodynamic potentials that are limited in one very important respect, and that is that they apply only to systems which do not change composition. Everything we have said is limited to constant-composition systems.

Such systems are indeed important, and they can include changes in composition in a limited sense. For example, we can use **G** to predict that, in a system having 1 gram of halite plus a kilogram of water, the water will change composition as the halite dissolves to an equilibrium composition. But the solution must have the same composition as the halite plus water that we started with. The system has not changed composition; components have just been redistributed between phases within the system, as in Figure 2.2(b).

The total Gibbs energy **G** depends on the mass of system we consider. The **G** of 2 kg of halite is twice the **G** of 1 kg of halite (Section 2.4.1). The total energy of a system depends on the mass or number of moles in the system, but it must also depend on what that mass consists of. Because we deal only with energy changes, we have no way of knowing whether a mole of halite has more or less energy than a mole of water or a mole of anything else, and we don't need to know. We just need to know how the energy of a system *changes* if we add halite or anything else to it. We know that, if we add halite to a system consisting of halite, the change is linear, but how does the energy change if we add halite to water?

Because we have a mathematical model of energy changes, we have a simple mathematical answer to this question. **G** is a state variable, so $d\mathbf{G}$ is an exact differential

(Section 1.4.1). This means that, among other things, we can write the total differential as in Equation (4.49), or in the total energy form as

$$dG = \left(\frac{\partial G}{\partial T}\right)_P dT + \left(\frac{\partial G}{\partial P}\right)_T dP \tag{4.68}$$

This shows, as long as we can integrate these derivatives, how G changes with changes in T and P. To see how it changes when we add n_1 moles of component 1 and n_2 moles of component 2, we just add more derivatives, so

$$dG = \left(\frac{\partial G}{\partial T}\right)_{P,n} dT + \left(\frac{\partial G}{\partial P}\right)_{T,n} dP + \left(\frac{\partial G}{\partial n_1}\right)_{T,P,n_2} dn_1 + \left(\frac{\partial G}{\partial n_2}\right)_{T,P,n_1} dn_2 \tag{4.69}$$

where n means (n_1, n_2), or in general, with c components,

$$dG = \left(\frac{\partial G}{\partial T}\right)_{P,n} dT + \left(\frac{\partial G}{\partial P}\right)_{T,n} dP + \sum_{i=1}^{c} \left(\frac{\partial G}{\partial n_i}\right)_{T,P,\hat{n}_i} dn_i \tag{4.70}$$

$$= -S\,dT + V\,dP + \sum_{i=1}^{c} \left(\frac{\partial G}{\partial n_i}\right)_{T,P,\hat{n}_i} dn_i \tag{4.71}$$

where n now means $n_1, n_2, \ldots, n_c$ (all components), n_i refers to any individual component i, and $\hat{n}_i$ refers to all components *except* i.

Our new derivative terms $(\partial G/\partial n_i)_{T,P,\hat{n}_i}$ are partial molar properties, introduced in Section 2.4.1. Partial molar properties allow us to deal with compositional changes, and as such they are one of the more important quantities in chemical thermodynamics. The partial molar Gibbs energy is given a name, chemical potential, and a symbol, μ, i.e.,

$$\mu_i = \left(\frac{\partial G}{\partial n_i}\right)_{T,P,\hat{n}_i}$$

We will have a lot more to say about them, but for the moment we will just get the equations we need for future reference. We can now write (4.71) as

$$dG = -S\,dT + V\,dP + \sum_{i}^{c} \mu_i\,dn_i \tag{4.72}$$

for c independent components which can be added to or subtracted from the system. This equation is not of much use as it is because we normally consider closed systems; we cannot add components to a closed system. However, it is useful in deriving other relationships, as we will see in the next section and in Chapter 13.

4.13.2   The Gibbs–Duhem Equation

An interesting result is obtained by integrating (4.72) at constant T and P, that is, by supposing that the quantity of system varies from zero up to some finite value. The result is[8]

---

[8] Equation (4.73) is also the result of Euler's theorem (see Topics in Mathematics C.7, Euler's Theorem for Homogeneous Functions, in the online resources) applied to G as a function homogeneous in the first degree in the masses of the components.

$$G_{T,P} = \sum_i \mu_i n_i \tag{4.73}$$

which shows that μ is not just some abstract partial derivative, but is the Gibbs energy per mole of a dissolved substance. That is, Equation (4.73) shows that the total Gibbs energy of a system is simply the sum of the number of moles of each component in the system (n_i) times the free energy per mole of that component (μ_i).

Our new "more complete" Fundamental Equation (4.72) is a bit inconvenient for some purposes, in that some of the differential terms (dT, dP) are intensive, but the others (dn_i) are extensive. We would like to have an equation which contains compositional terms, but which has differentials of intensive variables only. It could then be used for closed systems. We get this by first differentiating (4.73),

$$d\mathbf{G} = n_1\, d\mu_1 + \mu_1\, dn_1 + n_2\, d\mu_2 + \mu_2\, dn_2 + \cdots + n_c\, d\mu_c + \mu_c\, dn_c \tag{4.74}$$

and subtracting from this Equation (4.72). The result is

$$0 = \mathbf{S}\, dT - \mathbf{V}\, dP + \sum_{i=1}^{c} n_i\, d\mu_i \tag{4.75}$$

which is called the Gibbs–Duhem equation, or "Gibbs 97," because it is Equation 97 in Gibbs (1961a). One important application is in the derivation of the phase rule (Chapter 12).

The Integrated Form

Another application of this equation is to show that, even though the components in the system are compositionally independent of one another, their chemical potentials are not always independent. In a binary solution of components 1 and 2, you can change n_1 without changing n_2, but doing this will change *both* μ_1 and μ_2. Let's see how it works.

At constant T and P, Equation (4.75) becomes

$$\sum_{i=1}^{c} n_i\, d\mu_i = 0 \tag{4.76}$$

so for a binary system

$$n_1\, d\mu_1 + n_2\, d\mu_2 = 0$$

or, dividing by $n_1 + n_2$,

$$x_1\, d\mu_1 + x_2\, d\mu_2 = 0$$

so

$$d\mu_1 = -\frac{x_2}{x_1}\, d\mu_2 \tag{4.77}$$

$$\mu_1'' - \mu_1' = -\int_{x_1'}^{x_1''} \frac{x_2}{x_1}\, d\mu_2 \tag{4.78}$$

showing that, if you know how one potential changes as a result of a compositional change, you can calculate the change in the other. They are not independent. Because the chemical potential and the activity are closely related (e.g., Equation (8.18)) this is easily converted to calculation of the activity of one component of a binary solution if the activity of the other component is known. The result is

$$d \ln a_1 = -\frac{x_2}{x_1} d \ln a_2 \tag{4.79}$$

An example of the result of using this procedure is shown in Figure 8.3.

4.13.3   Other Kinds of Open Systems

The open system as a subsystem in an overall closed system (Figure 2.2(b)) is by far the most commonly used kind of open system in geochemistry. In any multiphase system, the compositions of the phases must be adjusted to achieve a Gibbs energy minimum, and this type of calculation is very common. But there are other kinds of open systems involving flow of material into and out of the system. This kind of system is common in mechanical engineering. These applications do not involve chemical reactions to nearly the same extent. All kinds of turbines, boilers, and so on can be modeled in terms of mass and energy balances. Conservation of mass and energy in the inputs and outputs, which is basically the first law, plus heat-to-work conversion, the second law, can account for much of thermodynamic modeling in these applications. Mechanical engineering texts may not even mention the concept of activity (Chapter 8), which is so central to geochemistry.

4.14   The Meaning of Entropy

As long as we are dealing with pure compounds, we have answered just about all our questions. We have an energy parameter, the Gibbs energy, which always decreases in spontaneous reactions at a given T and P, and we know how to measure this energy term – calorimetry. We know, however, that this energy term, ΔG, is made up partly of a fairly comprehensible term ΔH, which is just a heat flow term, and partly of another term ΔS, which is more mysterious. All we know about this one is that we defined the second law such that the entropy always increases in spontaneous reactions in isolated systems. The entropy is not itself an energy term, but the product of T and S, or T and ΔS, is an energy term.

If we had to rely on classical thermodynamics, we would know little more than we have already said about entropy. It is a parameter, with a method of measurement, which increases in spontaneous processes, *even when no energy changes are possible*, that is, in isolated systems. We would also notice, after measuring the entropy of many substances, that the entropies of gases are relatively large, those of solids relatively small, and those of liquids somewhere in between, but we would probably not have any mental picture of what entropy represents physically.

If you look at some processes that are quite irreversible but that involve little or no energy change, such as the mixing of two gases or the spreading of a colored dye in water, you observe a driving force for processes that is quite different from the energy-drop paradigm we have been pursuing (the ball rolling downhill). There is no energy drop when gases mix, but they do so invariably and irreversibly, and this shows that another driving force for reactions is *mixing*, or an increase in "mixed-upness," which will take place if it is possible. If you think about gases as collections of countless tiny molecules zipping around with the speed of rifle bullets, but with quite a lot of space between them, you realize that, if two different gases are brought together, it is no more difficult to understand why they will always mix together than it is to understand why a ball will roll downhill. But this mixing process involves no energy change (at least for ideal gases), so the first law of thermodynamics is powerless – it cannot be the basis for a thermodynamic explanation. The second law and entropy do provide it. Entropy can be thought of as a *degree of mixed-upness*, and increasing the randomness or mixed-upness of systems is *one* of the driving forces for spontaneous reactions.

The confusion arises because it is not the only one – the ball rolling down the hill (energy decrease) is also one. It is the two together that provide the complete answer. In some processes energy decrease is the dominant factor; in others, mixing or entropy increase is the dominant factor. The two are brought together in the Gibbs energy equation

$$dG_{T,P} = dH - T\,dS \tag{4.45}$$

which can also be written

$$\Delta G_{T,P} = \Delta H - T\,\Delta S$$

Here there are two factors that together determine whether $\Delta G_{T,P}$ will be positive or negative. One is ΔH ($= \Delta U + P\,\Delta V$), which is the energy change due to heat and work in the process represented by Δ, and the other is $T\,\Delta S$, the energy change due to the mixing factor. In many spontaneous reactions, ΔH is negative (the process is exothermic) and ΔS is positive (mixed-upness increases), so both factors are negative ($T\,\Delta S$ positive, $-T\,\Delta S$ negative) and ΔG is negative. In other reactions, ΔH is positive (the process is endothermic, e.g., melting ice), but the ΔS term is sufficiently large and positive that $T\,\Delta S > \Delta H$, and ΔG is negative in spite of the positive ΔH. This will be especially true at high temperatures, when T is large. There are all varieties of combinations; the point is that whether or not any particular process is spontaneous is the result of the two competing factors. Systems want to lower their energy content, but they also want to maximize their mixed-upness. The balance between these factors decides the issue.

4.14.1   But What Is Entropy, Really?

This question has been around since Clausius invented the term in a series of papers written during the years 1850–1865 (Clausius, 1867). A summary of this work in his own words appears in his book *The Mechanical Theory of Heat* (Clausius, 1879). The answer takes on many forms. Some follow the historical route, from steam engines, to Carnot, Clausius, Thompson, Joule, Rankine, and so on. A particularly lucid, concise account of this history is Purrington (1997). A central feature of this approach is the Carnot cycle, which was used by Clausius to deduce the existence of the entropy parameter. This approach is rather abstract, and needs some manipulation in order for it to be seen to be connected to thermodynamic potentials and chemical reactions. Others emphasize the impossibility of some processes, or the "availability" of energy, and some have a rather unique viewpoint, such as Reiss (1965), who considers entropy as the "degree of constraint."

Virtually since the beginning, however, a popular viewpoint has been to see entropy as a measure of disorder. Helmholtz used the word "Unordnung" (disorder) in 1882. This results from familiar relationships such as $S_{gases} > S_{liquids} > S_{solids}$, and the universally positive entropy of mixing. We used this relationship in the previous section when we spoke of "degree of mixed-upness." However, the "disorder" analogy can involve a serious fallacy, as made clear by Lambert (1999).

The rather subjective concepts of disorder and mixed-upness are useful analogies in certain situations, such as melting solids and mixing gases, but they fail completely in most other situations. You cannot tell whether α or β quartz is "more disordered" or has the higher entropy by looking at their structures. Shuffling a deck of cards perhaps increases its disorder, but it does not increase the entropy of the cards. To see this, just imagine cooling the cards down to near 0 K, measuring their heat capacity up to room temperature, and determining the entropy of the card deck from these measurements. The result will be the same no matter in what state of order the cards are. The same is true for the configuration of any macroscopic system. Checkers on a board are ordered at the start of a game, and become progressively disordered during a game, but the entropy of the checkers remains the same no matter what their arrangement.

In Chapter 1 we said that mixing processes can occur spontaneously with virtually no energy change, and in Chapter 2 we said "something is missing." That something turned out to be entropy. It also turns out that those two ideas are closely connected. Denbigh (1981, Section 1.17) says

> The irreversible process of temperature equalization may thus be regarded as a mixing of the available energy.... Here it is a question of the mixing or "spreading" of the total energy of the system over the whole range of quantized energy levels of the reactants and products. The occurrence of the reaction causes a larger number of these quantum states to become accessible, namely, those corresponding to the products.... There are thus two rather distinct types of mixing process; the first is the spreading of particles across positions in space, and the second is the sharing or spreading of the available energy of a system between the particles themselves.

So the central fact about entropy as used in science is that it involves the distribution of energy in a system. Energy tends to become "spread out," or delocalized, if that is not prevented from happening.

If you "really" want to understand entropy, you need to learn more than just equilibrium thermodynamics. In this book, we take the simple view that entropy is a parameter, having a clearly defined method of measurement, which enables us to define thermodynamic potentials in chemical systems. It is simply related to "disorder" in many simple situations, which is an intuitive aid, but this aid doesn't extend very far. Because of this resemblance to probability and disorder, entropy has been related to everything from shuffled cards to the fall of empires, but these connections for the most part have nothing to do with the second law of thermodynamics.

4.15   The End of the Road

We pause here to note that, in case you hadn't noticed, we have arrived at the answer to the question posed in Chapters 1 and 2. The question was as follows: what controls whether a reaction or a process will happen or not happen? Why does water freeze below $0\,°C$ and ice melt above $0\,°C$? What is the "chemical energy" term that always decreases to a minimum, like the ball rolling down the hill? In answering this question, we first had to define fairly carefully some terminology such as system, equilibrium, and process. We then noted (Chapter 3) that systems have fixed energy contents ($\mathbf{U}$, or U) at equilibrium, but this didn't help, because, although this energy is conserved, it doesn't distinguish at all between directions processes take (bricks could cool themselves, and use this energy to fly, as far as U is concerned).

The missing ingredient to understanding why reactions go one way and not the other is entropy. Entropy is defined as a state variable that always *increases* in spontaneous processes in isolated systems. But a parameter that is useful only in isolated systems is not of much practical use, so we defined another state variable, the Gibbs energy, that always decreases in spontaneous processes in systems at a given T and P (see Figure 4.10). This is the parameter we have been looking for. Figure 4.10 shows two different states of a system at the same temperature and pressure. A spontaneous process (the formation of gibbsite) occurs when the constraint keeping the reactants separated is removed (corundum and water are mixed together).

Similarly, in Figure 2.4, the "chemical energy" term is in fact the Gibbs energy. At $+5\,°C$, 1 bar, $G_{ice} > G_{water}$, and at $-5\,°C$, 1 bar, $G_{water} > G_{ice}$. In Figure 2.5 $G_{diamond} > G_{graphite}$, assuming both have the same P and T. Mathematically, the entropy and Gibbs energy potentials are two sides of the same coin – one implies the other, as shown in Figure 4.13. Looking at it in still another way, you see from Figure 4.5 that a single system can have $dU_{S,V} = 0$ and $dS_{U,V} = 0$ simultaneously. If we were not limited to three dimensions, we could show that the same system also has $dG_{T,P} = 0$. Each condition implies all the others.

In biochemistry, processes having a negative $\Delta_r G$ ($\Delta_r G < 0$) are termed *exergonic*, and those having a positive $\Delta_r G$ are termed *endergonic*. For some reason, these terms are not common in geochemistry.

The problem at the moment is that these new state variables S and G will have no "feeling of reality" for a reader new to the subject. That is, what *is* entropy or Gibbs energy, and how does one measure these things? Only by actually using these concepts will one become familiar with them. The next chapter is a first attempt at describing these variables in more familiar terms.

4.16   Summary

What you should know at this point is that we have defined a parameter, entropy, which can tell us which way reactions will go, but only in isolated systems. That is, only in

The Second Law:

WORDS:	There is a system property, entropy (**S**), which always increases in spontaneous reactions in isolated systems (those having constant **U** and **V**).
EQUATION:	$\Delta S_{U,V} \geq 0$

IMPLIES

WORDS:	There is a system property, Gibbs free energy (**G**), which always decreases in spontaneous reactions in constant-T, P systems.
EQUATION:	$\Delta G_{T,P} \leq 0$

Figure 4.13 All you need to know about the second law and the derivation of the Gibbs energy function. These statements and equations can be written equally well in their molar forms (using U, S, V, and G).

isolated systems if used by itself as a thermodynamic potential. Indirectly, i.e., combined with other state variables to define other thermodynamic potentials, it gives directionality parameters for any kind of equilibrium system. All thermodynamic potentials include entropy in some way.

The statement defining entropy is one way of stating the second law of thermodynamics. Combining entropy with the first law, we then defined another parameter, the Gibbs energy, which can tell us which way reactions go in systems at a given temperature and pressure. We also showed that the Gibbs energy is equal to the maximum amount of useful energy or work available from such reactions, but we have not yet seen how to measure any of these apparently useful quantities.

It is normal at this point for newcomers to this subject to be rather confused, or perhaps impatient. If we think about natural processes that we would like to understand, such as occur in living plants and animals, or even simpler inorganic processes such as occur in creating our weather patterns or in erupting volcanoes, we could be forgiven for wondering what earthly use the kind of material we have considered up to now can be. We seem to have restricted ourselves to ridiculously simple cases such as balls rolling in valleys, and, even though we have claimed that certain simple inorganic processes such as melting ice and polymorphic mineral changes are analogous, we haven't shown how to do anything remotely useful.

Not only is it not yet useful, but, even after restricting ourselves to simple cases and claiming to be dealing not with reality but with models of reality, we have introduced at least one concept (entropy) that is rather difficult to fully comprehend and have used a level of mathematics that, although not exceeding that taught in introductory calculus courses, has physical implications that are hard to grasp. The best remedy for this is to review the material to some extent, and to plunge ahead even if it is not entirely clear. After some familiarity with practical applications has been attained, some of the earlier material will become clearer, so the best approach is continuous review, in addition to assimilation of new material.

It may well seem that we have made no progress toward understanding complex processes, but this is not true. Most natural processes are so complex that we simply must start with the very simplest ones we can think of and define our terms very carefully. Our goal of finding the secret to why reactions go in one direction and not the other may seem overly simple, but it is in fact the basic concept necessary to build up an understanding of all the natural phenomena mentioned above. Of course, even when we have mastered thermodynamics, we will find that we don't have the answers to all our questions; in fact, we will find that the things thermodynamics can tell us are fairly limited. They have, however, a level of certainty which surpasses that of most other ways of looking at the same problems, and this makes the subject an absolutely essential element of all research into problems that involve energy transfers. You may wish to know much more than thermodynamics can tell you, but you need to know what it can tell you.

Exercises

E4.1 Both surfaces in Figure 4.11 slope down with increasing temperature, and up with increasing pressure. Why is that?

E4.2 Calculate the entropy of formation from the elements of the mineral anorthite ($CaAl_2Si_2O_8$). Combine this with the enthalpy of formation from the tables to calculate the Gibbs energy of anorthite. Compare your answer with the value in the tables.

E4.3 Calculate $\Delta_r H°$ and $\Delta_r G°$ for the reaction

$$NaAlSiO_4 (nepheline) + 2\,SiO_2 (quartz) = NaAlSi_3O_8 (low\ albite)$$

at 25 °C, 1 bar. Is the reaction endothermic or exothermic? Which way would the reaction go under standard conditions in the absence of kinetic barriers? What actually happens if you put quartz and nepheline together?

E4.4 (a) Calculate the value of R in $J\,mol^{-1}\,K^{-1}$ from the ideal gas equation ($PV = nRT$).

(b) Use the dimensions (Appendix A) of energy and pressure to show that $J\,bar^{-1}$ is a volume term, and calculate the conversion factor from $J\,bar^{-1}$ to cm^3.

E4.5 Equations for the two surfaces in Figure 4.11 along the 2.8 kbar isobar from 100 to 325 °C are shown in Table 4.1. Use molar volume data (assumed constant) from Appendix B to calculate and plot the position of the aragonite–calcite phase boundary.

Table 4.1. Equations giving $\Delta_f G°$ as a function of T for aragonite and calcite at $P = 2800$ bars. T is in °C.

	$\Delta_f G°$
Aragonite	$-1.10442 \times 10^{-4}\,T^2 - 8.64553 \times 10^{-2}\,T - 1.22259 \times 10^3$
Calcite	$-1.12440 \times 10^{-4}\,T^2 - 8.95055 \times 10^{-2}\,T - 1.22265 \times 10^3$

E4.6 Is magnesite ($MgCO_3$) or nesquehonite more stable in water?

E4.7 There are six naturally occurring oxides and hydroxides of aluminum listed in Appendix B, but complete data exist for only four of these (corundum, boehmite, diaspore, and gibbsite). Note that the compositions of these phases differ only by the number of H_2O, so it is relatively easy to write reactions between them. By writing balanced reactions between these four phases, determine which one is most stable in water.

E4.8 The origin of red-bed sandstones, in which the grains are coated with minute amounts of hematite, has long been controversial. A key question in the controversy is whether hematite is stable in water at low temperatures. Calculate whether goethite or hematite is stable in the presence of water at 25 °C.

Additional Problems

A4.1 Consider this cycle. You take 1 mole ($\approx$ 100 grams) of aragonite at 25 °C and put it in an oven at 35 °C. After a while you take it out and let it cool back to 25 °C. Show that this cycle is irreversible, despite the fact that the system (the aragonite) has recovered its original state and is completely unchanged. Section 4.10.1 should provide a clue.

A4.2 Calculate the heat input along the 600 °C isotherm (which extends from 10,000 bars to 300 bars) and output along the 500 °C isotherm, and the net heat in the Carnot cycle in Figure 4.11 In other words, calculate the area enclosed by the cycle. According to the equation of state for water in the program STEAM (Harvey *et al.*, 2000), the coordinates of the four corners of the cycle are as shown in Table 4.2.

Table 4.2. Data for Additional Problem A4.2.

T (°C)	P (bars)	V (cm^3 mol^{-1})	S (J mol^{-1} K^{-1})
600	10,000	18.66	70.24
600	300	206.2	112.4
500	3602	22.32	70.24
500	175.8	311.5	112.4

Program STEAM

The program STEAM (Harvey *et al.*, 2000) is an example of a highly sophisticated *equation of state*. A product of several years of development, it provides accurate data for many properties of H_2O over a very wide range of pressures and temperatures, and as such is valuable to geochemists modeling supercritical hydrothermal solutions. It is available from the National Institute of Standards and Technology (www.nist.gov/index.html), along with many useful databases and other publications. See Section 13.6.1, Water Substance, of Chapter 13 in the online resources.

A4.3 The P–V diagram for the Carnot cycle in Figure 4.11 is rather distorted because of the high pressure at one corner, which makes calculation of the area of the cycle very sensitive to the methods used. A more tractable cycle with which to compare the T–S and P–V representations is to use the 600 °C isotherm between 100 and 55 bars, and the 500 °C isotherm to complete the cycle. The four corners according to program STEAM (Harvey *et al.*, 2000) are as shown in Table 4.3.

Table 4.3. Data for Additional Problem A4.3.

T (°C)	P (bars)	V (cm^3 mol^{-1})	S (J mol^{-1} K^{-1})
600	100	691.39	124.39
600	55	1285.82	129.95
500	57.5	1067.6	124.39
500	31.4	1997.7	129.95

(a) Show that $q = -w$ for this H_2O Carnot cycle by calculating the areas in the T–S and P–V diagrams.

(b) Use the areas under the upper and lower isotherms on the T–S diagram to verify the theoretical conclusion that

$$\frac{q_2}{q_1} = \frac{T_2}{T_1}$$

where subscripts 2 and 1 refer to the high and low isotherms, respectively.

(c) Find the "efficiency" of the cycle, w/q_2.

A4.4 Write an equation for the surface in Figure 4.4.

5 Getting Data

Introduction

We have had quite enough theoretical discussion for now. Let's see how to get some numbers into our equations so as to be able to calculate something useful. Welcome to the world of experimental thermochemistry.

In this chapter we will have a look at a few of the ways in which the thermodynamic parameters we have derived are measured; i.e., where the numbers in the tables and databases come from. A deep knowledge of this subject is not necessary in order to use thermodynamics to model chemical, geological, or environmental systems, in the same sense that a knowledge of a composer's life and times is not necessary to enjoy his or her music. But it does enrich the experience, and in the case of using thermodynamics, such knowledge does serve to make the user conscious of the many reasons why his or her data might be incorrect.

Thermochemical data are produced for the most part by dedicated scientists, who devote a good part of their lives to tracking down elusive sources of error, and devising ever-improved methods for determining nature's fundamental parameters as defined by thermodynamic theory. When determined by independent methods and/or independent laboratories, the results are often satisfyingly in agreement, but almost as often they are not, meaning that there is some source of error, and identifying it can take a lot of discussion (perhaps argument would be a better term) and a long time.

Beginning with the establishment of the Geophysical Laboratory in Washington in 1905, scientists primarily interested in geological processes have contributed to our knowledge of these thermodynamic properties, especially, as might be expected, those of the rock and soil-forming minerals, liquid–solid phase relationships of these minerals, and aqueous solutions such as sea water and hydrothermal solutions. In addition to the measurement of mineral and solution properties, collecting these data into databases for use by computer programs and the testing of these databases for internal consistency have become increasingly important.

The experimental part and the self-consistent database part of using thermodynamics in the Earth sciences today have both become large and complex subjects, each worthy of separate study. Geologists and geochemists primarily interested in Earth processes, but who need to do some thermodynamic modeling in order to better understand these processes, can be forgiven for not wanting to be involved with either experiment or database development. Nevertheless, rigorous thermodynamic methods can produce nonsensical results, given incorrect data, so a blind faith in the database attached to some computer program can be a recipe for disaster.

In this chapter (and throughout this book) we cannot address these areas of study to any great extent, but understanding thermodynamics for most people is helped by having some knowledge of how fundamental data are determined. The aim of this chapter is thus to trace the connection between laboratory experiments and thermodynamic modeling. The idea is to impart not only some idea of how it is done, but also some idea of the great difficulties involved at every stage, and therefore some respect for data. Good data are a precious commodity; you should have some interest in where yours come from.

The discussion of methods in this chapter is very sketchy. Many details and variations are omitted, because all we want to do is illustrate what is involved. Do not conclude from the simplicity of the presentation that there is nothing much to this experimental business. It is not difficult to get data that are not much good; it is extremely difficult, time consuming, frustrating, and often expensive to get excellent data, ones that stand the test of time. An overview of many methods used by Earth science experimenters can be found in Ulmer and Barnes (1987).

5.2   What to Measure?

Imagine that you have an interesting field problem, and you want to do some thermodynamic modeling to better understand it, but for some reason there are no data for the mineral you are most interested in, gibbsite. You go to a fully equipped laboratory, but what do you do there?

5.2.1   What Not to Do

Perhaps in your field study you have concluded that gibbsite is being formed from some other mineral by some alteration process you have figured out, and you want to use thermodynamics to see whether it makes sense. Maybe thermodynamics will tell you that this process is not possible.

Your alteration process involves other minerals, as well as an aqueous solution. The most obvious thing to do then is to set up an experiment in which you try to duplicate this reaction. Perhaps you put some groundwater and some minerals together in some reaction vessel for a while, then look at the results. In other words, you try to duplicate nature. This approach has been tried many times, and generally the results are of very limited usefulness. The results are usually quite complex and difficult to interpret, and in any case they apply only to one specific set of starting materials – it is difficult to draw general conclusions from the results. Besides, the measurements you can make in such experiments, which are usually the compositions of some complex phases, have nothing to do with measuring heat and work, which as we have seen is the basis of all our thermodynamic parameters.

5.2.2   What to Do

If you wish to use thermodynamics, perhaps you should look at the equations to get a clue as to what to measure. Looking at the first law, $\Delta U = q + w$, you see that you will be involved in measuring heat and work. Looking at the second law, either $dS = (C_P/T)\,dT$ (4.32) or $dS = dH/T$ (4.31), you see that measuring heat at various temperatures is also involved. Measuring quantities associated with q and with w is therefore fundamental to experimental thermodynamics. There are quite a number of ways of doing this, but for the moment we can divide them into direct and indirect methods.

Direct Methods

Based on our discussion so far, it seems that we would like most to know $\Delta_r G$ for reactions of interest to us, because this will tell us which way the reactions will go, assuming that pressure and temperature are fixed. For example, if we were not sure which of the two forms of carbon was the stable form at 25 °C, 1 bar, we could measure $G_{\text{graphite}} - G_{\text{diamond}}$, which is $\Delta_r G$ for the reaction

C(diamond) = C(graphite)

and if this quantity was negative, then graphite would be stable, and if it was positive, diamond would be stable. The same reasoning would hold for any complex reaction involving gibbsite, as long as we know the Gibbs energy of every species in the reaction.

There are quite a number of ways of determining changes in Gibbs energy, but we will discuss only the most common one here. Others are associated with determining the equilibrium constant or cell voltages, as we will see in Chapters 9 and 10. To see how changes in G are measured, consider first Equations (4.62) and (3.30)

$$\Delta G_{T,P} = \Delta U - T\,\Delta S + P\,\Delta V \tag{4.62}$$

$$\Delta H_P = \Delta U + P\,\Delta V \tag{3.30}$$

Combining these, we have

$$\Delta G_{T,P} = \Delta H_P - T\,\Delta S \tag{5.1}$$

From this, we see that we can calculate ΔG for a process if we know ΔH_P and ΔS for that process. We know (from Section 3.6.1) that ΔH_P is simply the heat transferred to or from the closed system during a constant pressure process, so all we have to do is carry out some process (reaction) at some constant P, probably atmospheric P, and measure how much heat is evolved or absorbed.

This ΔH is also connected to ΔS, but only for a reversible process, as we have seen. This might be a little difficult to do, experimentally. But we also have an equation involving ΔS, C_P, and T (Equation (4.32)), and C_P is related to ΔH (Equation (3.40)), so it begins to look like measuring heat is pretty important, even if you are not concerned with heat flow in your field situation. Therefore, *calorimetry*, the art and science of measuring amounts of heat, is the secret to determining values of ΔG.

Indirect Methods

So far we have talked about using thermodynamics to determine phase relationships. But the opposite approach can also be used; phase relationships determined under strictly controlled conditions (meaning controlled by the phase rule, Chapter 12) can be used to deduce fundamental thermodynamic properties. For example, for gibbsite, you might conduct experiments to determine the temperature at which gibbsite changes to corundum,

$$2\,Al(OH)_3(s) = Al_2O_3(s) + 3\,H_2O(l) \tag{5.2}$$

or, you might determine the solubility of gibbsite in water at various temperatures and pH values. Both these relationships are interesting to Earth scientists even without thermodynamic manipulation, but they can also be used to determine the thermodynamic properties of gibbsite and related species. This is a subject for later chapters, but intuitively it would seem that, knowing the properties of corundum and water, you might be able to deduce those for gibbsite from the requirement that reaction (5.2) be at equilibrium.

Even more importantly, families of such relationships, that is, a number of relationships involving the same minerals, can be used to test the consistency of thermodynamic data. For example, suppose you have determined $\Delta_f H^\circ$ for each of gibbsite, corundum, and water from calorimetry. If you get a different result from studying reaction (5.2), then there is some error. Or you might have several different reactions involving gibbsite, so your data for gibbsite must fit all those reactions. In fact all the data for all the minerals in your database must fit all the reactions that you know about. Finding errors in these cases can be difficult, but the point is that phase relationships are a good way to examine how consistent your data are. For now, let's concentrate on calorimetry, the classical method for determining thermodynamic properties.

5.3   Solution Calorimetry

· ·

Heat flows can be measured in various ways. One way is to observe some process in which heat is liberated under controlled conditions, resulting in a rise in temperature, and then duplicate that temperature rise using an electrical heater. The energy used by the heater can be measured exactly, and will equal the energy released by the process considered. This is the principle used in the calorimeter in Figure 5.1.

This apparatus is used to measure how much heat is liberated when a known amount of solid material, such as a mineral, dissolves. Most minerals are notoriously insoluble in water and so an acid, such as hydrofluoric acid (HF), is used. The method is called *solution calorimetry*.

5.3.1   The Method

A measured number of grams (therefore a known number of moles) of crushed mineral are put into the *sample holder* and sealed with gold foil. The sample holder is then placed in the *reaction chamber*, which is then filled with acid. A long rod reaches from the top of the sample holder through various seals to the top of the apparatus. This assembly is then sealed and placed in a vacuum chamber, which goes into a water bath. The purpose of the vacuum and water bath is to minimize the loss of heat from the reaction chamber.

When everything has settled down, the mineral sample and the acid are at the same temperature, but are separated. The long rod is then pushed down. This punctures the seal

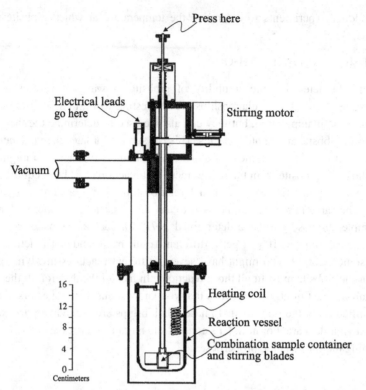

Figure 5.1 An adiabatic heat-of-solution calorimeter. The reaction vessel contains acid. Pushing down on the handle at the top punctures the upper seal and pushes out the bottom of the sample container, allowing the sample to dissolve. A thermometer is wound around the reaction vessel and records the change in temperature. No electrical leads are shown. (Simplified from Robie and Hemingway (1972).)

on the sample container, and the bottom also falls out, allowing the sample to mix with the acid and dissolve. The sample holder also has fins, and rotation stirs the solution and speeds up the dissolution process. The dissolution of the mineral releases heat, which raises the temperature of the acid, and the amount of temperature change is measured by a resistance thermometer, which is wrapped around the reaction vessel. The apparatus is calibrated by using an *electrical heating coil* to raise the temperature in a different experiment, but using exactly the same setup. The voltage drop across the electrical heater and the current flowing through it are known, and so the amount of heat required to raise the temperature of the calorimeter by any given amount is known exactly by turning on the heater for a short time and observing the temperature increase. By comparing the temperature change caused by the heating coil to that caused by the mineral dissolution, the heat liberated by the mineral dissolution can be determined quite precisely. A number of small corrections must be made for various heat losses in the apparatus, plus a correction to the heat measured over the temperature interval to what would have been observed if the process had occurred at a constant temperature of 25 °C. These calculations require a knowledge of the *heat capacity* of the calorimeter (see below). Because we know the mass of mineral grains used, the heat of solution per mole of mineral at 25 °C then can be calculated. The whole process is exacting and painstaking.

5.3.2   The Interpretation

What are the meaning and use of this heat of solution? In terms of the processes we have been discussing, we have observed an irreversible reaction between a metastable state (pure acid and mineral grains, separated, at T_1) and a stable state (mineral dissolved in acid at T_2), made some measurements, and calculated from this the heat that would be released in the reaction

mineral, HF separated $\rightarrow$ mineral dissolved in HF $+\, q_{\text{dissolution}}$

at 25 °C. If the calorimeter is open to the atmosphere, then the mineral dissolution process happens at a constant pressure, and by Equation (3.31), the heat measured, $q_{\text{dissolution}}$, is equal to the change in enthalpy of the system, ΔH. This change is illustrated in Figure 5.2.

Of course, a heat of solution is not exactly what we wanted, although it is a ΔH. Well, strictly speaking it is a $\Delta \mathbf{H}$ in Joules, which we convert to a ΔH, J mol^{-1}, using the number of moles used in the experiment (see below, Section 5.3.3). We want a ΔH that is the difference between products and reactants of reactions of all kinds, such as our corundum–water–gibbsite and diamond–graphite reactions, and innumerable others. But the heat of solution technique allows us to do this. Note that that is because, in any balanced chemical reaction, the total or bulk composition of the reactants must be exactly the same as that of the products. That's what "balanced" means – all the atoms on the left side of the reaction must appear also on the right. Therefore, if in separate experiments we dissolve the reactants and the products in the same kind of acid, we will get identical solutions. We will, however, measure different heats of solution, because the products and reactants have different structures and different energy contents. Therefore, the difference in the heats of solution must be equal to the difference in enthalpies of the products and reactants themselves.

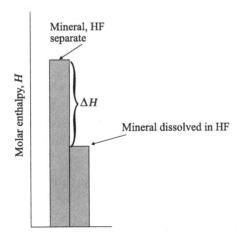

Figure 5.2 Enthalpy is a state variable of fixed but unknown value in the beginning and final equilibrium states. ΔH is obtained by measuring the heat liberated in the reaction at constant pressure.

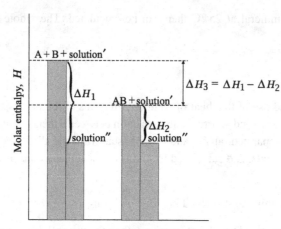

Figure 5.3 Compounds A + B together have heat of solution ΔH_1. Compound AB has heat of solution ΔH_2. Both processes result in solution″, so the heat of reaction for A + B → AB, which cannot be carried out in a calorimeter, is $\Delta H_1 - \Delta H_2$, or $\Delta H_2 - \Delta H_1$ if the reaction A + B → AB is exothermic.

To put this argument in formal terms, suppose our reaction is

A + B = AB

for example, $SiO_2 + Al_2O_3 = Al_2SiO_5$. First we dissolve the reactants, and then, in a separate experiment, we dissolve the products:

A + B + solution′ → solution″ + heat (ΔH_1)

AB + solution′ → solution″ + heat (ΔH_2)

As long as both solution′ and solution″ have the same composition in both reactions, the reactions may be subtracted, giving

A + B → AB + ΔH_3

where $\Delta H_3 = \Delta H_1 - \Delta H_2$ and is the heat of reaction ($\Delta_r H$) of the reaction A + B = AB, as shown in Figure 5.3. A, B, and AB can also refer to complex organic compounds of any kind. We need only be able to separate them into their pure forms, so as to be able to work with them.

Because of practical difficulties, the determination of $\Delta_f H°$ of a compound is rarely the sum of only two heats of solution, as in Figure 5.3. Quite often 10 or 15 solution reactions may have to be carried out to determine one $\Delta_f H°$, and the whole process may take several weeks.

5.3.3 A Real Example

The enthalpy of formation from the elements has been determined for gibbsite most recently by Hemingway and Robie (1977), and we look at their results here. The reactions they used, at a calorimeter temperature of 303.5 K, were as follows:

$$Al(s) + 3\,HF(aq) = AlF_3(aq) + \tfrac{3}{2}H_2(g) \qquad \Delta_r H_1° = -595{,}195 \pm 1192\,J\,mol^{-1} \tag{5.3}$$

$$Al(OH)_3(s) + 3\,HF(aq) = AlF_3(aq) + 3\,H_2O(l) \qquad \Delta_r H_2° = -2046 \pm 3\,J\,mol^{-1} \tag{5.4}$$

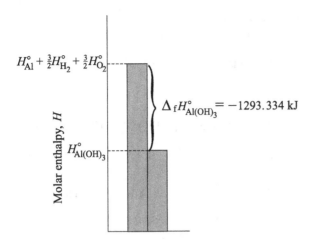

Figure 5.4 The meaning of $\Delta_f H^\circ_{\text{gibbsite}}$.

$$3\,H_2(g) + \tfrac{3}{2}O_2(g) = 3\,H_2O(l) \qquad \Delta_r H^\circ_3 = -857{,}490 \pm 75\,\text{J mol}^{-1} \qquad (5.5)$$

The enthalpy of formation from the elements for gibbsite is the reaction

$$\text{Al}(s) + \tfrac{3}{2}H_2(g) + \tfrac{3}{2}O_2(g) = \text{Al(OH)}_3(s) \qquad \Delta_f H^\circ_{\text{gibbsite}} \qquad (5.6)$$

so *apparently* we get this from the sum $\Delta_r H^\circ_1 - \Delta_r H^\circ_2 + \Delta_r H^\circ_3$.[1] Actually, we do not. We said that the aqueous solution would cancel out if it was identical in all cases. But in reaction (5.3) one mole of Al dissolves in HF solution and H_2 gas is evolved, which leaves the calorimeter. But in reaction (5.4), a mole of Al dissolves in the same kind of HF solution, but it brings with it some H_2O from the gibbsite, which dilutes the acid solution. So the solutions after reactions (5.3) and (5.4) do not have the same composition. This dilution of the acid solution is not a neutral process, but evolves heat, and this must be measured. But this raises another question – how do you write a dilution reaction? And what is the composition of the solutions in those reactions, anyway? The way we usually write reactions gives no clue as to the composition of the solutions after the dissolution reactions. The calorimetry people actually write their reactions in a much more explicit way, as shown in Box 5.1. A graphical idea of the meaning of $\Delta_f H^\circ_{\text{gibbsite}}$ is shown in Figure 5.4.

Box 5.1 The Enthalpy of Gibbsite

The gibbsite reactions in Section 5.3.3 were carried out by dissolving 0.005 moles of Al (0.1349 g Al; 0.3900 g Al(OH)$_3$) in 920.0 g of 20.1% HF solution. This HF solution has 184.92 g of HF (9.243 moles), and 735.08 g H_2O (40.8029 moles). Measurements give values of *q* in joules. To report measurements per mole of Al, these amounts of heat are multiplied by $1/0.005 = 200$, resulting in values of q_P or ΔH J mol^{-1}. Therefore the

---

[1] Remember that to subtract a reaction the best way is to *reverse* it, in this case reaction (5.4), change the sign of $\Delta_f H^\circ_2$, and then add the reactions and the $\Delta_f H^\circ$ values.

starting HF solution for the experiments, per mole of Al, has $9.243 \times 200 = 1849$ moles HF and $40.8029 \times 200 = 8163$ moles H_2O, so the dissolution reactions are written

$$Al(s) + [1849\,HF + 8263\,H_2O](aq) = [AlF_3 + 1846\,HF + 8163\,H_2O](aq) + \tfrac{3}{2}H_2(g)$$

and

$$Al(OH)_3(s) + [1849\,HF + 8263\,H_2O](aq) = [AlF_3 + 1846\,HF + 8166\,H_2O](aq)$$

The dilution reaction was performed by dissolving water into the solution produced by reaction (5.3), thus

$$3\,H_2O(l) + [AlF_3 + 1846\,HF + 8163\,H_2O](aq) = [AlF_3 + 1846\,HF + 8166\,H_2O](aq)$$

This results in a correction of $-622\,J\,mol^{-1}$. In addition, the heat capacities of $Al(OH)_3$, Al, H_2O, and H_2 were required to correct the measurement from 303.5 K to 298.15 K, $(+615\,J\,mol^{-1})$.

Finally, note that the actual calorimetric measurement of reaction (5.3) was not $-595,195\,J\,mol^{-1}$, but $-592,952\,J\,mol^{-1}$. The problem is that the H_2 gas which is evolved during the experiment, and which escapes from the calorimeter, carries with it some HF vapor and some H_2O vapor, and this causes some cooling due to evaporation. Correcting for this requires knowledge of the enthalpy of vaporization of HF and H_2O from the experimental solution. The final result is $\Delta_f H^\circ_{\text{gibbsite}} = -1,293,130\,J\,mol^{-1}$.

The $\Delta_f H^\circ$ of gibbsite in Appendix B, from the NBS tables of Wagman et al. (1982) is very slightly different, $-2586.67\,kJ\,mol^{-1}$ for $Al_2O_3 \cdot 3H_2O$, or $-1,293,334\,J\,mol^{-1}$ for $Al(OH)_3$, and is from Hemingway et al. (1982). The difference is due to the use of different data fitting techniques and is not statistically significant.

Thermodynamic data are not written in stone, they are written in blood, so to speak. Many obscure errors are possible, but you should not change data to suit your latest theory.

5.4   The Third Law

Now we must consider how to measure entropy. Differences in entropy between two equilibrium states can be measured using the heat capacity, as discussed in Section 4.10, possibly including provision for a changing heat capacity as in Section 3.6.6. By now you are probably accustomed to being told that we cannot know the absolute values of thermodynamic parameters, only differences. But this applies only to the internal energy, U, and any parameters that contain U, such as H and G. Entropy is different in that we *can* get absolute values, by virtue of the third law of thermodynamics.

5.4.1   The Third Law – Historical Aspects

Lacking an absolute value in some state, entropy is in the same boat as enthalpy and Gibbs energy, having only differences rather than absolute values. Differences in S can be determined from Equation (4.34), but this means that the only way to determine the difference in entropy for any chemical reaction is for there to be some *equilibrium* path between the products and reactants. For example, to determine ΔS between rhombic and monoclinic sulfur at 298.15 K and 1 bar, you would need to measure heat capacities from 25 to 95 °C (the equilibrium phase transition temperature) for both phases, then integrate (4.34) up to 95 °C for rhombic S and back to 25 °C for monoclinic S. To determine the ΔS between calcite and aragonite, you would need to determine the effect of pressure on the entropy of calcite up to the aragonite–calcite equilibrium pressure, then back down to the starting pressure using the effect of pressure on the entropy of aragonite.

This is but one of a host of difficulties you would have in finding an equilibrium path between states you were interested in. In some cases an equilibrium path exists but kinetic or other factors make the experimental determinations difficult. In other cases, such as with virtually all organic compounds, no equilibrium path is possible, so using thermodynamics would be greatly inhibited. Until the development and general acceptance of the third law, this was in fact the case. Contributions by Lewis (1899), Richards (1902), van 't Hoff (1904), Haber (1905), Nernst (1906), and Planck (1912), not to mention Einstein's (1907) fundamental work on heat capacities and Boltzmann's development of an atomistic approach (references in Lewis and Randell (1923, Chapter 31)), led to the general belief that the heat capacity, and perhaps entropy, became zero at absolute zero temperature. However, convincing calorimetric data supporting this idea, as well as confirming the limitations, developed only during the 1920s and 1930s. W. F. Giauque received the Nobel Prize in Chemistry in 1949 for his lifelong contributions to our understanding of the third law.

The physics of materials at low temperatures is now a large and important topic, and a complete understanding of the third law requires some knowledge of statistical mechanics and even some quantum mechanics. A fairly brief overview is Wilks (1961). However, for those whose interests lie at the other end of Earth's temperature spectrum and are mainly interested in having accurate thermochemical data, the only important aspect of the third law is that it provides an absolute reference point for entropy data.

The structure of thermodynamics is based on the first and second laws, but it is the third law which allows the structure to be useful for chemical reactions. By far the most data on Gibbs energy differences and equilibrium constants has been obtained through use of third law entropies.

5.4.2   Statement of the Third Law

The statement of the third law by Lewis and Randell (1923) is still useful:

If the entropy of each element in some crystalline state be taken as zero at the absolute zero of temperature: *every substance has a finite positive entropy, but at the absolute zero of temperature the entropy may become zero, and does so become in the case of perfect crystalline substances.*

The reference to perfect crystalline substances means that a non-zero entropy may be "frozen-in" at low temperatures, and is so in the case of glasses, gels, and various other substances having some configurational disorder. The most important case for geologists is that of solid solutions, in which two or more atoms occupying a crystal lattice site may be disordered. This disorder is undoubtedly still present when the crystal is cooled down to cryogenic temperatures for heat capacity measurements, so the entropy does not approach zero at 0 K. This "residual entropy" must be calculated and added to the entropy evaluated from Equation (5.7). Also note the condition that the entropy of the (perfectly crystalline) elements is assumed to be zero. The heat capacity is certainly zero at 0 K, but all we can really say about entropy is that the entropy of all perfectly crystalline substances becomes the same at 0 K, and is called zero by convention (Melrose, 1970).

If we let T_1 be absolute zero in Equation (4.33), the entropy of minerals at any temperature, say our standard temperature of 298.15 K, is (assuming no residual entropy)

$$S_{298} = \int_{T=0}^{T=298} \frac{C_P}{T}\, dT \tag{5.7}$$

and all that is required to determine "absolute" values for the entropy of minerals is to measure their heat capacity at a series of temperatures between zero and 298.15 K and to evaluate the integral. This gives rise to another kind of calorimetry, *cryogenic*, or low-temperature calorimetry.

5.4.3 Cryogenic Calorimetry

A cryogenic calorimeter (Figure 5.5) is an apparatus designed for the determination of heat capacities at very low temperatures. The procedure is to cool the sample down to a temperature within a few degrees of absolute zero (a temperature of absolute zero itself

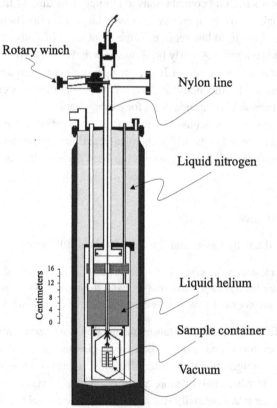

Rotary winch

Nylon line

Liquid nitrogen

Liquid helium

Sample container

Vacuum

Centimeters — 16, 12, 8, 4, 0

Figure 5.5 A cryogenic or low-temperature calorimeter. The sample container can be raised by the rotary winch so as to be in contact with the liquid helium reservoir for cooling to 4.2 K, or lowered into the vacuum for heating. The re-entrant well in the sample container contains a heating coil. (Simplified from Robie and Hemingway (1972).)

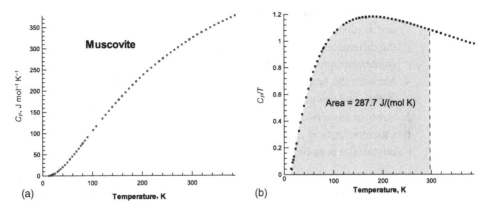

Figure 5.6 (a) Measured heat capacity of muscovite as a function of temperature (Robie *et al.*, 1976). (b) C_P/T vs. T for the same data. Integration gives the shaded area under the curve, which is equal to the entropy at the upper limit of integration, in this case, $S^{\circ}_{298.15} = 287.7 \, \mathrm{J \, mol^{-1}}$.

is actually impossible to achieve, a fact actually implicit in the third law), introduce a known quantity of heat using an electrical heating coil, and observe the resulting increase in temperature (usually a few degrees). The quantity of heat is equal to ΔH, and this divided by the temperature difference gives an approximate value of C_P at the midpoint of the temperature range. Corrections are then made to compensate for heat leaks, for the heat absorbed by the calorimeter, and to get exact C_P values from the approximate ones. The integration of C_P/T values to obtain the entropy at 298.15 K is illustrated in Figure 5.6. A much more detailed description of the calorimeter and its operation is in Robie (1987).

5.5 The Problem Resolved

We now know how to tell which way reactions will go, not just in a theoretical way (they will decrease their Gibbs energy at T, P), but in a practical way (how do we get this ΔG?). For example, SiO_2 comes in several crystalline varieties (polymorphs), such as the minerals quartz and cristobalite. They have the same composition SiO_2, but different crystallographic structures and energy contents, and one is stable and one is metastable at 25 °C, 1 bar. The problem is analogous to the diamond–graphite problem, and the reaction is

$$SiO_2^{\text{cristobalite}} = SiO_2^{\text{quartz}}$$

Which way does this reaction go at 25 °C, 1 bar? You could answer this question thermodynamically as follows.

1. Dissolve quartz in a solvent (HF acid) and measure the heat released.
2. Dissolve the same amount of cristobalite in the same amount of the same solvent and measure the heat released.

3. Because the solution after dissolution in the two cases has exactly the same composition and is identical in all respects, the difference in the two measured heat terms must be the difference in enthalpy between quartz and cristobalite, $\Delta_r H = H_{298}^{\text{quartz}} - H_{298}^{\text{cristobalite}}$ (remember, products minus reactants).

4. Measure the heat capacities of both quartz and cristobalite from near absolute zero to 298 K, and calculate S_{298}^{quartz} and $S_{298}^{\text{cristobalite}}$.

5. Subtract these two entropies to give $\Delta_r S = S_{298}^{\text{quartz}} - S_{298}^{\text{cristobalite}}$.

6. Calculate $\Delta_r G = \Delta_r H - 298.15 \cdot \Delta_r S$. If this is negative, quartz is stable; if it is positive, cristobalite is stable.

You are no doubt quite sure that you don't have to do this incredible amount of work to answer such a simple question; there must be an easier way. Well, there is, but only because other people have already done this incredible amount of work, and lots more like it. In other words, you can look up the data in tables. However, you do not look up heats of solution.

The trick we have just used to get the difference in enthalpy between two minerals (that is, to dissolve them both and subtract the heats of solution) is a very useful way of determining heats of reaction, because many reactions proceed very slowly, or not at all, so you cannot measure the heat of reaction directly in a calorimeter. You cannot measure the heat of reaction as cristobalite changes directly into quartz at 25 °C, because it never does – it is a truly metastable form of SiO_2. However, most minerals will dissolve fairly rapidly in some kind of solvent, providing an indirect means of getting their enthalpy differences.

To have tables of data that enable you to calculate $\Delta_r H$ for any reaction, it would seem that all you need to do is tabulate heats of solution. But you would soon find that although fine in theory, this would not work well in practice. For one thing, you would need to tabulate heats of solution for all combinations of substances that might be of interest. That is, the heats of solution of gibbsite, corundum, and water separately are not enough – you need the heat of solution of corundum + water in a 1:1 ratio. But for other reactions, you would need corundum + water in other proportions. Then you would need to be sure that the solution compositions were identical, and, given the variety and concentrations of solvents used, your database would soon become very large and unmanageable. This problem is resolved, of course, by using "formation from the elements" properties, as discussed in Chapter 3. This enables us to tabulate a single number for each property for each compound, and makes calculation of reaction deltas easy, at least for standard conditions.

5.5.1 Gypsum–Anhydrite Example

The change in enthalpy for any reaction between compounds for which there are formation-from-the-element data is given by a simple algebraic addition of these $\Delta_f H°$ terms, because in balanced reactions, the elements always cancel out. To take another example, consider reaction (3.34) again:

$$CaSO_4(s) \text{ (anhydrite)} + 2\,H_2O(l) = CaSO_4 \cdot 2H_2O(s) \text{ (gypsum)} \tag{5.8}$$

5.5 The Problem Resolved

Both gypsum and anhydrite occur at the Earth's surface, and it is not always clear which is the stable phase. To determine the enthalpy change in this reaction, we consider the reactions in which each phase in the reaction is formed from its elements:

$$Ca(s) + S(s) + 2O_2(g) = CaSO_4(s); \qquad \Delta_f H^\circ_{anhydrite} = -1434.11 \, kJ \, mol^{-1}$$

$$H_2(g) + \tfrac{1}{2}O_2(g) = H_2O(l); \qquad \Delta_f H^\circ_{water} = -285.830 \, kJ \, mol^{-1}$$

$$Ca(s) + S(s) + 3O_2(g) + 2H_2(g) = CaSO_4 \cdot 2H_2O(s); \qquad \Delta_f H^\circ_{gypsum} = -2022.63 \, kJ \, mol^{-1}$$

So for reaction (5.8) we have

$$\Delta_r H^\circ = \Delta_f H^\circ_{gypsum} - \Delta_f H^\circ_{anhydrite} - 2\,\Delta_f H^\circ_{water}$$

$$= -2022.63 - (-1434.11) - 2\,(-285.830)$$

$$= -16.86 \, kJ \, mol^{-1}$$

from which we see that the reaction between anhydrite and water to form gypsum is exothermic; that is, 16.86 kJ of heat would be released for every mole of anhydrite reacted.

As mentioned in Chapter 3, it is important to realize that this heat of reaction, $\Delta_r H^\circ$, is equal to the difference in the *absolute* enthalpies of the reactants and products – the enthalpies of the elements have nothing to do with it, because they all cancel out. Thus

$$\Delta_r H^\circ = \Delta_f H^\circ_{gypsum} - \Delta_f H^\circ_{anhydrite} - 2\,\Delta_f H^\circ_{water}$$

$$= H^\circ_{CaSO_4 \cdot 2H_2O} - H^\circ_{Ca} - H^\circ_{S} - 2H^\circ_{H_2} - 3H^\circ_{O_2}$$

$$- (H^\circ_{CaSO_4} - H^\circ_{Ca} - H^\circ_{S} - 2H^\circ_{O_2}) \qquad (5.9)$$

$$- 2(H^\circ_{H_2O} - H^\circ_{H_2} - \tfrac{1}{2}H^\circ_{O_2})$$

$$= H^\circ_{CaSO_4 \cdot 2H_2O} - H^\circ_{CaSO_4} - 2\,H^\circ_{H_2O}$$

$$= -16.86 \, kJ \, mol^{-1}$$

If you look carefully, you'll see that all the H° terms for the elements cancel out. But if you think before you look, you'll realize that they *must* cancel out if the reaction is balanced. Otherwise there is a mistake somewhere. Despite many statements to the contrary, the absolute enthalpies of the elements ($H^\circ_{Ca}, H^\circ_{O_2}$, etc.) are not assumed to be zero. There is no need to do so, because they all cancel out in balanced reactions.

As you might expect, all the hard work of cryogenic calorimetry has already been done too, for most common substances, and the results are obtainable in tables and compilations of data. From Appendix B, we find that

$$S^\circ_{CaSO_4} \, (anhydrite) = 106.7 \, J \, mol^{-1} \, K^{-1}$$

$$S^\circ_{CaSO_4 \cdot 2H_2O} \, (gypsum) = 194.1 \, J \, mol^{-1} \, K^{-1}$$

$$S^\circ_{H_2O} \, (water) = 69.91 \, J \, mol^{-1} \, K^{-1}$$

and the entropy of reaction for (5.8) is

$$\Delta_r S^\circ = S^\circ_{gypsum} - S^\circ_{anhydrite} - 2\,S^\circ_{water}$$
$$= 194.1 - 106.7 - 2(69.91)$$
$$= -52.42\,\mathrm{J\,mol^{-1}\,K^{-1}}$$

So Is Gypsum or Anhydrite Stable?

Now that we have numerical values for both $\Delta_r H^\circ$ and $\Delta_r S^\circ$ for reaction (5.8), it is a simple matter to calculate $\Delta_r G^\circ$ to see which way the reaction goes. Our number for enthalpy is in kJ and that for entropy is in J, so we must convert one of them to be consistent. Converting kJ to J, we have

$$\Delta_r G^\circ = \Delta_r H^\circ - T\,\Delta_r S^\circ$$
$$= -16{,}860 - 298.15(-52.42)$$
$$= -1231\,\mathrm{J\,mol^{-1}}$$

which is negative; therefore, gypsum is more stable than anhydrite in the presence of water at Earth surface conditions. We repeat that what we have found is that the *assemblage* of anhydrite plus water is metastable with respect to gypsum at 25 °C, 1 bar. Anhydrite by itself is not metastable, as there is no other form of $CaSO_4$ that has a lower energy.

Gibbs Energy from Tables

Although we must do the calorimetry experiments in order to calculate free energy differences, there is usually no need to use $\Delta_r H^\circ$ and $\Delta_r S^\circ$ values from tables to calculate $\Delta_r G^\circ$. Values of $\Delta_f G^\circ$ for most compounds have been calculated and are also to be found in the same tables of data, and so we can use these values directly, instead of going through the $\Delta_r H^\circ - T\,\Delta_r S^\circ$ calculation.

For example, $\Delta_f G^\circ$ for anhydrite can be calculated from

$$\Delta_f G^\circ_{CaSO_4} = \Delta_f H^\circ_{CaSO_4} - T\,\Delta_f S^\circ_{CaSO_4}$$

where $\Delta_f S^\circ_{CaSO_4}$ is

$$\Delta_f S^\circ_{CaSO_4} = S^\circ_{CaSO_4} - S^\circ_{Ca} - S^\circ_{S} - 2\,S^\circ_{O_2}$$

Don't forget that absolute entropies are obtainable for the elements just as well as for compounds, and these numbers are available in tables of data, such as Appendix B. These numbers are given in Table 5.1

So

$$\Delta_f S^\circ_{CaSO_4} = 106.7 - 41.42 - 31.80 - 2 \times 205.138$$
$$= -376.796\,\mathrm{J\,mol^{-1}\,K^{-1}}$$

Table 5.1. Absolute entropies.

Substance	$S°$ ($\mathrm{J\,mol^{-1}\,K^{-1}}$)
$CaSO_4(s)$	106.7
$Ca(s)$	41.42
$S(s)$	31.80
$O_2(g)$	205.138

Therefore, the Gibbs energy of formation of anhydrite is

$$\Delta_f G°_{CaSO_4} = \Delta_f H°_{CaSO_4} - T\,\Delta_f S°_{CaSO_4}$$
$$= -1,434,110 - 298.15(-376.796)$$
$$= -1,321,768\,\mathrm{J\,mol^{-1}}$$
$$= -1321.77\,\mathrm{kJ\,mol^{-1}}$$

which is the number for $\Delta_f G°$ in Appendix B ($-1321.79\,\mathrm{kJ\,mol^{-1}}$), within the limits of accuracy of the data.

The calculation for determining whether gypsum or anhydrite is stable is therefore a little easier – we just look up the $\Delta_f G°$ numbers instead of both the $\Delta_f H°$ and $S°$ numbers. Thus

$$\Delta_r G° = \Delta_f G°_{CaSO_4 \cdot 2H_2O} - \Delta_f G°_{CaSO_4} - 2\,\Delta_f G°_{H_2O}$$
$$= -1,797,280 - (-1,321,790) - 2(-237,129)$$
$$= -1232\,\mathrm{J\,mol^{-1}} \tag{5.10}$$

which is what we got before ($-1231\,\mathrm{J\,mol^{-1}}$), within the limits of accuracy of the tabulated data.

Again, although we use the Gibbs energies of formation, the G values for the elements all cancel out, and what we calculate is the difference between the absolute Gibbs energies of the compounds in the reaction. Free energy and enthalpy are similar in this respect.

5.5.2   An Aqueous Organic Example

To emphasize that our model is just as useful for organic or biochemical processes as for mineralogical ones, let's take another look at the reaction involving amino acids we considered in Section 2.6.1. Equations (2.1) and (2.3) are, to repeat,

$$C_8H_{16}N_2O_3(aq) + H_2O(l) = C_6H_{13}NO_2(aq) + C_2H_5NO_2(aq) \tag{5.11}$$

$$C_6H_{13}NO_2(aq) + C_2H_5NO_2(aq) = 2\,H_2(g) + 2\,NH_3(g) + 4\,H_2O(l) + 8\,C_{\mathrm{graphite}} \tag{5.12}$$

From the tables in Appendix C, we find the properties listed in Table 5.2.

Table 5.2. Properties of various substances.

Substance	Formula	$\Delta_f G°$ (J mol^{-1})
Leucine	$C_6H_{13}NO_2(aq)$	−343,088
Glycine	$C_2H_5NO_2(aq)$	−370,778
Leucylglycine	$C_8H_{16}N_2O_3(aq)$	−462,834
Hydrogen	$H_2(g)$	0
Ammonia	$NH_3(g)$	−16,450
Water	$H_2O(l)$	−237,129
Graphite	$C(s)$	0

Therefore, for reaction (5.11),

$$\Delta_r G° = \Delta_f G°_{leucine} + \Delta_f G°_{glycine} - \Delta_f G°_{leucylglycine} - \Delta_f G°_{water}$$
$$= -343,088 - 370,778 - (-462,834) - (-237,129)$$
$$= -13,903 \text{ J mol}^{-1}$$

and for reaction (5.12)

$$\Delta_r G° = 2\,\Delta_f G°_{H_2(g)} + 2\,\Delta_f G°_{NH_3(g)} + 4\,\Delta_f G°_{H_2O(l)} + 8\,\Delta_f G°_{C(s)}$$
$$\quad - \Delta_f G°_{leucine} - \Delta_f G°_{glycine}$$
$$= 2(0) + 2(-16,450) + 4(-237,129) + 8(0)$$
$$\quad - (-343,088) - (-370,778)$$
$$= -267,550 \text{ J mol}^{-1}$$

Thus we see that, as illustrated in Figure 2.9, both reactions have a negative "chemical energy," or $\Delta_r G°$. However, to say any more about these reactions, we must emphasize a factor we have not yet mentioned, and which we cannot develop fully until Chapter 9 (Section 9.7).

The Problem with Solutions

It matters not a bit whether the substances we consider are organic or inorganic, stable or metastable, as long as we have data for them. But it matters a great deal whether they are pure substances (such as gypsum, quartz, diamond, liquid water, etc.) or are dissolved in some solvent, as with all substances designated (aq) in the tables. The problem is that the Gibbs energy (and all other properties) of a pure substance is a fixed and known quantity, but the Gibbs energy of a substance in solution depends on its concentration. The tabulated values of $\Delta_f G°$ for (aq) substances are for one particular standard concentration. Therefore, although we have calculated a negative $\Delta_r G°$ for our two reactions above, they both involve at least some dissolved substances and, therefore, the conclusion that the reactions should proceed spontaneously applies only when all the (aq) substances have the standard concentrations. We look more carefully at these problems in Chapters 7, 8, and 9.

Enzymes as Catalysts

One more thing to note about chemical reactions is that living organisms have evolved mechanisms involving enzymes that overcome the energy barriers between reactants and products for reactions required by the organism. Such reactions, therefore, proceed easily and quickly, whereas in the inorganic world, diamond persists forever in its metastable state. No organism needs to change diamond to graphite, so no enzymes exist for this reaction. Living organisms also have mechanisms that drive some reactions "uphill," or against the Gibbs energy gradient. Thus peptide bonds are formed in organisms, as well as broken. The energy required to do this is obtained ultimately from the sun, but the exact mechanisms are complex. The study of such reactions forms a large part of the science of biochemistry.

5.6 Data at Higher Temperatures

Everything we have discussed so far is about determining data for "standard conditions," which usually means pure phases at 25 °C, 1 bar, although we will see later that it *can* mean something else. But as geologists we often deal with reactions at metamorphic and igneous temperatures of many hundreds of degrees. Specialized calorimeters can be used up to a few hundred degrees, but the experimental difficulties become great. Obviously other methods are needed. As usual, there are several, but we will mention just two.

5.6.1 Drop Calorimetry

The amount of heat required to raise the temperature of a mole of substance from T_r to T at constant pressure is simply $H_T - H_{T_r}$ (or $H_T^\circ - H_{T_r}^\circ$ for a standard reference substance); again, a difference between two unknown quantities. This quantity is conveniently determined by *cooling* the substance from T to T_r and measuring the amount of heat given up by the substance during this process. Drop calorimetry has actually been largely superseded by differential scanning calorimetry, but it illustrates the acquisition of high-temperature enthalpies and heat capacities more intuitively than does differential scanning calorimetry, and many of the data used nowadays were obtained by this method.

A calorimeter is placed directly under a furnace and the sample is dropped from the furnace where it has temperature T_1, into the calorimeter, where it gives up its heat and achieves temperature T_2 (Figure 5.7). The amount of heat given up by the sample is determined by using this heat to melt a working substance in the calorimeter (either H_2O or diphenyl ether $(C_6H_5)_2O$), and measuring the volume change of this substance by the displacement of mercury. The relationship between the volume change and the ΔH of the solid $\rightarrow$ liquid phase transition (T_2 in the calorimeter is 273.15 K for H_2O; 303.03 K for diphenyl ether) is accurately known, so this amount of heat equals $H_{T_1} - H_{T_2}$. Small corrections are then applied using heat capacities to adjust this ΔH to $H_T - H_{T_r}$, where T_r is invariably 298.15 K.

More details of the method are given by Robie (1987). Experimental results for muscovite are shown in Figure 5.8.

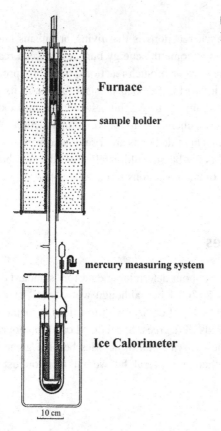

Figure 5.7 A drop calorimeter. Simplified from Douglas and King (1968).

Furnace

sample holder

mercury measuring system

Ice Calorimeter

10 cm

Values of $H_T - H_{T_r}$ can be combined to give $\Delta_f H°$ for substances at high temperatures. Thus for any substance

$$\Delta_f H_T° = \Delta_f H_{T_r}° + \Delta_f(H_T° - H_{T_r}°) \tag{5.13}$$

where Δ_f refers to the reaction in which the substance is formed from its elements. For example,

$$\Delta_f(H_T° - H_{T_r}°)_{SiO_2} = (H_T° - H_{T_r}°)_{SiO_2} - (H_T° - H_{T_r}°)_{Si} - (H_T° - H_{T_r}°)_{O_2}$$

$$= (H_{T,SiO_2}° - H_{T,Si}° - H_{T,O_2}°) - (H_{T_r,SiO_2}° - H_{T_r,Si}° - H_{T_r,O_2}°)$$

$$= \Delta_f H_{T,SiO_2}° - \Delta_f H_{T_r,SiO_2}° \tag{5.14}$$

and therefore

$$\Delta_f H_{T,SiO_2}° = \Delta_f H_{T_r,SiO_2}° + \Delta_f(H_T° - H_{T_r}°)_{SiO_2} \tag{5.15}$$

Having values of $\Delta_f H°$ for many compounds then allows calculation of the heat of reaction $\Delta_r H°$ for any reaction of interest.

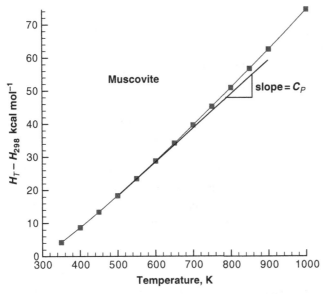

Figure 5.8 Values of $H_T - H_{298}$ for muscovite as measured in a drop calorimeter. The slope of the curve at any point equals the heat capacity at that temperature. Data from Pankratz (1964).

To get heat capacities from these measurements, the experimental values of $(H_T^\circ - H_{T_r}^\circ)$ for the substance and its elements are first fitted to a function, which is commonly

$$(H_T^\circ - H_{T_r}^\circ) = A + BT + CT^2 + DT^{-1} \tag{5.16}$$

Once the "best fit" values of A, B, C, and D are calculated, $(H_T^\circ - H_{T_r}^\circ)$ may be computed for any desired temperature. For example, the equation for the muscovite data in Figure 5.8 is

$$H_T^\circ - H_{298}^\circ = -38{,}793 + 97.65\,T + 13.19 \times 10^{-3}\,T^2 + 25.44 \times 10^5\,T^{-1}$$

The Heat Capacity

A knowledge of how the quantity $H_T^\circ - H_{298}^\circ$ varies with T is useful because the first derivative, or the slope of the curve, is the heat capacity, C_P. As we have said, $H_{T_r}^\circ$ is an unknown quantity, but it is certainly a constant, so that

$$\frac{d}{dT}(H_{T_r}^\circ) = 0$$

Therefore

$$\frac{d}{dT}(H_T^\circ - H_{T_r}^\circ) = \frac{d}{dT}(H_T^\circ)$$

$$= C_P^\circ$$

$$= B + 2CT - DT^{-2}$$

or

$$C_P^\circ = a + bT - cT^{-2} \tag{3.41}$$

where, by differentiating Equation (5.16), we see that

$a = B$, $b = 2C$, and $c = D$.

Thus the Maier–Kelley coefficients (Section 3.6.6) for the muscovite data mentioned above are $a = 97.65$, $b = 26.38 \times 10^{-3}$, and $c = 25.44 \times 10^5$, and these are the values for muscovite in Helgeson *et al.* (1978).

We have carried the superscript $^\circ$ throughout this derivation, so clearly we intend "standard conditions" to include high temperatures at times. In this case, it means simply that we are measuring some pure compound, rather than any arbitrary mixture, for which H and C_P would be more appropriate than H° and C_P°.

5.6.2   Entropies above 298 K

Just as with ΔH° values, a common method to calculate S_T° values, or more likely $\Delta_r S^\circ$ values, at elevated temperatures is by using an expression for the change in heat capacity as a function of temperature, such as the Maier–Kelley equation, (3.42). In other words, since

$$\frac{d}{dT}(S^\circ) = \frac{C_P^\circ}{T}$$

then

$$\frac{d}{dT}(\Delta_r S^\circ) = \frac{\Delta_r C_P^\circ}{T}$$

where $\Delta_r S^\circ$ refers to the entropy change of a balanced chemical reaction. Integrating,

$$\int_{T_r}^T d\Delta_r S^\circ = \int_{T_r}^T \frac{\Delta_r C_P^\circ}{T} dT \tag{5.17}$$

There are several functions available which give $\Delta_r C_P^\circ$ as a function of T, such as Equation (3.42), so this is easily integrated, giving values $\Delta_r S_T^\circ - \Delta_r S_{T_r}^\circ$, the change in entropy of a balanced chemical reaction from some temperature T_r, where ΔS° is known, to some other temperature T. Integrating,

$$\int_{T_r}^T d\Delta_r S^\circ = \int_{T_r}^T \frac{\Delta_r C_P^\circ}{T} dT \tag{5.18}$$

Combining this with the Maier–Kelley equation

$$\Delta C_P^\circ = \Delta a + \Delta bT - \Delta cT^{-2} \tag{3.42}$$

we have

$$\Delta_r S_T^\circ - \Delta_r S_{T_r}^\circ = \int_{T_r}^T \left(\frac{\Delta_r a}{T} + \Delta_r b - \frac{\Delta_r c}{T^3} \right) dT$$

or

$$\Delta_r S_T^\circ - \Delta_r S_{T_r}^\circ = \Delta_r a \ln\left(\frac{T}{T_r}\right) + \Delta_r b(T - T_r) + \frac{\Delta_r c}{2}\left(\frac{1}{T^2} - \frac{1}{T_r^2}\right) \tag{5.19}$$

In this equation $\Delta_r S_T^\circ$ refers to the entropy change of any balanced chemical reaction at temperature T. If the reaction is the formation of a compound from its elements, $\Delta_r S_T^\circ$ becomes $\Delta_f S_T^\circ$. Having high-temperature values of both enthalpy (Section 5.6.1) and entropy (Section 5.6.2) then allows calculation of high-temperature values of $\Delta_r G^\circ$, using Equation (5.1), or directly from the MK coefficients as

$$\begin{aligned}
\Delta_r G_T^\circ = {} & \Delta_r G_{T_r}^\circ - \Delta_r S_{T_r}^\circ (T - T_r) \\
& + \Delta_r a\left[T - T_r - T \ln\left(\frac{T}{T_r}\right)\right] \\
& + \frac{\Delta_r b}{2}\left(2TT_r - T^2 - T_r^2\right) \\
& + \frac{\Delta_r c\left(T^2 + T_r^2 - 2TT_r\right)}{2TT_r^2}
\end{aligned} \tag{5.20}$$

5.7   Data at Higher Pressures

In this section we will discuss only the effect of pressure on solid phases, i.e., minerals. The evaluation of the pressure integral is done in quite a different way for gases, water, and aqueous solutes, and will be treated in later chapters.

5.7.1   Effect of P on Gibbs Energy

As shown previously, the derivative of G with respect to P is V, i.e.

$$(\partial G_i / \partial P)_T = V_i \tag{4.51}$$

so that to calculate the effect of P on G_i we must know how V_i varies as a function of P.

Constant Volume

When substance i is a solid phase and thus has relatively small variation of V with both P and T (relative, that is, to liquids and gases), the errors introduced by the assumption that V is not affected by P or T tend to cancel one another out in reactions, and very little error is introduced by assuming that V_i is a constant at all P, T values. As a result, the assumption of constant V for solids is often adopted for minerals, and results in

$$\int_{P_r}^{P} (\partial G_i / \partial P)_T \, dP = G_{i,P} - G_{i,P_r}$$

$$= V_i(P - 1) \tag{5.21}$$

where V_i is the molar volume of the solid phase in $J\,bar^{-1}$, and $P_r = 1$ bar. Determination of the molar volume can be done experimentally, but the molar volume is more commonly calculated from X-ray determination of the mineral structure. If more accurate data are required, there are ways of calculating the change in V with pressure for solid phases.

5.7.2 Effect of P on Enthalpy and Entropy

For most geochemical modeling purposes, only the pressure effect on G is required. The effect of pressure on S is[2]

$$\left(\frac{\partial S}{\partial P}\right)_T = -\left(\frac{\partial V}{\partial T}\right)_P \tag{5.22}$$

showing that the effect of pressure on entropy can be obtained by measuring the effect of temperature on volume (or density), which is usually a much simpler task. Fairly simple manipulations then show the effect of pressure on enthalpy to be

$$\left(\frac{\partial H}{\partial P}\right)_T = V - T\left(\frac{\partial V}{\partial T}\right)_P \tag{5.23}$$

Therefore the effect of pressure on H and S using the constant molar volume assumption $((\partial V/\partial T)_P = 0)$ for solids is particularly simple. Equation (5.22) shows that there is no effect on S, and integration of Equation (5.23) with $(\partial V/\partial T)_P = 0$ shows that the pressure effect on H is the same as that on G, that is, $V(P - 1)$.

Equations (5.22) and (5.23) result in particularly simple expressions for an ideal gas, which are often useful. Substituting RT/P for V in Equation (5.22) leads to

$$\Delta S = S_{P_2} - S_{P_1}$$

$$= R \ln\left(\frac{P_1}{P_2}\right) \tag{5.24}$$

$$= R \ln\left(\frac{V_2}{V_1}\right) \tag{5.25}$$

and making the same substitution in (5.23) results in

$$\Delta H = H_{P_2} - H_{P_1}$$

$$= 0 \tag{5.26}$$

These results should make intuitive sense. Changing the pressure on an ideal gas does not change the fact that there are no interparticle forces, so there should be no effect of pressure on energy terms. However, it does change the ordering or arrangement of the particles, and hence the entropy.

5.8 Summary

. . . .

This chapter contains the transition from somewhat abstract theory to usable numbers. The Gibbs energy and enthalpy are forms of energy, closely related to the "energy in the

---

[2] This equation is obtained by applying the reciprocity relation (discussed in Topics in Mathematics, C.5 Exact and Inexact Differentials, in the online resources) to Equation (4.48).

deep pond," **U** (Figure 3.2). Energy can be transferred by heat and/or work, and assuming only mechanical ($P \Delta V$) work is involved, the fundamental properties we need to know in order to know the energy change are the thermal and volumetric properties ΔH (or more fundamentally C_P) and V.

Thermal properties are measured by some form of calorimetry, an exacting experimental procedure in which some kind of reaction is carried out, such as dissolution of a solid phase, and the heat (q) released or absorbed is measured. If the reaction occurred at constant pressure, the measured q is a $\Delta \mathbf{H}$ and is easily converted into a ΔH. Entropy can also be measured by calorimetry, though of a different type, and combining the enthalpy and entropy measurements gives ΔG numbers. Values of ΔG° can also be obtained by other methods, to be discussed in later chapters. All these quantities are related to the heat capacity, which turns out to be a very fundamental and important parameter. If pressure changes are important, then the volume or density is also required.

The use of these concepts in modern computer programs adds some complications. Although an understanding of these complications is not a prerequisite for understanding thermodynamics itself, they do need to be grasped in order to understand how the programs use data. These complications include the choice of algorithm to represent heat capacity as a function of temperature, how to represent the effect of pressure, and various conventions for "formation from the elements" quantities.

At this point, much of the theory and practice of chemical thermodynamics has been presented. It is worth pausing to reflect on just how it is that delicate measurements near absolute zero temperature, combined with a bunch of differential equations which refer to unattainable conditions, are essential in deciphering the origins of ore deposits, metamorphic rocks, and other geological phenomena.

Exercises

E5.1 In Section 3.6.3 we saw that calorimetry is used to determine the difference in enthalpy between a compound and its constituent elements under standard conditions, $\Delta_f H^\circ$. It is sometimes better or more convenient, at least for silicates, to determine the difference between the compound and its constituent oxides. For example, formation of andalusite (Al_2SiO_5) from its elements is

$$2\,Al(s) + Si(s) + 2.5\,O_2(g) = Al_2SiO_5(s)$$

for which $\Delta_f H^\circ$ is $-2590.27\,kJ\,mol^{-1}$ (Appendix B). The reaction in which andalusite is formed from its oxides (quartz and corundum) is

$$SiO_2 + Al_2O_3 = Al_2SiO_5$$

Holm and Kleppa (1966) made some calorimetric determinations of the heats of formation of kyanite, andalusite, and sillimanite from their constituent oxides in an oxide melt calorimeter at 968 K. For andalusite, the result was $\Delta_r H^\circ_{968} = -1.99\,kcal\,mol^{-1}$, and from this they calculated $\Delta_r H^\circ_{298.15} = -1.34\,kcal\,mol^{-1}$

using drop calorimetry data available at that time. Compare this result with the $\Delta_r H°$ obtained from data in Appendix B.

E5.2 On the other hand you could calculate Holm and Kleppa's result at 968 K. To do this we have to integrate Equation (3.40). We could assume that $\Delta C_P°$ is a constant, which would be silly, or we could use some formulation for the temperature dependence of the heat capacity. The only one we have shown is the Maier–Kelley equation, (3.42), and using this, integration gives

$$\int_{T_r}^{T} d\Delta_r H° = \int_{T_r}^{T} \Delta_r C_P° \, dT$$

$$\Delta_r H_T° - \Delta_r H_{T_r}° = \int_{T_r}^{T} (\Delta_r a + \Delta_r bT - \Delta_r cT^{-2}) dT$$

$$= \Delta_r a(T - T_r) + \frac{\Delta_r b}{2}(T^2 - T_r^2) + \Delta_r c \left(\frac{1}{T} - \frac{1}{T_r} \right) \quad (5.27)$$

Using this equation and the coefficients in Table 5.3, calculate $\Delta_r H°$ for the formation of andalusite from its oxides at 968 K (695 °C). Note that α and β quartz have different coefficients, and that the transition from α- to β-quartz at 573 °C involves an increase in enthalpy of $\Delta_{transition} H = 290 \, \text{cal mol}^{-1}$. However, remember that quartz is a reactant, not a product, so think carefully about the contribution of the transition enthalpy to the $\Delta_r H°$ of the overall reaction. When working with older data such as those of Holm and Kleppa it is more convenient to use calories rather than joules, because switching back and forth becomes tedious.

Table 5.3. Coefficients for the Maier–Kelley heat capacity equation. Using these coefficients in Equation (5.27) gives $C_P°$ values in $\text{cal mol}^{-1} \text{K}^{-1}$, not $\text{J mol}^{-1} \text{K}^{-1}$.

Mineral	a	b	c
Andalusite	41.311	0.006293	1,239,000
α-Quartz (25–573 °C)	11.22	0.00820	270,000
β-Quartz (573–1727 °C)	14.41	0.00194	0.0
Corundum	27.49	0.00282	838,000

Holm and Kleppa used their calorimetry data to calculate a phase diagram for the aluminum silicates. Look it up and compare their diagram with the one from question 7 in Chapter 6. Reading about their experiments will help you appreciate that good thermochemical data are not easily obtained.

Additional Problems

A5.1 The breakdown of spinels to their constituent oxides is thought to have possible tectonic significance. However, the pressures involved are very great (150 to

200 kbar). Therefore interest has focused on structural analogs of the minerals involved, which react at lower pressures but which may model the behavior of the higher-pressure reactions. These have been investigated both experimentally and calorimetrically. Two reactions of interest are

$$Mg_2SnO_4 = 2MgO + SnO_2$$

$$Co_2SnO_4 = 2CoO + SnO_2$$

Each of these compounds was dissolved in a molybdate salt melt at 986 K and the heat of solution measured (Navrotsky and Kasper, 1976).

(a) Calculate the enthalpy of formation of Mg_2SnO_4 and Co_2SnO_4 from their constituent oxides at 986 K (think carefully about the sign, + or −). See Table 5.4.

Table 5.4. Enthalpies of solution in $kcal\,mol^{-1}$ in $3NaO \cdot 4MoO_3$ at 986 K. The errors are standard deviations; the numbers in parentheses are the numbers of experiments performed.

Compound	$\Delta_{oxides}H°$
MgO	-8.564 ± 0.236 (4)
CaO	-5.038 ± 0.055 (4)
SnO_2	-0.214 ± 0.050 (3)
Mg_2SnO_4	-18.471 ± 0.345 (8)
Co_2SnO_4	-7.984 ± 0.264 (7)

(b) What information would you need to calculate $\Delta_f H°_{298}$ for these spinels?

(c) Calculate the volume change for each reaction. See Table 5.5.

Table 5.5. Molar volumes (Robie *et al.* 1967).

Compound	Molar volume	
	$cm^3\,mol^{-1}$	$cal\,bar^{-1}\,mol^{-1}$
CoO	11.64 ± 0.02	0.2782
MgO	11.248 ± 0.004	0.26889
SnO_2	21.55 ± 0.03	0.5151
Mg_2SnO_4	47.89	1.145
Co_2SnO_4	48.56	1.161

(d) The equilibrium pressure for the coexistence of Mg_2SnO_4 and its oxides at 1000 °C is 26 kb, and for Co_2SnO_4 is 12 kb. Calculate the entropy of formation of the spinels from their oxides at 1000 °C. What assumptions must you make here, and how would you improve on them?

Table 5.6. The $\Delta_f G°$ of halite from two different sources. Data in kJ mol^{-1}. Upper series from Robie *et al.* (1978). Lower series from Johnson *et al.* (1992).

T (K)	298.15	400	500	600	700	800	900
$\Delta_f G°$	−384.212	−374.837	−365.229	−355.730	−346.356	−337.136	−328.064
T (K)	273.15	373.15	473.15	573.15	673.15	773.15	873.15
$\Delta_f G°$	−382.376	−389.978	−398.995	−409.153	−420.279	−432.249	−444.981

(e) Calculate the *P–T* slopes of the univariant reaction curves at 1000 °C. (The experimental values are 40 ± 10 bar/deg and 0 ± 7 bar/deg for Mg_2SnO_4 and Co_2SnO_4, respectively).

A5.2 The values of $\Delta_f G°$ for the mineral halite (NaCl) from two different sources are shown in Table 5.6. Plot both sets of values on the same graph. You may get a surprise. There are no significant errors in either set of data.

6 Some Simple Applications

6.1 Introduction

We now know how to determine in which direction any chemical reaction will proceed at a given temperature and pressure, at least when all the products and reactants are pure phases. When even one of the products or reactants is a solute, that is, part of a solution, we would be stuck because we haven't considered how to obtain or use such data. We will start considering this problem in the next chapter. Before going on, however, we should explore some relationships using the concepts we have defined so far, so as to make sure we fully understand them. Naturally, we will only be able to consider some simple properties of pure phases, and reactions between pure phases.

6.2 Simple Phase Diagrams

The reason we are interested in knowing $\Delta_r G$ for reactions is that we can then tell which way the reaction will go, or which side is more stable at one particular T and P. If we know how $\Delta_r G$ varies with T and P, we might find that under some conditions $\Delta_r G$ changes sign, so that the other side is more stable. This implies that there is a boundary between regions of T and P, with one side of the reaction stable on one side of the boundary, and the other side of the reaction stable on the other side of the boundary. A phase diagram shows which phases are stable as a function of T, P, composition, or other variables.

For example, calcium carbonate ($CaCO_3$) has two polymorphs, namely calcite and aragonite. Their properties (from Appendix B) are shown in Table 6.1. Because $\Delta_f G^\circ_{calcite} < \Delta_f G^\circ_{aragonite}$, we conclude immediately that calcite is the stable form of $CaCO_3$ at 25 °C, 1 bar, and that aragonite is a metastable form. But what about other temperatures and pressures? Is aragonite stable at high temperature? At high pressure? How can we tell?

6.2.1 Le Chatelier's Principle

When looking at thermodynamic data, or the results of some thermodynamic calculation, it is always a good idea to ask yourself if it makes sense, if it is reasonable. To some extent this is a matter of experience, but in another way, "making sense" means obeying Le Chatelier's principle. This simply says that *if a change is made to a system, the system will respond in such a way as to absorb the force causing the change*. For example, if the pressure on a system is raised, the system will respond by lowering its volume, that is,

Table 6.1. Thermodynamic properties of calcite and aragonite, from Appendix B.

Formula	Form	$\Delta_f H°$ (kJ mol^{-1})	$\Delta_f G°$ (kJ mol^{-1})	$S°$ (J mol^{-1} K^{-1})	$V°$ (cm^3 mol^{-1})
CaCO$_3$	calcite	−1206.92	−1128.79	92.9	36.934
CaCO$_3$	aragonite	−1207.13	−1127.75	88.7	34.150

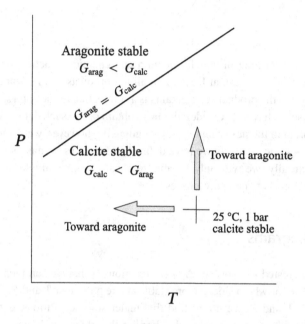

Figure 6.1 The form of the calcite–aragonite phase diagram deduced from Le Chatelier's principle.

by being compressed. Systems never expand as a result of increased pressure. The result of a change in temperature is less obvious, though equally certain. If the temperature of a system is raised, the enthalpy and the entropy of the system will both increase. This is because of Equations (3.39) and (4.32), which show that the temperature derivative of each is a simple function of C_P, the heat capacity, which is always positive for pure compounds.

Therefore, by looking at $V°$ and $\Delta_f H°$ or $S°$ for calcite and aragonite, and assuming that the relative magnitudes of these properties do not change much with T and P, we can tell something about their relative positions on the phase diagram. We note that $V_{\text{aragonite}} < V_{\text{calcite}}$; therefore, increasing the pressure on calcite should favor the formation of aragonite. Also, $\Delta_f H°$ and $S°$ for calcite are greater than the values for aragonite, and so raising the temperature of calcite will *not* favor the formation of aragonite. In other words, *lowering* the temperature of calcite should favor the formation of aragonite. If the stability field of aragonite lies somewhere at higher pressure and lower temperature than 25 °C, 1 bar, the boundary between the two phases must have a positive slope, as shown in Figure 6.1. This is the common case for phase boundaries; it is normal for the high-pressure, lower-volume side to be the lower-enthalpy, lower-entropy side. The most common exception to this is the ice–water transition, as shown in Figure 3.1.

In Figure 6.1 we see that a phase diagram is a kind of free energy map – it shows a T–P region where calcite is stable ($G_{calcite} < G_{aragonite}$), and another where aragonite is stable ($G_{aragonite} < G_{calcite}$). These two regions are necessarily separated by a line where $G_{aragonite} = G_{calcite}$, the phase boundary. We have a lot more to say about phase diagrams in Chapter 12.

6.2.2 The Effect of Pressure on $\Delta_r G°$

Having figured out the relationship between calcite and aragonite qualitatively, the next step is to define the stability field of aragonite, that is, to calculate the position of the phase boundary. This should be possible, because we know that

$$\partial G/\partial P = V \tag{4.51}$$

and thus

$$\partial \Delta G/\partial P = \Delta V$$

ΔG and ΔV refer to the difference in G and V between any two equilibrium states. In this case we are dealing with a chemical reaction between two compounds in their pure states, so we can also write

$$\partial \Delta_r G°/\partial P = \Delta_r V°$$

Integrating this equation between 1 bar and some higher pressure P, we have

$$\Delta_r G°_P - \Delta_r G°_{1bar} = \int_{1bar}^{P} \Delta_r V° \, dP \tag{6.1}$$

and if we assume that $\Delta_r V°$ is a constant, this becomes

$$\Delta_r G°_P - \Delta_r G°_{1bar} = \Delta_r V° \int_{1bar}^{P} dP$$
$$= \Delta_r V°(P - 1)$$

We could use this to evaluate $\Delta_r G°_P$ at any chosen value of P. However, we are particularly interested in a value of $\Delta_r G°_P = \Delta_r G°_{P_{eqbm}} = 0$, that is, on the phase boundary. We know the values of

$$\Delta_r G°_{1bar} = \Delta_f G°_{aragonite} - \Delta_f G°_{calcite}$$
$$= -1127.75 - (-1128.79)$$
$$= 1.04 \text{ kJ mol}^{-1}$$
$$= 1040 \text{ J mol}^{-1}$$

and

$$\Delta_r V° = V°_{aragonite} - V°_{calcite}$$
$$= 34.150 - 36.934$$
$$= -2.784 \text{ cm}^3 \text{ mol}^{-1}$$

So we can solve the equation for P_{eqbm}, the pressure of the calcite–aragonite equilibrium at 25°C.

However, there is one little problem.

The Units of Volume

Volumes are generally measured in cubic centimeters, milliliters, liters, and so on. But if you look at an equation such as

$$w = -P \Delta V$$

you see that we have a problem with our units. Work (w) and $P \Delta V$ are obviously energy terms (J mol^{-1}), but the product of P in bars and ΔV in $\text{cm}^3 \text{ mol}^{-1}$ is not joules. We must always convert our volumes to joules bar^{-1}, so that the product of P and V or ΔV is J mol^{-1}. The conversion factor (Appendix A) is

$$1 \text{ cm}^3 = 0.10 \text{ J bar}^{-1}$$

so now our $\Delta_r V°$ is $-2.784 \times 0.1 = -0.2784 \text{ J bar}^{-1}$.

Now we can solve for pressure P_{eqbm}:

$$\Delta_r G°_{P_{\text{eqbm}}} - \Delta_r G°_{1 \text{ bar}} = \Delta_r V°(P_{\text{eqbm}} - 1)$$

$$0 - 1040 = -0.2784(P_{\text{eqbm}} - 1)$$

$$P_{\text{eqbm}} = 3737 \text{ bar}$$

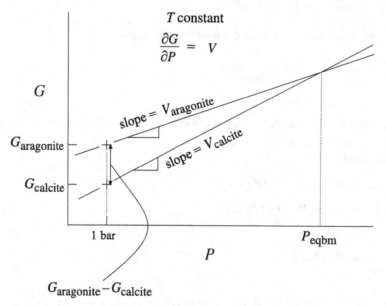

Figure 6.2 The relationship between G and P. Note that we don't know individual G values, so there are no numbers on the y-axis. We do know $G_{\text{calcite}} - G_{\text{aragonite}}$ and the slopes of the lines (the molar volumes), and this is sufficient to solve for P_{eqbm}. At P_{eqbm}, $G_{\text{calcite}} = G_{\text{aragonite}}$, the two phases can coexist, and we have a phase boundary.

The relationship between G and P in this calculation is shown in Figure 6.2. This gives us one point on the calcite–aragonite phase boundary. We also know that the boundary has a positive slope, and so we could sketch a diagram that would be approximately right, but we really need one more piece of information – either another point on the boundary or its slope.

6.3   The Slope of Phase Boundaries

The phase boundary is the locus of T and P conditions where $\Delta_r G = 0$, i.e., where

$$G_{calcite} = G_{aragonite} \tag{6.2}$$

It follows that on the boundary,

$$dG_{calcite} = dG_{aragonite} \tag{6.3}$$

This simply says that as you move along the boundary, the change in $G_{calcite}$ has to be the same as the change in $G_{aragonite}$; otherwise you won't stay on the boundary. From Equation (4.48) we have

$$dG = -S\,dT + V\,dP \tag{4.48}$$

This applies to each mineral, and combining with (6.3) gives

$$-S_{calcite}\,dT + V_{calcite}\,dP = -S_{aragonite}\,dT + V_{aragonite}\,dP$$

Rearranging this gives

$$\frac{dP}{dT} = \frac{S_{calcite} - S_{aragonite}}{V_{calcite} - V_{aragonite}}$$

or, for any reaction,

$$\frac{dP}{dT} = \frac{\Delta_r S}{\Delta_r V} \tag{6.4}$$

which gives the slope of an equilibrium phase boundary in terms of the entropy and volume changes between the phases involved in the reaction. This is called the Clapeyron equation.

Equation (5.1) says

$$\Delta G_{T,P} = \Delta H - T\,\Delta S \tag{5.1}$$

This applies to any change between two equilibrium states at the same T and P. If those two equilibrium states have the same value of G, such as calcite and aragonite do on their phase boundary (6.2), then $\Delta G_{T,P} = 0$, and

$$\Delta H = T\,\Delta S \tag{6.5}$$

or

$$\frac{\Delta H}{T} = \Delta S \tag{6.6}$$

This is a useful relationship for any phase boundary,[1] which is the usual place to find $\Delta G_{T,P} = 0$. This gives an alternative form of the Clapeyron equation,

$$\frac{dP}{dT} = \frac{\Delta H}{T\,\Delta V} \tag{6.7}$$

6.3.1   The Slope of the Calcite–Aragonite Boundary

We have one point on the calcite–aragonite boundary at 3737 bar, 25 °C. If we assume that the $\Delta_r S$ and the $\Delta_r V$ at this P and T are the same as those at 1 bar, 25 °C, we can calculate the slope from the data in our tables. Thus

$$\Delta_r S = S_{\text{aragonite}} - S_{\text{calcite}}$$
$$= 88.7 - 92.9$$
$$= -4.2 \,\text{J}\,\text{mol}^{-1}\,\text{K}^{-1} \tag{6.8}$$

and

$$\Delta_r V = V_{\text{aragonite}} - V_{\text{calcite}}$$
$$= 34.150 - 36.934$$
$$= -2.784 \,\text{cm}^3\,\text{mol}^{-1}$$
$$= -0.2784 \,\text{J}\,\text{bar}^{-1}$$

Therefore

$$\frac{dP}{dT} = \frac{\Delta_r S}{\Delta_r V}$$
$$= \frac{-4.2}{-0.2784}$$
$$= 15.09 \,\text{bar/°C}$$

Therefore, to get another point on the calcite–aragonite phase boundary, we simply choose an arbitrary temperature increment, say 100 °C, calculate the corresponding pressure increment, $100 \times 15.09 = 1509$ bar, and add these increments to our first point. We now have a second point at 125 °C, $3737 + 1509 = 5246$ bar, and we can plot the boundary as in Figure 6.3.

Keep in mind that we have assumed that the $\Delta_r S$ and $\Delta_r V$ from the tables are unchanged at all temperatures and pressures, that is, that they are constants. This is quite a good approximation for a reaction involving only solid phases such as this one, but you would not use it for reactions involving liquids, gases, or solutes. In general, all thermodynamic parameters do vary with T and P, so phase boundaries are in principle curved and not straight as we have assumed. However, the amount of curvature is quite small in some cases, such as this one.

---

[1] That is, any phase boundary in a one-component system. With two or more components, the relationship is in principle the same but becomes more complicated, and less useful.

Table 6.2. Experimental results for the system CaCO₃ from Crawford and Hoersh (1972).

Temperature (°C)	Pressure (bars)	Experimental result	Duration days
128	5180	A	21
132	5180	A	21
153	4830	C	35
76	4480	A	3
90	4140	C	28
93	4140	C	17
56	4140	A	28
70	4140	A	17
70	3690	C	8
81	3520	C	36

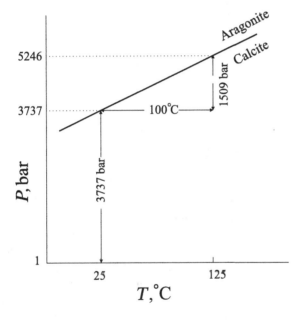

Figure 6.3 Calculation of the calcite–aragonite phase diagram.

6.3.2   Comparison with Experimental Results

Table 6.2 shows the results of some experiments on the stability of CaCO₃ at elevated temperatures and pressures. A mixture of calcite and aragonite was held at the indicated *T* and *P* for the length of time shown, then quenched and examined. The stable phase is shown as C (calcite) or A (aragonite). These points are plotted in Figure 6.4.

Also shown in this figure are the two points we have just calculated at 25 and 125 °C, plus results using data from Helgeson *et al.* (1978), and a line showing the experimenter's best estimate of the phase boundary. As you see, the calculated results using data from Appendix B are a little high, and the Helgeson *et al.* results are a bit low. Thermodynamic data have many possible sources of error, but then so do experimental data.

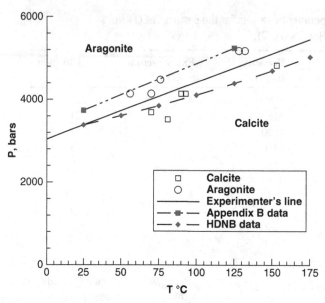

Figure 6.4 Comparison of experiment and calculation of the calcite–aragonite phase diagram.

6.4   Another Example

6.4.1   The Effect of Temperature on $\Delta_r G°$

To illustrate the effect of temperature on $\Delta_r G°$, we could continue with the calcite–aragonite case and try to calculate the temperature where the phase boundary crosses the 1 bar pressure line (Figure 6.3). Unfortunately, this turns out to be close to absolute zero, so it is not a very useful example. As another case let's consider the polymorphs of Al_2SiO_5. There are three of these, kyanite, andalusite, and sillimanite. Therefore there are three two-phase boundaries, and these three boundaries meet at a single point, where $G_{kyanite} = G_{andalusite} = G_{sillimanite}$ as shown in Figure 6.5. These minerals, which form quite commonly in rocks subjected to high temperatures and pressures in the Earth's crust, are of special interest to geologists who study these rocks because the "triple point," the point where the three phase boundaries meet, is in the middle of a rather common range of T–P conditions. If a rock contains one of these minerals, the geologist immediately has a general idea of the T and P conditions at the time the rock formed. It is only a "general idea" because it is not safe to assume that rocks reach chemical equilibrium at some P and T, and then remain unaltered as they are exhumed and are exposed at the Earth's surface. Many complications can occur, which are not within the subject of thermodynamics.

According to Figure 6.5, the kyanite–andalusite boundary crosses the 1 bar line at some elevated temperature. We should be able to calculate what this is by methods perfectly

6.4 Another Example

Table 6.3. Thermodynamic data for the Al_2SiO_5 minerals, from Appendix B.

Formula	Form	$\Delta_f H°$ (kJ mol^{-1})	$\Delta_f G°$ (kJ mol^{-1})	$S°$ (J mol^{-1} K^{-1})	$V°$ (cm^3 mol^{-1})
Al_2SiO_5	Kyanite	−2594.29	−2443.88	83.81	44.09
Al_2SiO_5	Andalusite	−2590.27	−2442.66	93.22	51.53

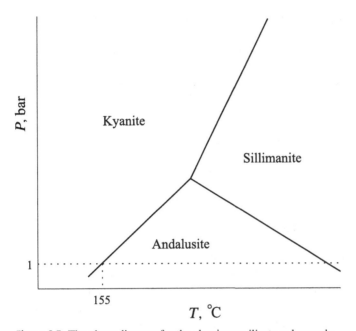

Figure 6.5 The phase diagram for the aluminum silicate polymorphs.

analogous to those we used for calcite–aragonite. The data we need (from Appendix B) are shown in Table 6.3.

First, we note that all seems to be well to start with, in that there is no conflict between the data and Figure 6.5. Kyanite has the lower value of $\Delta_f G°$, and so it should be the stable phase at 25 °C, 1 bar, as shown in the diagram. The values of $\Delta_r S°$ and $\Delta_r V°$ would indicate that kyanite is the high-pressure phase, and that the kyanite–andalusite boundary has a positive slope, also as shown by the diagram.

To calculate the temperature of the kyanite–andalusite boundary at 1 bar, we start with Equation (4.50),

$$(\partial G/\partial T)_P = -S \tag{4.50}$$

from which we can write immediately[2]

$$(\partial \Delta_r G^\circ / \partial T)_P = -\Delta_r S^\circ$$

Integrating this from 298.15 K to some higher temperature T, we get

$$\Delta_r G_T^\circ - \Delta_r G_{298}^\circ = \int_{298}^{T} -\Delta_r S^\circ dT \tag{6.9}$$

and if we assume that $\Delta_r S^\circ$ is a constant, this becomes

$$\Delta_r G_T^\circ - \Delta_r G_{298}^\circ = -\Delta_r S_{298}^\circ \int_{298}^{T} dT$$

$$= -\Delta_r S_{298}^\circ (T - 298.15) \tag{6.10}$$

Now if we let $\Delta_r G_T^\circ = 0$, T becomes T_{eqbm}, and we can solve for this. From the tables, $\Delta_r G_{298}^\circ = 1220\,\mathrm{J\,mol^{-1}}$ and $\Delta_r S_{298}^\circ = 9.41\,\mathrm{J\,mol^{-1}\,K^{-1}}$, so

$$\Delta_r G_T^\circ - \Delta_r G_{298}^\circ = -\Delta_r S_{298}^\circ \int_{298}^{T_{eqbm}} dT$$

$$0 - 1220 = -9.41(T_{eqbm} - 298.15)$$

$$T_{eqbm} = 427.8\ \mathrm{K}$$

$$= 154.6\,^\circ\mathrm{C}$$

The relationship between G and T is shown in Figure 6.6. Note that in Figure 6.2, the slope of G vs. P is positive, whereas in Figure 6.6, the slope of G vs. T is negative. This is because for pure substances V is always positive, and S is always positive by virtue of the third law. This is only true in general for pure substances; for differences (i.e., $\Delta_r V$, $\Delta_r S$) or for solutes, these quantities may be negative, as we will see.

---

[2] It is not immediately clear to many students why, if $(\partial G/\partial T)_P = -S$, we can "write immediately" $(\partial \Delta_r G^\circ / \partial T)_P = -\Delta_r S^\circ$, that is, why we can just stick in a Δ whenever we wish. It is because the derivative relationship can be applied to all terms of any balanced reaction. For example, if the reaction is A + 2B = C (e.g., reaction (5.8)),

$$\Delta_r G^\circ = \Delta_f G_C^\circ - \Delta_f G_A^\circ - 2\,\Delta_f G_B^\circ$$

$$= G_C^\circ - G_A^\circ - 2\,G_B^\circ$$

so the derivative with respect to T is

$$(\partial \Delta_r G^\circ / \partial T) = (\partial G_C^\circ / \partial T) - (\partial G_A^\circ / \partial T) - 2\,(\partial G_B^\circ / \partial T)$$

$$= -S_C^\circ + S_A^\circ + 2\,S_B^\circ$$

$$= -(S_C^\circ - S_A^\circ - 2\,S_B^\circ)$$

$$= -\Delta_r S^\circ$$

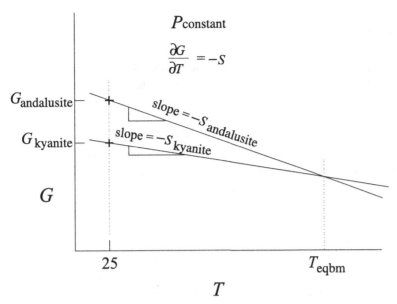

Figure 6.6 The relationship between G and T. Note the general similarity to Figure 6.2, with the exception that the slopes are negative.

6.4.2 A Different Formula for $\Delta_r G°_T$

Another useful way of expressing the effect of temperature on G is given by expanding (6.10). Thus

$$\Delta_r G_T^\circ - \Delta_r G_{298}^\circ = -\Delta_r S^\circ (T - 298.15)$$

$$\Delta_r G_T^\circ - (\Delta_r H_{298}^\circ - 298.15\,\Delta_r S_{298}^\circ) = -T\,\Delta_r S_{298}^\circ + 298.15\,\Delta_r S_{298}^\circ$$

Collecting and rearranging terms gives

$$\Delta_r G_T^\circ = \Delta_r H_{298}^\circ - T\,\Delta_r S_{298}^\circ \tag{6.11}$$

In other words, you can calculate $\Delta_r G^\circ$ at some temperature T using the values of $\Delta_r H^\circ$ and $\Delta_r S^\circ$ at 298.15 K. However, this is subject to the same restriction as before, that both $\Delta_r H^\circ$ and $\Delta_r S^\circ$ are not functions of temperature. Of course, both these terms always *are* functions of temperature, but often this can be neglected without introducing much error, especially if T is not very different from 298 K.

Both (6.10) and (6.11) are therefore approximations, to be used only over a small temperature interval, or in cases where the result need only be approximate. More accurate formulae involve the heat capacity as a function of T, as discussed in Chapter 5.

6.5   Summary

The main idea in this chapter is to illustrate the uses of our thermodynamic variables, using only pure phases. We did this by calculating simple phase diagrams. In phase diagrams, the condition $\Delta_r G = 0$ becomes a central concern, and for pure phases, this is the same as $\Delta_r G° = 0$.

Many reactions involving only pure phases also involve water, but calculating the Gibbs energy of fluids at high T and P is more difficult than for minerals, because we cannot assume that they are incompressible. How to handle this, as well as how to deal with solutions, is an important topic in later chapters.

Exercises

E6.1 At what pressure will graphite be converted to diamond at 25 °C? At 100 °C?

E6.2 Calculate the pressure at which jadeite is in equilibrium with low albite and nepheline at 300 °C.

E6.3 At what temperature does gibbsite break down (dehydrate) to form corundum and water? Assume a pressure of 1 bar. Actually, because the temperature is greater than 100 °C and liquid water is present, the pressure will be somewhat greater than 1 bar, but this will not greatly affect the Gibbs energies.

E6.4 Show that dolomite is stable relative to calcite plus magnesite at 25 °C and 1 bar.

E6.5 Consider the data in Table 6.4 for the two polymorphic compounds, α and β. Assume these data don't change much with changes in T or P.

Table 6.4. Data for two polymorphic compounds.

	$\Delta_f H°$ (kJ mol^{-1})	$\Delta_f G°$ (kJ mol^{-1})	$S°$ (J mol^{-1} deg^{-1})	$V°$ (cm^3 mol^{-1})
α (s)	−2600	−2440	90.0	50.0
β (s)	−2598	−2437	93.4	49.0
α (aq)		−2415		

(a) Which polymorph is more stable at 25 °C, 1 bar?

(b) Which is the high-temperature form?

(c) Which is the high-pressure form?

(d) Calculate the slope of the phase boundary in bars/degree.

(e) Calculate the equilibrium pressure at 25 °C.

(f) Sketch the P–T phase diagram showing the location of 25 °C, 1 bar, and the phase boundary between α and β.

(g) Sketch a G–T and a G–P section through the phase boundary.

(h) What is the solubility of α?

$$\alpha(s) = \alpha(aq)$$

(i) Does β have a greater or smaller solubility? Explain.

Additional Problems

A6.1 Using the data in Appendix B, calculate a more accurate version of Figure 6.5.

A6.2 Find the pressure and temperature of the triple point. You can do this

(a) graphically, by finding the location of at least two boundaries at at least two different pressures; or,

(b) even better, by obtaining an equation for at least two of the boundaries in the form $y = ax + b$, and solving for P and T.

7 Solutions

7.1

Introduction

If the world were made of pure substances, our development of the thermodynamic model would now be complete. We have developed a method, based on measurements of heat flow, that enables predictions to be made about which way reactions will go in given circumstances. But one of the reasons why the world is so complex is that pure substances are relatively rare, and strictly speaking they are nonexistent (even "pure" substances contain impurities in trace quantities). Most natural substances are composed of several components, and the result is called a *solution*. Therefore, we need to develop a way to deal with components in solution in the same way that we can now deal with pure substances – we have to be able to get numerical values for the Gibbs energies, enthalpies, and entropies of components in solutions. We will then be able to predict the outcome of reactions that take place entirely in solution, such as the ionization of acids and bases, and reactions that involve solids and gases as well as dissolved components, such as whether minerals will dissolve or precipitate. Our thermodynamic model will then be complete.

In this chapter we have a look at how to deal with dissolved substances – solutes. When we mix two substances together, sometimes they dissolve into one another, like sugar into coffee or alcohol into water, and sometimes they do not, like oil and water. In the former case, if we thought about it at all, we would probably expect that the properties of the mixture or solution would be some kind of average of the properties of the two separate substances. This is more or less true for some properties, but decidedly not true for the most important one, Gibbs energy.

After making sure we understand how to express the composition of solutions, we begin by considering properties of *ideal solutions*, which are, as you might expect, the simplest possible properties that solutions might have. As you might also expect, no real solutions are in fact ideal, although some come fairly close.

7.2

Measures of Concentration

A number of concentration terms are used in describing solutions, and it is naturally important to be able to change from one to another.

Mole Fraction

Consider a solution containing a number of components, n_1 moles of component 1, n_2 moles of component 2, and so on. If it is an aqueous solution, then water is one of the components, normally the major component. The mole fraction of any one component i is defined as

$$x_i = \frac{n_i}{\sum_i n_i} \tag{7.1}$$

where $\sum_i n_i$ is the total number of moles of components, $n_1 + n_2 + n_3 + \cdots$.

The mole fraction is very commonly used, especially in theoretical discussions, because it is perfectly general, and it can cover the entire range of compositions from dilute solutions to pure components. It is inconvenient for aqueous solutions because these are usually quite dilute on the mole fraction scale; that is, water is by far the dominant component, and the mole fractions of the solutes are numerically very small.

The mole fraction is a simple concept, but there is one important thing to note. In any mole fraction the question is, n_1, n_2, etc., are moles of *what*? This is not as simple as it might seem. Let's say you have a solution containing 1 mole ($\approx 58\,g$) of NaCl, 1 mole ($\approx 75\,g$) of KCl, and 50 moles ($\approx 900\,g$) of water. What is the mole fraction of NaCl? Is it

$$
\begin{aligned}
x_{NaCl} &= \frac{n_{NaCl}}{n_{NaCl} + n_{KCl} + n_{H_2O}} \\
&= \frac{1}{1 + 1 + 50} \\
&= 0.0192
\end{aligned}
$$

or is it

$$
\begin{aligned}
x_{NaCl} &= \frac{n_{Na^+} + n_{Cl^-}}{n_{Na^+} + n_{K^+} + n_{Cl^-} + n_{H_2O}} \\
&= \frac{2}{1 + 1 + 2 + 50} \\
&= 0.0370
\end{aligned}
$$

Completely dissociated electrolytes under ambient conditions have long been a major topic in solution chemistry, so the second option is possible. However, the situation is not as clear under conditions of high T and P, where even "strong" electrolytes dissociate to a variable extent, and may hardly dissociate at all. So the message is, if you use mole fractions, make sure you know how they are defined.

Molality

The molality (m_i) of component i is the number of moles of i (n_i) per kilogram of pure solvent, usually water. Even if the aqueous solution contains several solutes, the molality is the number of moles of one of them in 1000 g of *pure* water. It is less general than mole fraction in the sense that you cannot express the composition of the pure solute in molal units, because m_i becomes infinite for pure i.

The use of molality is virtually universal for aqueous solutions because it is independent of the temperature and pressure of the solution, and equations in molality are usually simpler than equations in mole fractions.

Molarity

The molarity (M_i) of component i is the number of moles of i in 1 liter of solution (not a liter of pure solvent). This is a convenient unit in the laboratory, where solutions are prepared in volumetric flasks. It has the disadvantage that as temperature or pressure changes, the volume of the solution changes but the definition of the liter does not, and the molarity is therefore a function of temperature and pressure. The conversion to molality requires a knowledge of the density of the solution, which is readily available in handbooks for binary solutions at 25 °C, but usually not available for natural solutions. For dilute solutions at ambient conditions, m and M are about the same. The relationship for NaCl is shown in Table 7.1.

Weight Percent

Measurement in weight %, the grams of solute per 100 grams of solution, is used in the metallurgical literature, and in some areas of geochemistry. Natural solutions found in fluid inclusions, basinal brines, and evaporitic environments can reach concentrations of several molal. To convert to or from molality you need to know the molecular weight (more formally, relative molecular mass) of the component. The relationship between NaCl weight % and molality is shown in Table 7.1.

Parts per Million

For trace components a million grams of solution rather than 100 grams may be used, giving parts per million (ppm). For example, an aqueous solution that is 10^{-4} molal in

Table 7.1. Relationship between NaCl molality, weight %, and molality at 25 °C. Density data from Pitzer *et al.* (1984).

Molality (mol kg^{-1})	Weight % (g/100 g)	Density (g cm^3)	Molarity (mol L^{-1})
0.01	0.058	0.99746	0.0100
0.10	0.58	1.00117	0.0995
0.25	1.44	1.00722	0.2482
0.50	2.84	1.01710	0.4941
0.75	4.20	1.02676	0.7377
1.0	5.52	1.03623	0.9790
2.0	10.46	1.07228	1.9201
3.0	14.92	1.10577	2.8225
4.0	18.95	1.13705	3.6864
5.0	22.61	1.16644	4.5133
6.0	25.96	1.19423	5.3051

Table 7.2. Zinc concentration units.

Molality	ppm	mg/kg
0.0001	6.5	6.5
0.001	65	65
0.005	327	327
0.010	653	654
0.050	3258	3269
0.100	6495	6537
0.150	9710	9806
0.200	12,905	13,074

Zn contains $0.0001 \times 65.37 = 0.006537$ grams of NaCl in $(1000 + 0.006537)$ grams of solution, or about 10^3 grams of solution. Therefore there would be 6.537 grams of Zn in 10^6 grams of solution, or 6.5 ppm. If the solution contains a number of other solutes, they should all be included in the denominator, but it is common practice to ignore all components except the solute of interest and water.

Roughly equivalent and perhaps more common units are milligrams per liter and milligrams per kilogram of solvent. Being a volumetric unit, conversion of mg/L should involve the density, but for dilute solutions, mg/L, mg/kg, and ppm are about the same. Table 7.2 gives a comparison for Zn.

7.3   Properties of Ideal Solutions

What are the properties of true ideal solutions and why do real solutions not behave this way? The picture differs for gases, liquids, and solids. Before developing the equations, it will help to have a mental picture of what an ideal solution is.

7.3.1   Ideal Gaseous Solutions

Taking the simplest case first, an ideal gas consists of hypothetical, vanishingly small particles that do not interact in any way with each other. They are unaware of the existence of the other particles and there are no forces or energies of attraction or repulsion. An ideal gas obeys the ideal gas law, $PV = nRT$, where n is the number of moles, T is related to the movement and individual energies of the particles, V is the volume occupied by the particles, and P comes from the only interaction allowed in the system – particles bouncing off the walls or boundaries. A *solution* of two ideal gases will also obey the ideal gas law since the particles of the different constituents remain unaware of all other particles, just as with an ideal single-component gas. You might say that molecules in an ideal gas, whether pure or a solution, think they are in a perfect vacuum. Of course, real gases do interact at the molecular scale and can only be expected to approach ideal behavior at very low densities and pressures, or in the limit as $P \to 0$.

Equations of State

The ideal gas equation $PV = nRT$ or $PV = RT$ is an equation of state; an equation relating the volume of an ideal gas to its pressure and temperature. For real gases there have been many modifications of this equation, and there are other equations of state which interrelate not only P, V, and T, but also thermodynamic parameters such as enthalpy, Gibbs energy, and entropy. See Chapter 13 Equations of State, in the online material.

7.3.2   Ideal Liquid Solutions

Liquids are necessarily more complicated than gases. To start with, they have much greater cohesiveness than gases; for example, a liquid equilibrated with its gaseous vapor develops a meniscus. This boundary has a measurable surface tension caused by the fact that particle interactions in the liquid are stronger than those in the vapor. A liquid must have significant interaction among its particles – if it did not, it would disperse and become a gas.

In an ideal liquid solution, the forces of interaction among all the molecules, whether of one type or another, are exactly the same. For example in a liquid solution of constituents A and B, interactions A–A, A–B, and B–B must be identical. This means that all constituents A, B, ..., must have the same molecular properties (size, charge, polarity, bonding characteristics). This is never the case, of course, but mixtures of some organic compounds come fairly close.

Given this uniformity of intermolecular forces in the ideal liquid solution (as opposed to the *absence* of such forces in the ideal gas), it follows that many properties of the solution are very simply related to the properties of the pure compounds. Thus the volume of the solution is the sum of the volumes of the pure components before mixing, and no heat is absorbed or given off when the solution is prepared (because such effects are caused by changes in the particle interactions, which we have just ruled out).

7.3.3   Ideal Solid Solutions

A solid has a rigid structure, and its component molecules, ions, or atoms are confined to specific structural sites. The regularity of the structure varies, of course, from glassy to fully crystalline materials, but whatever the degree of ordering, the positions of the particles are fixed. Whereas ideal gases and gaseous solutions have a complete absence of interparticle forces and ideal liquid solutions have a complete uniformity, solids must have highly specific interactions between different constituents. We speak of specific sites in crystals, such as tetrahedral silicon–oxygen bonds and octahedral aluminum sites, and the same is true (although to a lesser extent) of glassy solids. The Si–Si, Si–O, Si–Al, O–O, and Al–O interactions in an aluminosilicate are all quite different. However, within the framework of a perfectly crystalline compound it is frequently possible to substitute one element for another. This substitution and the corresponding solid solution would be ideal *if the two substituting elements or species were completely indistinguishable.* The closest approximation to an ideal solid solution would be the substitution of two isotopes of the same element on the same crystal site. Like ideal gaseous and liquid solutions, there would be no heat evolved on mixing the components and the total volume of the solution must simply be the sum of the volumes of the pure constituents before mixing.

7.3.4   Two Kinds of Ideal Solution

There is only one kind of ideal gas solution, as discussed above, but there are two kinds of ideal liquid and solid solutions.

Our discussion of liquid and solid solution ideality above makes no provision for the possibility that there might be more than one definition of ideality; that a solution might act ideally in one way but not in another. If in a liquid solution of A and B there are A–A and A–B interactions but no B–B interactions, we have another kind of ideal solution, in which changing the concentration of B results in a perfectly linear change in the properties of B, and has no effect on the properties of A. In order for there to be no B–B interaction, particles of B must be quite widely separated – the concentration of B in A must be very small. This leads to the concept of the "infinitely dilute solution," in which there is only one particle of B in a sea of A, and therefore there is interaction between the particle of B and the surrounding A, and of course A–A interaction, but, there being no other B particles, there is no B–B interaction. We can't deal with a single molecule of B, so we need to think of a mole of B particles and have so much A that no B particle is influenced by another B particle – perhaps a roomful of A, as in the room analogy on page 171.

These two kinds of ideality permeate discussions of liquid and solid solution properties, and are formalized by two ideal solution laws – Raoult's law and Henry's law.

7.4   Ideal Solution Laws

These relationships or laws were discovered in the nineteenth century by investigations of gas or vapor pressures associated with solutions of known composition. Because the gas or vapor was at a fairly low pressure, it acted as an ideal gas, and because it was in equilibrium with the solution, it provided information on the nature of the liquid solution. Today, the original connection with an associated vapor or gas phase is a secondary concern. The relationship between the ideal solution components themselves proves to be more useful, a subject to be discussed in terms of *activities*, an important topic introduced here and treated more fully in Chapter 8. Before discussing these relationships, we look first at solutions of ideal gases.

7.4.1   Dalton's Law

The simplest imaginable system other than a vacuum is undoubtedly an ideal gas. One mole of ideal gas occupies 22.41 liters at 0 °C, 1 atm, so that (from the ideal gas law) one mole of ideal gas occupying one liter at 0 °C would have a pressure of 22.41 bars.

It was an early discovery (by Dalton, in 1811) that mixtures of gases would exert a pressure equal to the sum of the pressures that each of the species gases would have if each alone occupied the same volume. This was established using gases at relatively low pressures where they behave close to ideally, and in fact it is only strictly true for mixtures of ideal gases, which are also then ideal gases. Thus, for each species gas 1, 2, 3, etc.

$$P_1\mathbf{V} = n_1 RT$$

$$P_2\mathbf{V} = n_2 RT$$

etc ...

and for the gas mixture

$$P_{\text{total}}\mathbf{V} = \sum_i n_i RT$$

Thus

$$\frac{P_1}{P_{\text{total}}} = \frac{n_1}{\sum_i n_i} = x_1$$

$$\frac{P_2}{P_{\text{total}}} = \frac{n_2}{\sum_i n_i} = x_2$$

etc ...

or

$$\left.\begin{aligned} P_1 &= x_1 \cdot P_{\text{total}} \\ P_2 &= x_2 \cdot P_{\text{total}} \\ &\ \vdots \qquad\quad \vdots \\ \text{etc} &\dots \end{aligned}\right\} \qquad (7.2)$$

P_1, P_2, etc. are called the *partial pressures* of the solution gases and equations (7.2) are now normally used as the definition of partial pressure, even though in real, non-ideal solutions they give a quantity that is not equivalent to the original meaning, i.e., the pressure a gas would exert if it alone occupied the total volume.[1]

7.4.2   Henry's Law

Henry's law in its original form stated that the solubility of a gas in a liquid is proportional to the pressure on the gas. In Figure 7.1 is shown an apparatus for controlling the pressure

---

[1] There are several definitions of partial pressure. de Heer (1986, Section 23.4) says there are five, and explains the three most common. In my experience, only the definition in Equations (7.2) is ever used in geochemistry.

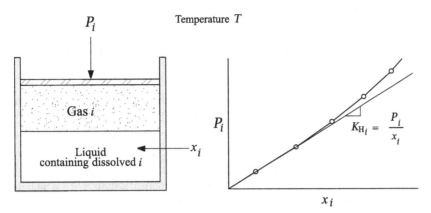

Figure 7.1 Illustration of Henry's law. As the pressure P_i on the gas i increases, more of it goes into solution in the liquid, increasing x_i.

on a gas i in contact with a liquid. As the pressure on the gas P_i increases, more of it dissolves in the liquid, and so x_i increases. When x_i is sufficiently small, it is directly proportional to P_i, and the constant of proportionality is called the Henry's-law constant, K_{H_i}. As x_i gets larger, there is inevitably some deviation from strict proportionality, as shown.

Henry's law is expressed mathematically as

$$P_i = K_{H_i} x_i \tag{7.3}$$

where P_i is the pressure or partial pressure of some component i, x_i is its mole fraction in solution, and K_{H_i} is a constant, specific for component i, the Henry's-law constant. Actually, it is more often used in terms of molality,

$$P_i = K_{H_i} m_i \tag{7.4}$$

where m_i is the molality of i in solution. K_{H_i} will have a different numerical value in the two cases. It follows that in the concentration range where Henry's law is obeyed,

$$\frac{\partial P_i}{\partial m_i} = K_{H_i}$$

$$= \frac{P_i}{m_i} \tag{7.5}$$

a result we will use later (Section 7.5.3).

In this experimental situation, it will be noted that the total pressure is not, strictly speaking, P_i, because some of the liquid solvent will evaporate into gas i, so that the piston is supported partly by gas i and partly by vaporized liquid. In other words, there are always at least two partial pressures in a gas in contact with a liquid. However, if the vapor pressure of the liquid is small compared with the gas pressure, it can be neglected, and the pressure on the piston equated with P_i. This was the case in the early experiments of Henry and others.

Henry's law results from a lack of interaction between the solute particles, and represents the limiting behavior as solute concentrations approach zero. It has been generalized to refer not only to gas concentrations and pressures, but to any linear proportionality between the activity and concentration.

Finally, it is important to be aware of the physical meaning of the tangent in Figure 7.1. As $x_i \to 0$, it represents the values of P_i in equilibrium with the solution in which a particle of i interacts with the solvent but is unaware of any other particles of i; i is "infinitely dilute." As we go out along the tangent to larger values of x_i, the value of P_i given by the tangent at any x_i of course deviates from the actual measured P_i, but still represents the P_i which *would be* in equilibrium with i in solution, *if i continued to fail to be aware of other i particles*. It represents the P_i for a hypothetical solution of i which displays dilute solution behavior at all concentrations. This rather esoteric sounding situation proves to be surprisingly useful when we generalize P_i to the activity of dilute species (Section 8.3.2).

7.4.3   Raoult's Law

Raoult's law originally concerned the composition of a vapor phase in equilibrium with a solution of two or more components. This sounds quite different from the Henry's law situation, but the two are intimately related. In fact, Raoult's law can be considered to be just a special case of Henry's law. Many combinations of components A and B (e.g., water and alcohol, or two organic liquids) were dissolved into one another in various proportions, and the composition and pressure of the coexisting vapor phase were measured (Figure 7.2). The results of these measurements varied widely, but a very few systems showed a particularly simple relationship. When the two liquids A and B were very similar, the vapor pressure of their mixture was a simple function of the vapor pressures of the pure liquids,

$$P_{\text{mixture}} = x_A P_A^\circ + x_B P_B^\circ \tag{7.6}$$

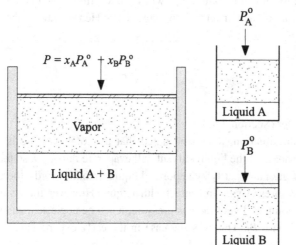

Figure 7.2 The vapor pressure of a solution of A and B that obeys Raoult's law.

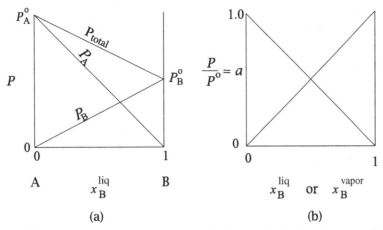

Figure 7.3 (a) The vapor pressure (P_{total}) of a binary solution that obeys Raoult's law is $P = P_A + P_B = x_A P_A^\circ + x_B P_B^\circ$. The partial pressure of each component is given by the diagonal lines, e.g., between 0 at $x_B = 0$ and P_B° at $x_B = 1$. (b) The partial pressure of each component divided by the vapor pressure of the pure component.

and the partial pressures of A and B in the vapor were found to be directly proportional to their concentration in the liquid (Figure 7.3(a)).

$$P_A = x_A^{liquid} P_A^\circ$$

$$P_B = x_B^{liquid} P_B^\circ$$

The only way that these simple relationships can hold is for the liquid solution to be ideal, in the sense discussed in Section 7.3.2. That is, the A–A, B–B, and A–B intermolecular forces must be identical, so that a molecule of A behaves exactly the same way whether it is surrounded mostly by A or mostly by B. Thus the relationship

$$P_i = x_i^{liquid} P_i^\circ \tag{7.7}$$

can be taken as a statement of Raoult's law, which means the solution is ideal. In Figure 7.3(b) we see that in an ideal solution of this type, P/P° for both components (i.e., P_A/P_A° and P_B/P_B°) are represented by diagonal lines, so that if we define *activity* as $a = P/P^\circ$, then the activity equals mole fraction ($a_A = x_A$; $a_B = x_B$) in Raoultian systems. The concept of activity is actually a bit more complicated. We will get to that in Chapter 8.

Raoult's law has therefore been generalized to refer not only to the partial pressures of gases, but to any solution, including solid solutions, in which component activities equal their mole fractions.

Although some authors use different terminology to distinguish between Raoultian and Henryan ideality, many do not. When referring to an ideal solution, we must always be clear whether we are referring to Raoultian or Henryan ideality.

7.5   Ideal Solution Equations

7.5.1   Volume, Enthalpy, Heat Capacity

These terms are simply additive in ideal solutions. Volume is the most intuitive of these terms, and we illustrate the additivity of volumes in ideal solutions in Section 7.7.1. This leads to Equation (7.38), the generalized form of which is

$$V_{\text{ideal solution}} = \sum_i x_i V_i^\circ \tag{7.8}$$

where V is the molar volume of an ideal solution, and V_i° is the molar volume of the pure component i.

The form of Equation (7.8) applies also to any thermodynamic parameter which does not contain entropy in its definition. The important ones are enthalpy and heat capacity, so that

$$H_{\text{ideal soltuion}} = \sum_i x_i H_i^\circ \tag{7.9}$$

and

$$C_{P\text{ideal solution}} = \sum_i x_i C_{p_i}^\circ \tag{7.10}$$

The difference between the property of a solution and the combined properties of the pure components is called a *mixing property*. For an ideal solution, from Equations (7.8)–(7.10)

$$\Delta_{\text{mix}} V_{\text{ideal solution}} = V_{\text{ideal solution}} - \sum_i x_i V_i^\circ \tag{7.11}$$

$$= \left(\frac{\partial \Delta_{\text{mix}} G}{\partial P} \right)_T$$

$$= 0$$

$$\Delta_{\text{mix}} H_{\text{ideal solution}} = H_{\text{ideal solution}} - \sum_i x_i H_i^\circ \tag{7.12}$$

$$= \left(\frac{\partial \Delta_{\text{mix}} G/T}{\partial (1/T)} \right)_P$$

$$= 0$$

$$\Delta_{\text{mix}} C_{P\text{ideal solution}} = C_{P\text{ideal solution}} - \sum_i x_i C_{p_i}^\circ \tag{7.13}$$

$$= \left(\frac{\partial \Delta_{\text{mix}} H}{\partial T} \right)_P$$

$$= 0$$

Entropy, Gibbs Energy

Entropy and other potential quantities which contain entropy (such as G) are specifically defined so as to change in spontaneous processes, and two or more substances dissolving into one another is a perfect example of a spontaneous process.

Consider two ideal gases which are allowed to mix at constant P and T. In the final mixture, which is also an ideal gas, the partial pressure of gas 1 is $P_1 = x_1 P$, and of gas 2 is $P_2 = x_2 P$, where x_1 and x_2 are the mole fractions. The change in entropy on mixing, $\Delta_{mix}S$, is equal to the ΔS involved in expanding each gas from its initial pressure P to its partial pressure in the gas mixture. From Equation (5.24), this process is, for each gas

$$\Delta S_1 = R\ln(P/P_1)$$

and

$$\Delta S_2 = R\ln(P/P_2)$$

and the total change in entropy is

$$
\begin{aligned}
\Delta_{mix}S_{\text{ideal solution}} &= x_1 R\ln(P/P_1) + x_2 R\ln(P/P_2) \\
&= x_1 R\ln(1/x_1) + x_2 R\ln(1/x_2) \\
&= -R(x_1 \ln x_1 + x_2 \ln x_2)
\end{aligned}
\tag{7.14}
$$

or, in general terms,

$$\Delta_{mix}S_{\text{ideal solution}} = S_{\text{ideal solution}} - \sum_i x_i S_i^\circ \tag{7.15}$$

$$= -R \sum_i x_i \ln x_i \tag{7.16}$$

$$= -\left(\frac{\partial \Delta_{mix}G}{\partial T}\right)_P \tag{7.17}$$

Because the x terms are fractional, $\Delta_{mix}S_{\text{ideal solution}}$ is inherently positive. Equation (7.16) can also be derived from reasonably simple statistical considerations which have nothing to do with the physical state of the particles. In other words, it applies equally to ideal gas, liquid, and solid solutions.

It is important to note the fundamental difference between the ideal mixing of volumes and other terms not containing entropy [Equations (7.8)–(7.10)], which are just linear combinations of the pure end-member terms, and the ideal entropy of mixing, Equations (7.15) and (7.16), which are nonlinear, and result in all mixtures having a higher entropy than points on the $\sum_i x_i S_i^\circ$ line or plane. It is this property which gives the entropy, Gibbs energy and other thermodynamic potentials (all of which contain an entropy term, either as part of the definition or as a constraint) their ability to predict energy differences, and hence reaction directions.

It follows from Equations (5.1) and (7.15), and the fact that $\Delta_{mix}H_{\text{ideal solution}} = 0$ (Equation (7.12)), that the Gibbs energy of (ideal) mixing is

$$\Delta_{mix}G_{\text{ideal solution}} = \Delta_{mix}H_{\text{ideal solution}} - T\Delta_{mix}S_{\text{ideal solution}}$$

$$= G_{\text{ideal solution}} - \sum_i x_i G_i^\circ \tag{7.18}$$

$$= RT \sum_i x_i \ln x_i \tag{7.19}$$

The $\sum_i x_i G_i^\circ$ term defines a straight line (or plane surface) between points representing end-member components, and the $\sum_i x_i \ln x_i$ term (which is inherently negative) describes how far *below* this line or plane the surface representing the G of the (ideal) solution is.

Changes in Gibbs energy are useful for individual reactions and for whole systems. Both cases are important. To get the extensive form of Equation (7.19), consider that each mole fraction is

$$x_i = \frac{n_i}{\sum_i n_i}$$

for the mole numbers n_i, so $\sum_i n_i$ represents the whole system. Multiplying both sides of (7.19) by $\sum_i n_i$ then gives

$$\Delta_{mix}\mathbf{G}_{\text{ideal solution}} = RT \sum_i n_i \ln x_i \tag{7.20}$$

Figure 7.4 shows $\Delta_{mix}H$, $\Delta_{mix}S$, and $\Delta_{mix}G$ for an ideal solution. There are several things to note in this diagram.

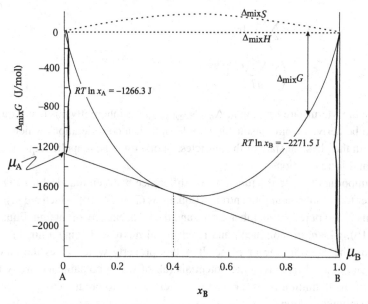

Figure 7.4 Gibbs energy of ideal mixing, from Equation (7.19).

1. Because the mixing of A and B takes place at a constant T and P and is a spontaneous process, $\Delta_{mix}G$ must be negative. The curve shown is an expression of Equation (7.19). No experimental data are required; just mole fraction numbers from 0 to 1.

2. The $\Delta_{mix}S$ shown is an expression of Equation (7.15), but because it is a much smaller quantity (reaching a maximum of $5.76\,\mathrm{J\,mol^{-1}\,K^{-1}}$ at $x_B = 0.5$) it is exaggerated in the diagram.

3. $\Delta_{mix}H$ is from Equation (7.12) and is of course zero at all x_B. $\Delta_{mix}G$ is therefore a mirror reflection of $T\,\Delta_{mix}S$, because $\Delta_{mix}G = \Delta_{mix}H - T\,\Delta_{mix}S$. The $T\,\Delta_{mix}S$ curve is not shown.

4. We don't have absolute values for G, so we must always measure the difference in G from some other state at the same T and P. In Figure 7.4 this other state is pure A $(x_A = 1)$ and pure B $(x_B = 1)$.

5. Despite appearances, the molar Gibbs energy curve in Figure 7.4 is actually asymptotic to each vertical axis, whereas the volume curve in Figure 7.6 on p. 170 is not. This is not of great importance by itself, but it is connected to the fact that we cannot use the infinite dilution standard state for Gibbs energy. We leave this important topic to Chapter 8.

Chemical Potentials

Figure 7.4 also introduces us to a new problem. Because the mixing curve is concave downward, the Gibbs energies of components A and B in the solution are necessarily less than the corresponding values of G_A° and G_B°, the molar Gibbs energies of the pure compounds. The fact that this is so provides the (thermodynamic) reason why A and B form a solution. If we mix n_A moles of A and n_B moles of B the reaction is

$$n_A\,\mathrm{A} + n_B\,\mathrm{B} = (n_A\,\mathrm{A}, n_B\,\mathrm{B})_{solution} \qquad (7.21)$$

and the Gibbs energy change for this reaction is $\Delta_{mix}G$, which is negative. During this reaction, the Gibbs energies of both A and B become lower. If the mixing line lies *above* the straight line joining G_A° and G_B°, then ΔG_{mix} would be positive, the dissolution reaction (7.21) would not be spontaneous, and no solution would form – A and B would be immiscible, like oil and water.

So $G_A^{solution} < G_A^\circ$, and $G_B^{solution} < G_B^\circ$. But this raises a few questions, like what are these quantities $G_A^{solution}$ and $G_B^{solution}$? Where are they on the diagram? How can you separate ΔG_{mix} into these two separate quantities? If we knew that, then perhaps we could evaluate $G_A^\circ - G_A^{solution}$ and $G_A^\circ - G_A^{solution}$.

You can see by simple inspection of Figure 7.4 that at any point on the $\Delta_{mix}G$ curve, the tangent at that point allows calculation of the numerical value of that point on the curve by combining two points on the tangent in a linear combination. In Figure 7.4 we show the tangent to the curve at $x_B = 0.4$, and identify the intersections of the tangent with the two ordinate axes as μ_A and μ_B. In Equation (7.18) we see that if G_A° and G_B° are zero, then $\Delta_{mix}G = G_{solution}$, and in drawing Figure 7.4, this is exactly what we have assumed. We have made $\Delta_{mix}G$ the difference between the G of the solution and zero. Evidently then,

$$G^{solution} = x_A\mu_A + x_B\mu_B \qquad (7.22)$$

But what is the physical meaning of μ? There are two ways of answering this.

Euler's Answer. Mathematically minded people simply invoke Euler's theorem for homogeneous functions. In plain language, this says that for any extensive (total) function such as $\mathbf{V}$ or in this case $\mathbf{G}$, having n_A moles of component A and n_B moles of component B

$$\mathbf{G} = n_A \left(\frac{\partial \mathbf{G}}{\partial n_A} \right)_{n_B} + n_B \left(\frac{\partial \mathbf{G}}{\partial n_B} \right)_{n_A} \tag{7.23}$$

or, dividing by $(n_A + n_B)$,

$$G = x_A \left(\frac{\partial \mathbf{G}}{\partial n_A} \right)_{n_B} + x_B \left(\frac{\partial \mathbf{G}}{\partial n_B} \right)_{n_A} \tag{7.24}$$

Comparing (7.22) and (7.24), we have

$$\left. \begin{aligned} \mu_A &= \left(\frac{\partial \mathbf{G}}{\partial n_A} \right)_{T,P,n_B} \\ \mu_B &= \left(\frac{\partial \mathbf{G}}{\partial n_B} \right)_{T,P,n_A} \end{aligned} \right\} \tag{7.25}$$

so the tangent intersections are in fact the partial derivatives of the total Gibbs energy of the solution $\mathbf{G}$ with respect to n_A and n_B. These are in fact our first examples of *partial molar* terms, which were introduced in Sections 2.4.1 and 4.13.1. μ is a partial molar Gibbs energy, and it is our answer to the question, what is the Gibbs energy per mole of a dissolved substance? It is important to develop an intuitive understanding of partial molar terms, so we devote more discussion to this in Section 7.7.2. Notice that it is the derivative of $\mathbf{G}$, not G, that results in the partial molar variable μ. This will be seen in physical terms when we get to the room analogy (Section 7.7.2).

Our Answer. But we don't have to use Euler's theorem. We can simply expand our definition of $\mathbf{G}$, which so far is restricted to closed (constant composition) systems. If we exclude chemical work, which means we deal only with systems at complete stable equilibrium, we know from Equation (4.72)

$$d\mathbf{G} = -\mathbf{S}\,dT + \mathbf{V}\,dP + \sum_i \mu_i\,dn_i$$

At constant T and P (by implication a condition of Equations (7.23) and (7.24)) the first two terms on the right drop out, and integrating the other terms from an increment of solution up to the whole system, we get, for two components 1 and 2,

$$\mathbf{G} = n_1 \left(\frac{\partial \mathbf{G}}{\partial n_1} \right)_{T,P,\hat{n}_1} + n_2 \left(\frac{\partial \mathbf{G}}{\partial n_2} \right)_{T,P,\hat{n}_2}$$

which is Equation (7.23), and from there we get Equations (7.25) again.

So that answers one question – how do we split G^{solution} into G_A^{solution} and G_B^{solution}, and where are these things on the diagram? The next question is, how do we evaluate $G_A^\circ - G_A^{\text{solution}}$ and $G_B^\circ - G_B^{\text{solution}}$?

7.5.3   ## The Relation Between Composition and Gibbs Energy

Figure 7.4 also shows a very important relationship between Gibbs energy and composition, namely, that

$$\left.\begin{array}{l} \mu_A - G_A^\circ = RT \ln x_A \\[2mm] \mu_B - G_B^\circ = RT \ln x_B \end{array}\right\} \tag{7.26}$$

Equations (7.26) provide a relationship between the concentration of an ideal solution component and its Gibbs energy. This is an important milestone. The equations

$$(\partial G/\partial T)_P = -S \tag{4.50}$$

$$(\partial G/\partial P)_T = V \tag{4.51}$$

in Chapter 4 showed how Gibbs energy varies with T and P, respectively (expanded upon in Figures 6.2 and 6.6); now we can see how Gibbs energy varies with concentration of something in solution. If we can calculate the Gibbs energy of solids, liquids, gases, and solutes over a range of T, P, and composition (x), we have just about solved all our problems, in principle. Basically, from here on we will be amplifying and coming to grips with practical matters, such as the fact that Equation (7.26) applies only to Raoultian solutions.

But where do Equations (7.26) come from?

Gibbs Energy and Mole Fraction I

The most direct way to derive the relationship between Gibbs energy and mole fraction is to simply differentiate the total form of Equation (7.19) with respect to n_i. For two components, $G = \mathbf{G}/(n_1 + n_2)$, so multiplying both sides of (7.19) by $(n_1 + n_2)$, then differentiating, gives

$$\frac{\partial}{\partial n_1}\left(\mathbf{G}_{\text{solution}} - n_1 G_1^\circ - n_2 G_2^\circ\right) = \frac{\partial}{\partial n_1}\left[RT\left(n_1 \ln x_1 + n_2 \ln x_2\right)\right]$$

which, with n_2, G_1°, and G_2° constant, gives

$$\mu_1 - G_1^\circ = RT\left[\frac{\partial}{\partial n_1}\left(n_1 \ln x_1\right) + \frac{\partial}{\partial n_1}\left[n_2 \ln(1 - x_1)\right]\right]$$

$$= RT\left(\ln x_1 + x_2 - x_2\right)$$

$$= RT \ln x_1 \tag{7.27}$$

which is (7.26) for component 1.

Applying the same method to Equation (7.14), we get

$$\bar{S}_i - \bar{S}_i^\circ = -R \ln x_i \tag{7.28}$$

and recall that $\Delta_{mix}H = 0$, so

$$\overline{H}_i - \overline{H}_i^{\circ} = 0 \tag{7.29}$$

as well, for ideal solutions.

There is a complication in (7.27) and (7.28) which is actually very general but only becomes important in the study of solid solutions. We discuss it in more detail in Section 9.8.2, but briefly, the problem is that in Equation (7.27) the left side is a difference in energy *per mole*, which obviously depends on how the mole is defined, but the right side has a mole fraction, which is independent of the definition of the mole as long as all components of the solution are treated in the same way.

For example, normally we define a mole of nitrogen as Avogadro's number of N_2 in nitrogen gas, and oxygen as O_2. But we could define these as the same number of N and O, or of N_4 and O_4, without affecting the mole fraction. Whether these forms exist or not is not relevant. The mole fraction of N_2 in a solution of nitrogen and oxygen does not depend on which of these ways we use to define the mole, but the value of μ does. Avogadro's number of N_4 particles has twice the mass and twice the energy of the same number of N_2 particles, so

$$\mu_{N_4} - \mu_{N_4}^{\circ} = 2\left(\mu_{N_2} - \mu_{N_2}^{\circ}\right) \tag{7.30}$$

Therefore, if there is any question as to the size of the mole, which usually only arises in defining components in solid solutions, Equation (7.27) is generalized to

$$\mu_1 - G_1^{\circ} = n\,RT\ln x_1$$
$$= RT\ln x_1^n \tag{7.31}$$

where n is the factor relating the two definitions of the mole.

Gibbs Energy and Mole Fraction II

It will also prove useful to derive (7.27) another way, especially when we use molal units of concentration instead of mole fractions. So first we will re-derive (7.27), and then use the same method for molal units in Section 8.2.3.

Equation (7.27) expresses the relationship between μ and x, or concentration. To derive this, it would seem natural to find an expression for the derivative of μ with a concentration term, and then integrate. In other words, what is the value of $(\partial\mu_i/\partial n_i)_{\hat{n}_i}$? How does μ_i (the G of i in solution) vary with the amount of i in solution in the ideal case? This is a partial derivative, so if n_i is the number of moles of i, we need to keep the concentrations of all other components constant. We denote all other components by $\hat{n}_i$.

If we expand $(\partial\mu_i/\partial n_i)_{T,P,\hat{n}_i}$ by introducing P_i, the pressure on gaseous i which is, or might be, in equilibrium with solute i (whether or not there is such a gas phase is irrelevant), we get

$$\left(\frac{\partial\mu_i}{\partial n_i}\right)_{\hat{n}_i} = \frac{\partial\mu_i}{\partial P_i}\frac{\partial P_i}{\partial n_i} \tag{7.32}$$

where μ_i is the same in the solution and in the (perhaps hypothetical) gas phase, where it can be called G_i (the gas being assumed ideal), so that $(\partial \mu_i / \partial P_i) = (\partial G_i / \partial P_i) = V_i = RT/P_i$, and where $(\partial P_i / \partial n_i) = P_i / n_i$ is an expression of Henry's law (Equation (7.5)).[2] Combining all this, we get

$$\left(\frac{\partial \mu_i}{\partial n_i} \right)_{\hat{n}_i} = \frac{RT}{n_i} \tag{7.33}$$

for ideal solutions. Integrating this equation between two values of n_i, n_i', and n_i'', we get

$$\mu_i'' - \mu_i' = RT \ln \left(\frac{n_i''}{n_i'} \right)$$

$$= RT \ln \left(\frac{P_i''}{P_i'} \right) \quad \text{by Henry's law}$$

$$\mu_i'' - \mu_i^\circ = RT \ln \left(\frac{P_i''}{P_i^\circ} \right) \quad \text{if state ' is pure } i$$

and, from Equation (7.7),

$$\mu_i - \mu_i^\circ = RT \ln x_i \quad \text{(state '' no longer needs a superscript)} \tag{7.34}$$

which is Equation (7.26) for component i.

When $x_B = 0.4$, $x_A = 0.6$, $R = 8.31451 \, \text{J mol}^{-1} \text{K}^{-1}$, and $T = 298.15 \, \text{K}$, Equation (7.26) gives

$$\mu_A - G_A^\circ = -1266 \, \text{J mol}^{-1}$$

and

$$\mu_B - G_B^\circ = -2271 \, \text{J mol}^{-1}$$

This says that a mole of A has $1266 \, \text{J mol}^{-1}$ less in the dissolved state than in the pure state, and this is the "thermodynamic explanation" for why A dissolves in B.

7.6 Real Solutions

We have now considered both ideal solution behavior and deviations from it, but in a rather generalized way, using activity coefficients. We now have to start to consider how to measure these things, and doing this means we have to consider partial molar properties in much more detail.

We start with a fairly detailed look at the volumetric properties of solutions, because these are the most intuitive. Partial molar properties of the other state variables are the same in principle, but become more complicated in the case of enthalpy measurements because of the relative nature of enthalpy. The Gibbs energy is also a relative property,

---

[2] OK, Equation (7.5) shows the ratio of P_i to m_i, not n_i. But as m_i is just n_i per kilogram of solvent, the ratio is the same.

but is treated in quite a different way. Most of the material in this chapter is quite general, and can be applied to any kind of solution, although most of our examples are for aqueous solutions.

Real Solutions

The thermodynamics of solutions is actually a huge subject. The fundamental, most important variable as described here is the partial molar property, especially the partial molar Gibbs energy, but there are several other aspects that deserve mention, such as regular solutions, excess properties, the infinite dilution standard state, and the Margules equations. See Chapter 10, Real Solutions, in the online material.

7.7 Solution Volumes

All real solutions are of course non-ideal. Our discussion of their properties will be concerned for the most part with deviations from the properties of ideal solutions, whether Henryan or Raoultian.

7.7.1 Partial and Apparent Properties

The properties of a dissolved substance are described in terms of *partial*, *apparent*, and *excess* total or molar properties, so we begin by discussing these terms, using volume as an example.

The Volume of Mixing

If two substances are immiscible (they do not dissolve into one another to any appreciable extent, like oil and water), obviously the volume of the two together is simply the sum of the two volumes separately. But if they are completely miscible (they dissolve into one another completely, forming a solution), this may be more or less true, but probably not exactly true. Why?

If you mix white sand and black sand together, there is no interaction or chemical reaction at all between the two kinds of sand, and the volume of the mixture is the same as the two volumes separately. If the volume of the white sand is $\mathbf{V}_w$ and the volume of the black sand is $\mathbf{V}_b$, the total volume is

$$\mathbf{V} = \mathbf{V}_w + \mathbf{V}_b$$

It's somewhat like stacking boxes as in Figure 7.5. There is no change in total volume just because they are together.

However, using total volumes usually turns out to be inconvenient. If the volume per mole of white sand is V_w and that of black sand is V_b, then the total volume is

$$\mathbf{V} = n_w V_w + n_b V_b \tag{7.35}$$

7.7 Solution Volumes

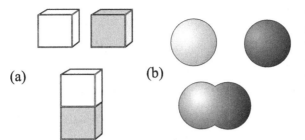

Figure 7.5 (a) There is no volume change when boxes are stacked together – they do not interact. (b) When molecules are mixed together, they may occupy less volume than they did separately.

where n_w and n_b are the numbers of moles of white and black sand in the mixture. The molar volume is defined as the total volume divided by the number of moles of all components in the system (i.e., the molar volume of pure white sand is therefore $\mathbf{V}_w/n_w$); so if the mixture contains n_w moles of white sand and n_b moles of black sand, the total number of moles in the mixture is $n_w + n_b$. Dividing both sides of Equation (7.35) by $n_w + n_b$, we get

$$V = x_w V_w + x_b V_b \tag{7.36}$$

Here, V is the *molar volume* of the mixture and x is the *mole fraction*, where

$$\begin{aligned} x_w &= \frac{n_w}{\sum n} \\ &= \frac{n_w}{n_w + n_b} \end{aligned} \tag{7.37}$$

and similarly for x_b. This equation simply says that the volume of the mixture is the same as the volume of the two things separately. The introduction of n and x is just to determine how much of each is used. If we plot molar volume against mole fraction of either component sand, we get a straight line (Figure 7.6), called the *ideal mixing* line.

Clearly these relations do not depend on the grain size of the sands;[3] they depend on the fact that the sands do not react in any way with each other. Each grain of white sand is indifferent to what kind of sand is next to it. Now imagine that the grain size of the sands gets smaller and smaller. Soon they get so small that you can no longer distinguish the colors – the mixture becomes gray. Imagine the grain size continuing to get smaller and smaller – right down to atomic proportions, so that instead of having a mechanical mixture of black and white sand, we have a true solution of black and white atoms. If the black and white atoms continue to have no attraction, repulsion, or chemical reaction with one another, the volume of the two together will continue to be exactly the same as the sum of the two separately. Actually, we have oversimplified a bit – normally the white molecules interact with each other even in the pure state, and similarly with the black molecules. If these interactions are very similar in nature, then when they are mixed together the molecules will continue to interact with each other in the same way, and the volumes will be additive. In other words, it is not necessary for there to be no molecular interactions

---

[3] Actually, only as long as the grain sizes of the black and white sands are the same.

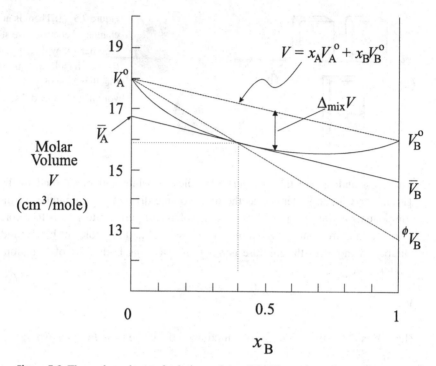

Figure 7.6 The molar volume of solutions of A and B. The molar volume of pure A (V_A^o) is 18.0 cm^3 mol^{-1} and that of pure B (V_B^o) is 16.0 cm^3 mol^{-1}. The molar volume of an ideal solution having $x_B = 0.4$ is $0.6 \times 18.0 + 0.4 \times 16.0 = 17.2$ cm^3 mol^{-1}. The molar volume of a real solution having $x_B = 0.4$ is actually 15.9 cm^3 mol^{-1}. It may be calculated in the same way, but using $\overline{V}_A$ and $\overline{V}_B$ instead of V_A^o and V_B^o. The difference between the real and the ideal molar volumes is the change in V on mixing $\Delta_{mix}V$. There is also something called the apparent molar volume of B ($^\phi V$) in a solution, which is the intercept on the $x_B = 1$ axis of a line joining the molar volume of pure A and the molar volume of the solution, in this case $^\phi V = 12.75$ cm^3 mol^{-1}.

for ideal mixing, only that white molecules react with black molecules in exactly the same way that they do with other white molecules.

But suppose that at this molecular size, white (w) and black (b) particles are attracted to one another more than to others of the same kind, perhaps even forming a new kind of particle (wb). Because of this attraction, the particles will be closer together than they would otherwise be, and the total volume of the mixture will be smaller, as shown in Figure 7.5(b), and instead of getting a straight mixing line as in Figure 7.6, the line is curved downward as in Figure 7.6. Alternatively, if the white and black particles repel each other, the total volume will be greater, and in Figure 7.6(b) the curved line for the molar volume of the mixture will lie above the straight line that represents no interaction. The volume change on mixing ($\Delta_{mix}V$, Figure 7.6) caused by the attraction between A and B is the difference between the straight line and the curved line. The straight line

$$V = x_A V_A^o + x_B V_B^o \tag{7.38}$$

is called *ideal mixing* and is rarely observed. The curved line represents non-ideal mixing, the general case. The difference between the ideal mixing line and the actual molar volume V is called the change in volume on mixing, $\Delta_{mix}V$. Thus

$$\Delta_{mix}V = V - (x_A V_A^\circ + x_B V_B^\circ) \tag{7.39}$$

7.7.2   Partial Molar Volumes

Now suppose in our mixture of white and black particles that attract each other, that we are not satisfied to have the total volume or the molar volume of the mixture as a whole. We would like to know the volume of each component in the mixture, not just the combined volume. But how can this be done, when each is dispersed at the molecular level and is interacting strongly with another component? Simple. Just draw the tangent to the molar volume curve at the composition you are interested in. The intercepts of this tangent give the volumes of each component in the solution, called *partial molar volumes*, which are combined to give the total molar volume in exactly the same way as the black and white sands in Equation (7.36) and Figure 7.6.

Looking at partial molar volumes in this way, they seem to be just a sort of geometrical construct. They are defined such that they can be substituted for V_A and V_B in Equation (7.36) in cases where mixing results in a curved line for the molar volumes; thus

$$V = x_A \overline{V}_A + x_B \overline{V}_B \tag{7.40}$$

(Figure 7.6) or, multiplying both sides by $(n_A + n_B)$,

$$\mathbf{V} = n_A \overline{V}_A + n_B \overline{V}_B \tag{7.41}$$

In Figure 7.6 we have shown a case where A and B are attracted to each other, and their partial molar volumes are both *less* than the volumes of the pure components ($\overline{V}_A < V_A^\circ$). If A and B repelled one another, the mixing line would lie above the straight line and the partial molar volumes would be *larger* than the pure volumes. There is no general rule for the shapes and positions of these mixing curves; they must be measured experimentally. This would be done by density measurements in the case of volume, and calorimetry in the case of enthalpy and entropy. It is quite possible for the mixing curve to be shaped such that in a certain range of composition one of the tangent intercepts is at less than zero volume – a negative partial molar volume. This is why some of the tabulated thermodynamic parameters in Appendix B are negative for some solute components. It is, of course, not possible for pure components to have a negative volume.

The Room Analogy

But there is another way of looking at partial molar volumes which shows that they really are the volume of a mole of each component in solution. Just for a change we will switch from components A and B to a solution of salt (NaCl) in water. Consider an extremely large quantity of water – say enough to fill a large room (Figure 7.7). Now let's add enough salt to make the concentration exactly 1 molal, and adjust the volume of the solution so that the room is full and a little excess solution sticks up into a calibrated tube inserted into the

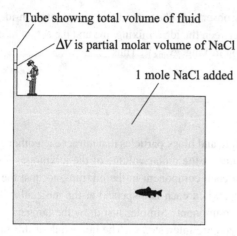

Tube showing total volume of fluid

ΔV is partial molar volume of NaCl

1 mole NaCl added

Figure 7.7 A roomful of 1 molal salt solution. The observer sees the change in volume caused by adding one mole of salt, which is the partial molar volume of salt in the 1 molal solution.

ceiling. By observing changes in the level of solution in the tube, we can accurately record changes in the **V** of the solution in the room.

Now, when we add a mole of NaCl (58.5 g of NaCl occupying $27 \, cm^3$) to the solution, the change in concentration is very small. In fact, if we can detect *any* change in concentration by the finest analytical techniques available, then our room is too small, and we must find and inundate a larger one. Eventually, we will fill a sufficiently large room with salt solution that on adding 58.5 g of NaCl we are unable to detect any change in concentration – it remains at 1.000 mole NaCl/kg H_2O. But although the concentration remains unchanged, the volume of course does not. The salt added cannot disappear without a trace. The level in the tube in the ceiling changes, and the Δ**V** seen there is evidently the volume occupied by 1 mole of NaCl in a 1 molal NaCl solution, in this case about $19.47 \, cm^3 \, mol^{-1}$ of NaCl. This is, in quite a real sense, the volume occupied by a mole of salt in that salt solution and has a right to be thought of as a *molar volume* (just as much as $27 \, cm^3 \, mol^{-1}$ is the molar volume of crystalline salt) rather than as an arbitrary mathematical construct. It is referred to as the *partial* molar volume of NaCl in the salt solution, $\overline{V}_{NaCl}$.

Some readers will have difficulty in seeing how, on adding our salt, the concentration does not change but the total volume does. If this is the case, think of the room as containing not a solution, but nine million white tennis balls and one million black tennis balls, all mixed together. The room is full, the balls are arranged so that no space is available for another ball, and a few balls overflow into the tube in the ceiling. The total volume is the volume of ten million tennis balls. Now we add one more black tennis ball, somewhere in the middle of the room. The fractional concentration of black balls changes from $10^6/10^7$ to $(10^6 + 1)/(10^7 + 1)$, or from 0.1 to 0.10000009, a change so small it is completely negligible.[4] But the total volume has changed by the volume of one tennis ball, and this change must be reflected by the level of the balls in the tube, which will rise by the volume of one ball. We can even extend the analogy by imagining that the balls in the room are

---

[4] If you don't find it negligible, just imagine a bigger room and more tennis balls, until the change *is* negligible.

compressed by the pressure, so that when we add another ball, it becomes compressed too, and the level in the tube rises by the volume of a compressed tennis ball, not a normal (standard state) tennis ball.

The Formula for Partial Molar Properties

The partial molar concept is applied to most thermodynamic properties, not just volume. The mathematical expression, introduced in Section 2.4.1 on page 22, is

$$\left(\frac{\partial \mathbf{Z}}{\partial n_i}\right)_{\hat{n}_i} = \overline{Z}_i \tag{7.42}$$

where $\mathbf{Z}$ is a thermodynamic parameter such as $\mathbf{V}$, $\mathbf{S}$, $\mathbf{G}$, etc., n_i is the moles of component i, and $\hat{n}_i$ is the moles of all solution components *except* i. It is important to note that the derivative is taken of the *total* quantity, $\mathbf{Z}$, not the molar property, Z. It is the change in the *total* volume of the solution in the room that is measured, not the molar volume.

Put in this partial differential form, partial molar properties look somewhat obscure. However, it is important to have an intuitive grasp of their meaning, and you will be well advised to think of them in the sense of the room analogy, as *molar* properties of solutes in solutions of particular compositions, rather than in terms of Equation (7.42).

7.8   Next Step – The Activity

We have now developed the relationship between the Gibbs energy of a component of a solution and the concentration of that component (Equations (7.26), (7.27), and (7.34)). However, it applies only to ideal solutions, and only for concentrations in mole fractions. Obviously we need to expand the range of applicability of this relationship.

Doing this gets complicated, because we have gaseous, liquid, and solid solutions, a variety of concentration scales, non-ideal solutions, and several different standard states that μ° refers to. That is, the quantity $\mu_i - \mu_i^\circ$ need not always refer to the difference between i in solution and i in its pure state. At the same time, the form of Equation (7.34) is very convenient, and we want to retain it for all these conditions. We do this by defining the *activity*, already mentioned in Section 7.4.3, as

$$\mu_i - \mu_i^\circ = RT \ln a_i \tag{7.43}$$

All the complications are accommodated by this parameter, and we will try to sort it all out in Chapter 8.

7.9   Summary

In Chapter 4 (Section 4.11.1) we saw that

$$(\partial G/\partial T)_P = -S \tag{4.50}$$

and

$$(\partial G/\partial P)_T = V \tag{4.51}$$

We didn't bother to write, though it is equally true, that

$$(\partial \mathbf{G}/\partial T)_P = -\mathbf{S} \tag{7.44}$$

and

$$(\partial \mathbf{G}/\partial P)_T = \mathbf{V} \tag{7.45}$$

and, differentiating by n_i, that

$$(\partial \mathbf{S}/\partial n_i)_{T,P} = \overline{S}_i \tag{7.46}$$

and

$$(\partial \mathbf{V}/\partial n_i)_{T,P} = \overline{V}_i \tag{7.47}$$

meaning that any solution property, not just Gibbs energy, can be split up into the contributions of the individual components. Finally, because

$$\frac{\partial}{\partial T}\left(\frac{\partial \mathbf{G}}{\partial n_i}\right) = \frac{\partial}{\partial n_i}\left(\frac{\partial \mathbf{G}}{\partial T}\right)$$

and similarly for P, then

$$\frac{\partial \mu_i}{\partial T} = -\overline{S}_i$$

and

$$\frac{\partial \mu_i}{\partial P} = \overline{V}_i$$

So it turns out that

$$\left.\begin{aligned}
(\partial \mu_i/\partial T)_{P,n} &= -\overline{S}_i \\[4pt]
(\partial \mu_i/\partial P)_{T,n} &= \overline{V}_i \\[4pt]
(\partial \mu/\partial n_i)_{T,P,\hat{n}_i} &= RT/n_i
\end{aligned}\right\} \tag{7.48}$$

are the relationships we need to be able to evaluate (by integration) to know the Gibbs energy of any substance as a function of T, P, and composition. But we're not finished yet. The first two equations apply to any solution, but the $(\partial \mu/\partial n_i)$ equation deals only with ideal solutions. We need to find ways to evaluate the partial molar properties in the first two equations, and how to modify the third one so that it works for any solution. A central concept in all of this is the activity, which we take up in Chapter 8.

Exercises

E7.1 Calculate $\Delta_{\mathrm{mix}}G$ for $x_B = 0.4$, and compare with Figure 7.4.

E7.2 If the molality of solute A is 0.05 and its activity coefficient (γ_H) is 0.8, what can you say about the chemical potential of the solute?

E7.3 The composition of the air we breathe is shown in Table 7.3.

Table 7.3. Gases in the air.

Gas	Percent by volume
N_2	78.084
O_2	20.946
Ar	0.934
CO_2	0.035
CH_4	0.00017
H_2	0.00005

Calculate the partial pressure and fugacity of each gas, assuming the atmosphere is an ideal gas. Atomic weights are not required.

E7.4 If a sample of air was compressed to 100 bar, what would be the fugacity of methane in the gas? Assume it behaves as an ideal gas ($\gamma_f = 1.0$).

E7.5 If the fugacity coefficient of methane at 100 bar is actually 0.95, what is its fugacity?

Additional Problems

A7.1 The equation for $\Delta_{mix}G$ in Figure 7.4 is

$$\Delta_{mix}G = RT(x_A \ln x_A + x_B \ln x_B)$$

Use calculus to find the minimum in this function.

Table 7.4. Densities of Nacl solutions.

m_{NaCl} (moles NaCl/kg H_2O)	Density, ρ $(g\,cm^{-3})$
0.00	0.99710
0.10	1.00117
0.25	1.00722
0.50	1.01710
0.75	1.02676
1.00	1.03623
2.00	1.07228
3.00	1.10577
4.00	1.13705
5.00	1.16644
6.00	1.19423

A7.2 The densities of NaCl solutions, taken from Pitzer *et al.* (1984), are shown in Table 7.4. All other necessary data (molecular weights, molar volumes of NaCl(s), H_2O) are in Appendix B.

From these data, calculate for each molality

(a) the mass of the solution containing 1000 g water.

(b) the volume occupied by this solution.

(c) the mole fraction of NaCl (x_{NaCl}).

(d) the solution molar volume.

(e) the ideal mixing molar volume.

(f) the ΔV of mixing ($\Delta_{mix}V$).

Present the results in tabular form.

A7.3 Plot the ideal mixing line and the solution molar volume vs. x_{NaCl}, where x_{NaCl} extends from 0.0 to the highest concentration given.

A7.4 Plot the same data against x_{NaCl}, where x_{NaCl} extends from 0.0 to 1.0.

A7.5 Estimate the partial molar volume of NaCl in pure water ($\overline{V}^{\circ}_{NaCl}$). This requires drawing a tangent to the molar volume curve at $x_{NaCl} = 0$, which is difficult to do accurately. An approximation is obtained by finding the difference in solution molar volume between 0.0 and 0.10 molal (you should find a slight decrease in V between 0.0 and 0.10 molal), then assuming that this change in volume per 0.10 change in m_{NaCl} is continued all the way over to $x_{NaCl} = 1.0$. Plot this tangent on the diagram having x_{NaCl} from 0.0 to 1.0. The tangent intercept at $x_{NaCl} = 1.0$ is $\overline{V}^{\circ}_{NaCl}$. The accepted value of $\overline{V}^{\circ}_{NaCl}$ is $16.80 \, cm^3 \, mol^{-1}$. (Note that partial molar volumes are not actually measured by this graphical method. More accurate methods are available.)

A7.6 Explain in your own words what a partial molar volume of NaCl in pure water of $16.80 \, cm^3 \, mol^{-1}$ means. Sketch a diagram to show how you might obtain the partial molar volume of NaCl in water saturated with NaCl.

8 Fugacity and Activity

8.1 Fugacity

. . . .

8.1.1 Introduction

The fugacity was introduced by G. N. Lewis in 1901, and became widely used after the appearance of *Thermodynamics*, a very influential textbook by Lewis and Randall in 1923. Lewis describes the need for such a function in terms of an analogy with temperature in the attainment of equilibrium between phases. Just as equilibrium requires that heat must flow such that temperature is the same in all parts of the system, so matter must flow such that chemical potentials are also equalized. He referred to the flow of matter from one phase to another as an "escaping tendency," such as a liquid escaping to the gas form to achieve an equilibrium vapor pressure. He pointed out that in fact vapor pressure is equilibrated between phases under many conditions, and could serve as a good measure of escaping tendency if it behaved always as an ideal gas.

The chemical potential is of course another measure of "escaping tendency," but Lewis pointed out that there are "certain respects in which this function is awkward." This refers to the fact that because activity is defined as $\mu - \mu° = RT \ln a$, $\mu \rightarrow -\infty$ as $a \rightarrow 0$. Lewis defined a function which would be much like a vapor pressure, which would be equilibrated between phases at equilibrium, even in non-ideal cases, and even if no vapor phase actually existed. It is a system property, in the sense that it does not depend on a standard state, and it has units of pressure. Lewis and Randall (1923) called it a kind of "ideal or corrected vapor pressure."

Fugacity has proved useful in a number of ways. One way is to provide a relatively simple way to evaluate the integral $\int VdP$. In Section 5.7.1 we saw one way to do this. That is, for solids, we often assume that the molar volume is constant, making the integration very simple. Another way, for gases, is to assume the ideal gas law (see below). This is actually a special case of the most general method, which is to develop an "equation of state" for the system, from which you can generate all its thermodynamic properties.

Lacking an equation of state, how do we evaluate the integral $\int V dP$ for a fluid such as H_2O or CO_2, either in the pure form, mixed with other fluid components, or reacting with solid phases? A possible way to proceed would be to express V as a function of P in some sort of power series, just as we did for C_P as a function of T (Equation (3.41)). $\int V dP$ could then be integrated, and we could experimentally determine the values of the power series coefficients for each gas or fluid and tabulate them as we do for the Maier–Kelley coefficients.

8.1.2   Definition of Fugacity

Fortunately, thanks to the insight of Lewis, we can proceed in a simpler and completely different fashion. To see how the inspiration for such a function might have arisen, consider the form of the volume integral $\int V \, dP$ for an ideal gas. Starting with (4.51),

$$dG = V \, dP$$

$$= \frac{RT}{P} dP$$

$$= RT d \ln P \tag{8.1}$$

$$\int_{P_1}^{P_2} dG = \int_{P_1}^{P_2} \frac{RT}{P} dP$$

$$G_{P_2} - G_{P_1} = RT \ln \left(\frac{P_2}{P_1} \right) \tag{8.2}$$

If P_1 is 1 bar and this is designated a standard or reference state denoted by a superscript $°$, then P_2 becomes simply P, and

$$G - G^° = RT \ln \left(\frac{P}{P^°} \right)$$

$$= RT \ln P \qquad \text{since } P^° = 1 \text{ bar} \tag{8.3}$$

Thus for ideal gases $RT \ln P$ all by itself gives the value of $\int_{P=1}^{P} dG$, and thus the difference in Gibbs energy of an ideal gas at pressure P and at 1 bar. Unfortunately, this doesn't work for real gases, although it's not a bad approximation at low pressures and high temperatures where real gases approach ideal behavior. However, the *form* of the relationship

$$dG = V \, dP = RT \, d \ln P$$

(Equation (8.1)) is sufficient to suggest that we could define a function such that the relationship *would* hold true for real gases. This function is the fugacity, f, where

$$dG = V \, dP = RT \, d \ln f \tag{8.4}$$

and

$$\int_{P_1}^{P_2} dG = \int_{P_1}^{P_2} V \, dP$$

$$G_{P_2} - G_{P_1} = \int_{P_1}^{P_2} RT \, d \ln f$$

$$= RT \ln \left(\frac{f_{P_2}}{f_{P_1}} \right) \tag{8.5}$$

Because f appears as a ratio in (8.5), this equation cannot serve as a full definition of f. To complete the definition, i.e., to be able to get real numbers for fugacity, we need to know the fugacity at some pressure (at temperature T) to use as f_{P_1}. The integral $\int V \, dP$ will then give us numbers for f_{P_2} at some fixed T. Gases behave more or less ideally at low values of P, so we accomplish this by stipulating that

$$\lim_{P \to 0} \left(\frac{f}{P} \right) = 1 \tag{8.6}$$

This means that for an ideal gas, $f = P$, and for real gases at low pressures, $f \approx P$. So the fugacity of a gas, say CO_2 or H_2O at very high T and P, can be determined by using measured volumes V (actually $V - RT/P$ works better) to integrate $\int V\, dP$, starting from some very low pressure where f_{P_1} has a known value. Equations (8.6) plus (8.4) or (8.5) make up the definition of fugacity.

Fugacity Coefficient

The ratio f/P is called the fugacity coefficient, γ_f. Thus

$$f_i = \gamma_{f_i} P \tag{8.7}$$

where P is the pressure of a pure fluid compound, or the *partial pressure* of a compound in a solution. The partial pressure is a measure of concentration, so the fugacity coefficient is a kind of *activity coefficient*, discussed in Section 8.3.

Box 8.1 Fugacity Standard States

The fact that fugacities often appear in the ratio f/f° has led to a common confusion about "fugacity standard states." For example, in one textbook the authors say

The fugacity is a relative function because its numerical value is always relative to that of an ideal gas at unit fugacity; in other words, the standard state fugacity f_i° in [... the equation $\mu_i - \mu_i^\circ = RT \ln (f_i/f_i^\circ)$...] is arbitrarily set equal to some fixed value, usually 1 bar.

This is a bit misleading. The numerical value of f_i is not relative to anything, but $\mu - \mu^\circ$ is. Fugacity is calculated from measured densities or molar volumes, and the fact that the lower limit of integration is some very low pressure does not change the fact that fugacity is a system property, while activity is not. Fugacity is independent of whatever one chooses as f°, but the choice of f° of course governs the corresponding value of $\mu - \mu^\circ$.

The expression $RT \ln f_i$ gives the difference between μ_i at T, P and μ_i of ideal gas i at T, 1 bar, just as $RT \ln (m_i \gamma_i)$ gives the difference between μ_i at T, P and μ_i in an ideal one molal solution at T, P. f_i is no more a relative value than is $m_i \gamma_i$. So fugacities do not have standard states any more than corrected concentrations have standard states.

In other words, $a_i = 0.01$ is meaningless unless the standard state is known, but $f_i = 0.01$ bars is unambiguous.

8.2 Activity

8.2.1 Introduction

In Chapter 7 (Section 7.8) we introduced the *activity*, and we said that in the form $RT \ln a_i$, it gives the quantity $\mu_i - \mu_i^\circ$. If you think about this, you will realize that there are not many more important concepts related to using thermodynamics in chemical systems. The goal of finding the minimum value of our thermodynamic potential G (or μ) in chemical systems is made complicated by the variety and complexity of our systems, and the fact that we use a variety of standard states in calculating our difference in Gibbs energy. In a sense, all these complexities are transferred to a single quantity, the activity, a dimensionless number which is directly related to $\mu_i - \mu_i^\circ$ for any component i in any system under any conditions. How to calculate a_i for various kinds of components (pure phases, associated and dissociated solutes etc.) in various kinds of systems (multiphase solid, liquid and gaseous solutions) is therefore an important topic.

Let us first summarize our development of the concept of the fugacity, f. Starting with the definition

$$dG = RTd\ln f \tag{8.4}$$

we found

$$G_{P_2} - G_{P_1} = RT \ln \left(\frac{f_{P_2}}{f_{P_1}} \right) \tag{8.5}$$

which expresses the relationship between the Gibbs energy and fugacity of a gas at two different pressures at the same T. The other part of the definition,

$$\lim_{P \to 0} (f/P) = 1 \tag{8.6}$$

allows us to experimentally derive numbers for the fugacity. However, changing the pressure on a pure phase is not the only way of changing the fugacity. Because fugacity approximates partial pressure, we might, for instance, simply introduce other components at the same P and T, which will also change the fugacity. Dealing with a solution rather than a pure phase, though, means we should use μ rather than G. So generalizing from a single gas to a gas i in a mixture of gases, and from two states at different pressures to any two states $'$ and $''$ at the same temperature, this becomes

$$\mu_i'' - \mu_i' = RT \ln \left(\frac{f_i''}{f_i'} \right) \tag{8.8}$$

One implication of this is that the fugacity of i is the same in any two states or phases that are in mutual equilibrium, because if $\mu' = \mu''$ then $f' = f''$. This of course was Lewis's intention in defining the fugacity in the first place. Because in principle any substance or species has a fugacity, Equation (8.8) seems to offer a general method for determining Gibbs energy differences. The problem with that is that the fugacities ($\approx$ vapor pressures) of substances other than gases are far too small to measure, and are mostly unknown. However, even in systems where species fugacities are unknown, the *ratio* of a species fugacity to its fugacity in some other state is quite often a measurable and useful quantity, and, comparing (8.8) with Equation (7.43), we see that this ratio is in fact a way of expressing the *activity*.

8.2.2   Definition I: Gases

Rewriting (8.8) so that state $''$ is any (unsuperscripted) state and state $'$ is a standard state designated by superscript $°$, we have

$$\mu_i - \mu_i^° = RT \ln \left(\frac{f_i}{f_i^°} \right) \tag{8.9}$$

This is a simple generalization of (8.8), and hence a direct result of the definition of fugacity. We now define the activity of species i as

$$a_i = \frac{f_i}{f_i^°} \tag{8.10}$$

where f_i and $f_i^°$ are the fugacities of i in the particular solution or state of interest to us and in some reference state at the same temperature. Thus

$$\mu_i - \mu_i^° = RT \ln a_i \tag{8.11}$$

which is of course Equation (7.43), arrived at in a different way. We begin now to see why using the activity can be confusing. In Chapter 7 (Equation (7.34)) the state that $\mu_i^°$ refers to is i as a pure liquid or solid, and in this case (8.8) $\mu_i^°$ refers to i existing as a gas or perhaps fluid in some as yet undefined state, which might be, and is in fact, completely arbitrary. It will be interesting to see how it is that we can use μ_i in a multicomponent, multiphase system, where at equilibrium μ_i must be the same in every phase, while limited by the fact that we can only know μ_i as the difference between it in whatever state it is and some other, arbitrary state which will be different for each kind of phase. We will try to do this in the remainder of this chapter and the next chapter, where activities become part of the equilibrium constant.

8.2.3   Definition II: Solutes

We use the same method we used in Section 7.5.3. We need an expression for the derivative of μ with a concentration term, which we can integrate. The derivative of μ_i with respect to the molality of i, m_i, is $(\partial \mu_i / \partial m_i)_{T,P,\hat{m}_i}$, where $\hat{m}_i$ means the molality of all solution components *except* i.

If, as before (in Equation (7.32)), we expand $(\partial \mu_i / \partial m_i)_{T,P,\hat{m}_i}$ by introducing P_i, the pressure on gaseous i which could be in equilibrium with solute i, we get

$$\left(\frac{\partial \mu_i}{\partial m_i} \right)_{\hat{m}_i} = \frac{\partial \mu_i}{\partial P_i} \frac{\partial P_i}{\partial m_i} \tag{8.12}$$

where μ_i is the same in the solution and in the vapor phase, where it can be called G_i (the vapor being assumed an ideal gas), so that $(\partial \mu_i / \partial P_i) = (\partial G_i / \partial P_i) = V_i = RT/P_i$, and where $(\partial P_i / \partial m_i) = P_i / m_i$ is an expression of Henry's law (Section 7.4.2), as mentioned earlier. Combining all this we get

$$\left(\frac{\partial \mu_i}{\partial m_i} \right)_{\hat{m}_i} = \frac{RT}{m_i} \qquad (8.13)$$

for ideal (Henryan) solutions. Integrating this equation between two values of molality, m_i' and m_i'', we get

$$\mu_i'' - \mu_i' = RT \ln \left(\frac{m_i''}{m_i'} \right) \qquad (8.14)$$

showing the effect of changing solute concentration on the chemical potential, as we wanted. However, it is limited to ideal (Henryan) solutions. The relationship is generalized to any kind of solution by introducing a correction factor at each concentration. Thus

$$\mu_i'' - \mu_i' = RT \ln \left(\frac{\gamma_H'' m_i''}{\gamma_H' m_i'} \right) \qquad (8.15)$$

where γ_H is the Henryan activity coefficient, and Equation (8.15) now refers to any real solution at a given temperature in which species i changes concentration, all other species remaining unchanged.[1] Henryan and Raoultian activity coefficients are discussed in Section 8.3.

Equation (8.15) can be generalized and so made more useful by choosing a single concentration m_i' for all solutes. In choosing this concentration, we should realize that

(1) γ_H in the denominator will be different for all different solutes unless we choose some idealized state, and
(2) it would be convenient to have the denominator $(\gamma_H' m')$ disappear, i.e., be unity.

The only state which satisfies these conditions and is equal to one molal for all solutes is the ideal (Henryan) one molal solution, and this is universally used as the standard state for solutes. Introducing superscript $\circ$ for the standard state, and dropping the now unnecessary superscript $''$, we get

$$\mu_i - \mu_i^\circ = RT \ln \left(\frac{\gamma_{H_i} m_i}{\gamma_{H_i}^\circ m_i^\circ} \right) \qquad (8.16)$$

and because $\gamma_{H_i}^\circ = 1$ and $m_i^\circ = 1$, this is usually written

$$\mu_i - \mu_i^\circ = RT \ln (\gamma_{H_i} m_i) \qquad (8.17)$$

The quantity $(\gamma_{H_i} m_i)/(\gamma_{H_i}^\circ m_i^\circ)$ is another definition of the *activity*, a_i, so

$$\mu_i - \mu_i^\circ = RT \ln a_i \qquad (8.18)$$

The activity thus allows calculation of the difference between the μ_i in a solution and μ_i in the ideal one molal standard state at the same T and P as the solution. This sounds like a

---

[1] Note that you can do this, i.e., change the concentration of i without changing any other concentrations, because molality is moles per kg of *solvent*. You cannot do it using mole fraction or molarity.

fairly esoteric thing to do, but because standard Gibbs energies of formation are determined for this ideal standard state (albeit at 25 °C, 1 bar), it is immensely useful, as we will see.

8.2.4   Definition III: Solids and Liquids

Now that we know about the fugacity, we can derive Equation (7.27) in still another way, because for an ideal gas f_i/f_i° is equal to P_i/P_i°, which is equal to the mole fraction, x_i (Equation (7.7)). So, for ideal gaseous and liquid solutions, and by extension, for any ideal (Raoultian) solution,

$$\mu_i - \mu_i^\circ = RT \ln x_i \tag{8.19}$$

For solutions covering a wide range of compositions, such as many solid and liquid solutions, this equation can be used by introducing another correction factor, the Raoultian activity coefficient, γ_R. Thus

$$\mu_i - \mu_i^\circ = RT \ln (x_i \gamma_{R_i}) \tag{8.20}$$

As before, we now define another activity term

$$a_i = x_i \gamma_{R_i} \tag{8.21}$$

which is useful for solutions covering a wide range of concentrations, and for which γ_R is known or can be estimated. In geochemistry, this tends to be for solid and gaseous solutions only, but it is widely used in metallurgy for liquids as well. The standard state, as before, is that state for which $a = 1$, in this case the pure liquid or solid ($x = 1$; $\gamma_R = 1$ in this state by definition).

8.2.5   Summary

Here are our various definitions of activity:

$$\left.\begin{aligned}
\mu_i - \mu_i^\circ &= RT \ln a_i \\[1ex]
\mu_i - \mu_i^\circ &= RT \ln \left(\frac{f_i}{f_i^\circ}\right) \\[1ex]
\mu_i - \mu_i^\circ &= RT \ln \left(\frac{\gamma_{H_i} m_i}{\gamma_{H_i}^\circ m_i^\circ}\right) \\[1ex]
\mu_i - \mu_i^\circ &= RT \ln (x_i \gamma_{R_i})
\end{aligned}\right\} \tag{8.22}$$

In any equilibrium state, both μ_i and μ_i° are absolute, finite quantities with a fixed difference between them. If the same standard state is chosen for each of these equations, then $\mu_i - \mu_i^\circ$ is the same in each equation, and the activity would be the same in all phases at equilibrium. This would be nice, but it would mean using a vapor pressure as the standard state for activity in solids, or an ideal one molal solution standard state for activities in a gas, or perhaps an ideal gas at one bar for an aqueous solute. This would be not only inconvenient, but impossible in many cases. So we accept the small inconvenience of having different activities for the same species in different phases.

In a multicomponent, multiphase system at equilibrium, μ_i is the same in every phase, but in most cases μ_i° (and therefore $\mu_i - \mu_i^\circ$) is different for solids, liquids, gases, and solutes (we know this without knowing the numerical value of either term). Thermodynamic properties are determined and tabulated for substances in these various standard states, and how they relate to one another in chemical reactions can be seen when we consider the equilibrium constant (Chapter 9).

Finally, note that fugacities have units of pressure (e.g., bars), but that activities and activity coefficients are always dimensionless.

8.3 Standard States and Activity Coefficients

The real usefulness of the ideal solution or ideal mixing concept is that it serves as a model with which real solutions are compared. Solute activities are compared with the activities they would have if the solution were ideal, and this ratio is called an *activity coefficient*. Care is required in using this number however, because Raoult's law and Henry's law describe two types of ideal solution behavior, and this results in two types of activity coefficients.

An activity coefficient of a constituent in a system is a number (always dimensionless) which when multiplied by the solute concentration (the mole fraction or molality) gives the solute activity. This is illustrated for the two types of ideal solution in Figure 8.1. In this binary system, B shows positive deviation from Raoultian behavior, so the Raoultian activity coefficient will be greater than one. Solutes that obey Raoult's law have $a_i = x_i$, so the Raoultian activity coefficient γ_R is defined as

$$a_i = x_i \gamma_R$$

In system A–B in Figure 8.1, when $x_B = 0.3$, $a_B = 0.5$, so

$$\gamma_R = 0.5/0.3$$
$$= 1.67$$

Note that the activity of B (0.5) is measured on the scale defined by Raoultian behavior, on the left-hand axis.

On the other hand B shows a negative deviation from Henry's law, and will therefore have a Henryan activity coefficient less than one. A Henryan scale of activities is created by extending the Henry's-law slope defined by dilute solutions of B in A right over to the pure B axis. This intercept is then called unit activity, and this defines a new scale of activities shown on the right-hand axis. This new Henryan activity scale also has $a_i = x_i$ for ideal systems, but the activity scale is different, so that now the real or measured activity at $x_B = 0.3$ is 0.2, measured on the scale defined by Henryan behavior, the right-hand axis. Thus the Henryan activity coefficient, γ_H, defined as

$$a_i = x_i \gamma_H$$

is in this case

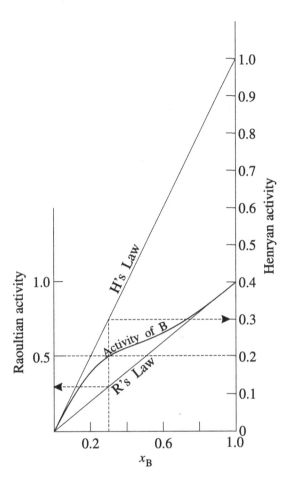

Figure 8.1 The behavior of component B in system A–B, showing the difference between the Raoultian and Henryan activity scales.

$$\gamma_H = 0.2/0.3$$
$$= 0.67$$

A major difference between the two kinds of activity coefficients which should be noted is that $\gamma_R \rightarrow 1$ as $x \rightarrow 1$, but $\gamma_H \rightarrow 1$ as $x \rightarrow 0$. Thus γ_H is generally more useful for constituents having low mole fractions, that is, solutes in dilute solutions.

8.3.1   A Real Example

Somewhat more important than the system A–B in geochemistry is the system CO_2–H_2O. At high temperatures and pressures these components are completely miscible, and form a solution that shows a small positive deviation from Raoult's law. Bowers and Helgeson (1983) present calculated fugacities of both H_2O and CO_2 at a number of temperatures and pressures, illustrated in Figure 8.2. The Henry's-law slope is inconveniently steep in this case, so the intersection with the pure CO_2 axis is not shown. Usually this is not required anyway, because in aqueous solution work, the region of interest is the lower left-hand corner, i.e., the dilute-solution region.

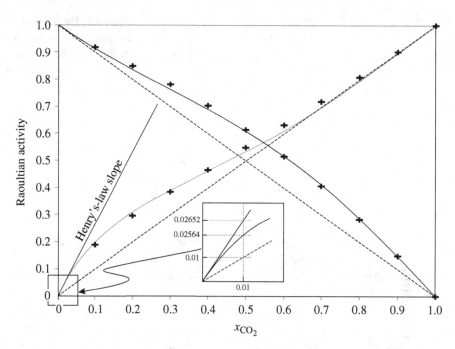

Figure 8.2 Raoultian activities of H_2O and CO_2 in the binary solution at 600 °C and 2 kbar. Data from Bowers and Helgeson (1983) and listed in Anderson and Crerar (1993, Table 11.1). The inset shows data at $x_{CO_2} = 0.01$, which are used in Additional Problem A8.1.

Solid Solutions

Figures 8.2 and 8.3 show that the thermodynamics of gaseous and solid solutions can be remarkably similar. Solid phases and solid solutions however have some distinctive aspects, such as the lack of any fugacity data, which make the determination of activities more difficult. They also have some advantages such as providing a means of controlling (buffering) the activity of components in real systems. See Chapter 14, Solid Solutions, in the online material.

8.3.2   The Ideal One Molal Standard State

Systems like H_2O-CO_2 (Figure 8.2) or mineral systems like olivine–forsterite or FeO–MgO (Figure 8.3) in which a large range of concentrations is of interest are best described using mole fractions and Raoultian activity coefficients. But aqueous systems involving mineral solubilities have concentrations confined to the dilute solution region of very small mole fractions, where concentrations in molality are better. The mineral becomes saturated in the dilute solution region and compositions beyond this do not exist. Extrapolation of the Henryan slope to a mole fraction of 1.0 can be done but it is not very useful.

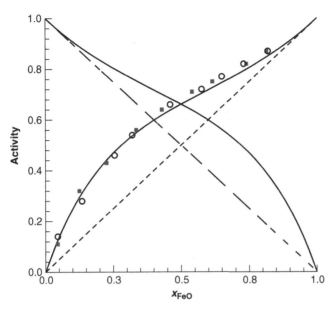

Figure 8.3 Raoultian activity of FeO and MgO in magnesiowüstite solid solutions. Data from Hahn and Muan (1962). Measured activities of FeO: squares, 1100 °C; and circles, 1300 °C. Activities of MgO are calculated from a version of the Gibbs–Duhem equation (Section 4.13.2).

But in fact the standard state can equally well be chosen anywhere on the Henry's-law slope, all points on which have the properties of the infinitely dilute solution. Therefore, as discussed in Section 8.2.3 a concentration of one molal on this slope is chosen as the system that real molal concentrations are compared with, namely the standard state for aqueous solutes as shown in Figure 8.4. Metallurgists sometimes use a weight percent axis, and define the standard state at 1% on the Henryan slope (Lupis, 1983, Chapter 7). In chemistry and geochemistry where aqueous solutions are used, a molality axis is best, and one molal on the Henryan slope becomes the standard state.

8.4   Activities and Standard States: An Overall View

We have now said everything necessary about activities and standard states, but the overall effect for the newcomer is often one of confusion at this stage. To try to draw the various threads together we consider in Figure 8.5 a hypothetical three-phase equilibrium at temperature T and pressure P.

A solid crystalline solution of B in A is in contact with an aqueous solution of A(aq) and B(aq), which is in turn in contact with a vapor phase containing A(v) and B(v) in addition to water vapor. We can suppose the dissolution of (A,B)(s) to be stoichiometric so that the ratio of A to B is the same in all three phases, but this is irrelevant to our development as we consider only component A. Let's say that for a solid solution composition of $x_A = 0.5$, $x_B = 0.5$, the concentration of A(aq) at equilibrium (m_A) is 10^{-2} molal, and the fugacity

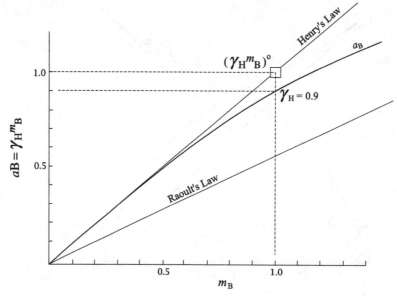

Figure 8.4 Activities of a slightly soluble component B based on the ideal one molal standard state. A concentration of B of 1 molal has an activity coefficient and hence an activity of 0.9.

of A in the vapor (f_A) is 10^{-5} bars. Assuming activity coefficients in the solid and liquid phases to be 1.0, the activity of A in the solid solution (using a standard state of pure crystalline A at T and P) is 0.5, the activity of A in the aqueous solution (using a standard state of the hypothetical ideal one molal solution of A at T and P) is 10^{-2}, and the activity of A in the vapor (using a standard state of pure ideal gaseous A at T and one bar) is 10^{-5}. Because the system is at equilibrium, the chemical potential of A (μ_A) is the same in each of the three phases, but because the three standard states are different, the standard chemical potential of A (μ_A°) is different for the three phases. The difference $(\mu_A - \mu_A^\circ)$ is calculable from the equations we have just derived. Thus, letting $T = 25\,^\circ$C,

$$(\mu_A - \mu_A^\circ)_{\text{solid}} = RT \ln x_A$$

$$= 8.31451 \times 298.15 \times \ln{(0.5)}$$

$$= -1.72\,\text{kJ}\,\text{mol}^{-1}$$

$$(\mu_A - \mu_A^\circ)_{\text{aq}} = RT \ln m_A$$

$$= 8.31451 \times 298.15 \times \ln(10^{-2})$$

$$= -11.4\,\text{kJ}\,\text{mol}^{-1}$$

$$(\mu_A - \mu_A^\circ)_{\text{gas}} = RT \ln f_A$$

$$= 8.31451 \times 298.15 \times \ln(10^{-5})$$

$$= -28.5\,\text{kJ}\,\text{mol}^{-1}$$

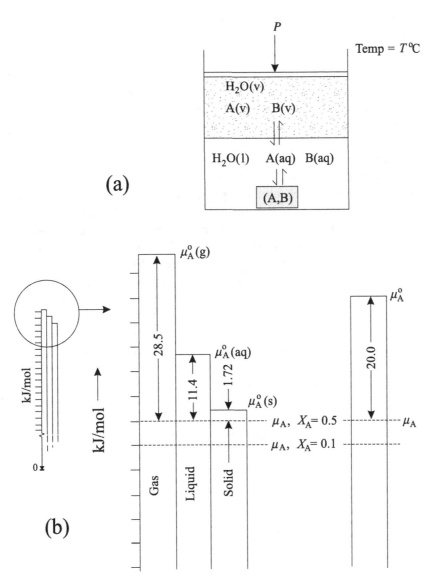

Figure 8.5 (a) A hypothetical three-phase system at equilibrium at pressure P and temperature T. (b) The top part of the histogram of chemical potentials in kJ mol^{-1}. The length of the bar for each phase is fixed when the standard state is chosen, and the chemical potential of A in the equilibrium system is represented by a line across the histogram at a level depending on the amount of B in the system. The lengths of the bars on the left represent traditional standard states, but any position for the top of the bars could be chosen, such as the one on the right, thus defining a new standard state.

It is instructive to consider these differences on a histogram (Figure 8.5) in which the ordinate is a scale of kJ mol^{-1}, on which we plot the *absolute* chemical potentials of A. These absolute potentials may be very large, so we look at only the tops of the bars in the histogram, and we unfortunately don't know the values of the absolute potentials individually, so we can't put an absolute scale on the ordinate. But we *can* plot the relative positions of the tops of the bars, and the position of the equilibrium chemical potential of A in the system.

If we now consider systems having more and more B in the solid solution (and hence in the other two phases), but always at equilibrium, the histogram bars stay where they are (because we are not changing standard states) but the level of the (absolute) chemical potential of A is lowered, increasing the distance between the top of the histogram bar for each phase and the level of μ_A, that is, increasing the (negative) value of $(\mu_A - \mu_A^\circ)$ as the activity of A is lowered.

This diagram is worth careful thought. It illustrates several things that are useful in understanding activities, chemical potentials, and standard states, such as the absolute nature of chemical potentials and the necessity of using differences, the equality of chemical potentials in each phase, and the arbitrary nature of the standard state.

This is all very nice, but wouldn't it be a lot simpler to have the same activity in every phase, just as we have the same chemical potential in every phase? This is worth some thought too. Having the same activity in each phase means having the same value of $\mu - \mu^\circ$ in each phase, and presumably we could choose whatever value we like for this. If we chose this to be $-20\,\text{kJ}\,\text{mol}^{-1}$, the histogram for every phase would look like that shown in Figure 8.5. Then, because the three standard states have the same value of μ°, they could all coexist at equilibrium (if they could exist at all). However, arbitrarily choosing a_i means arbitrarily choosing f_i° and m_i°, and this results in some standard states that are even more weird than the traditional ones. For example, if $\mu - \mu^\circ = -20\,\text{kJ}\,\text{mol}^{-1}$, then $a_i = 10^{-3.50}$, and if $m_i = 10^{-5}$ and $m_i/m_i^\circ = 10^{-3.5}$, then $m_i^\circ = 10^{1.5}$ or 31.9. So the standard state becomes a hypothetical ideal 31.9 molal solution.

So probably it is better to stick to having different standard states and different activities in each phase.

Units Again

A reminder.

- Activities and activity coefficients (a, γ) have no units, but fugacity (f) does.
- Activities have standard states, but fugacities do not.

8.5   Summary

It would be hard to overemphasize the importance of the concepts of fugacity and activity and their relationship to the chemical potential, at least in chemical thermodynamics. In thermodynamics for engineers or physicists, chemical reactions play a smaller role, but for anyone interested in processes involving chemical changes, this is a central topic. That is because the fugacity and activity are the parameters which relate the composition of a system to its Gibbs energy, governing how changes in phase compositions change the Gibbs energy of the system, and so determine the equilibrium phase compositions at a given T and P.

Fugacity and activity are basically compositional terms. In ideal solutions they are not necessary; pressure and various composition terms can be directly linked to the Gibbs energy. Real solutions have a variety of intermolecular forces, so that ideal solution models need correction factors. These corrections can be made either to the composition terms (fugacity and activity coefficients) or to the thermodynamic potentials (excess functions, not discussed here), and efforts to model these correction factors in mathematical terms have always been, and likely always will be, an important research field.

Because there are three main kinds of solutions and there are several commonly used methods of expressing concentrations, activity can take on several different forms requiring different kinds of correction factors, and because Gibbs energy is always a difference in energy between two states, there are several different standard or reference states in common use. This all adds up to possible confusion, and although it is possible to learn how things are done and to follow the rules, it is of course much better to understand the reasons for why things are done this way.

Exercises

E8.1 Consider crystalline A to be in equilibrium with water saturated with A at a pressure of 1234.0 bars and a temperature of 567.0 °C. A is very slightly soluble (0.001 moles/kg H_2O) so that $P_{H_2O} = P_{total}$. The vapor pressure of crystalline A varies with temperature according to the relation

$$\log P \text{ (bars)} = (-19{,}130/T \text{ (K)}) + 7.65$$

Its molar volume is 22.7 cm^3 mol^{-1} and is essentially constant in the pressure range under consideration. The fugacity coefficient for water at this P and T is 0.541.

(a) What is the activity of crystalline A, using the following standard states:

   (i) pure A(s) at T, P;
   (ii) pure A(s) at T and a pressure of 1 bar;
   (iii) pure A(s) at T and under its own vapor pressure;
   (iv) pure ideal gaseous A at T and one bar.

(b) What is the activity of water, using the following standard states:

   (i) pure water at T and P;
   (ii) pure water at T and a pressure of 500 bars (the fugacity coefficient is 0.747);
   (iii) ideal gaseous water at T and a pressure of 123.456 bars;
   (iv) hypothetical ice at T and a pressure of 1 bar. (Suggest how this could be done, without performing any calculations.)

(c) What is the activity of A(aq) (with γ_A(aq) assumed to be 1.0) using the following standard states:

   (i) an ideal one molal solution of A at T and P;
   (ii) a 17.2 molal solution of A having an activity coefficient of 0.123, at T and P.

(d) (i) Under what conditions is $\mu_{A(s)} = \mu_{A(aq)}$?

    (ii) Under what conditions is $\mu^\circ_{A(s)} = \mu^\circ_{A(aq)}$?

E8.2 Calculate the difference in chemical potential between dissolved A at T and P and dissolved A in the standard state (i) of Exercise E8.3(c) above.

E8.3 Experiments show that at 9.7 kbar, 1080 °C, the system SiO_2–H_2O shows a second critical end point at which quartz and a supercritical fluid of composition 75 wt. % SiO_2, 25 wt. % H_2O coexist at equilibrium. What is the activity of SiO_2 in this fluid, referred to a standard state of quartz at 1080 °C and one bar?

E8.4 How would you modify Equation (8.9) if you wished to have a constant T standard state?

Additional Problems

The next two problems involve changing the standard state.

A8.1 Consider the system H_2O–CO_2 at 600 °C, 200 bars, illustrated in Figure 8.2. Looking at the inset we see that the CO_2 activity according to Raoult's law at $x_{CO_2} = 0.01$ is 0.02564. But to do speciation in this solution you want a_{CO_2} using the ideal one molal standard state. Calculate γ_R and γ_H and the CO_2 activity at $x_{CO_2} = 0.01$ using the ideal one molal standard state. At $x_{CO_2} = 0.01$ the molality of CO_2 (m_{CO_2}) is 0.56.

A8.2 According to Wellman (1969), the fugacity of NaCl in equilibrium with nepheline ($NaAlSiO_4$) and sodalite ($3NaAlSiO_4 \cdot NaCl$) at 600 °C, 1 bar, is $10^{-10.566}$ bars, and according to Zimm and Mayer (1944) the fugacity of pure halite at the same T, P is $10^{-5.743}$ bars.

(a) What is the activity of NaCl at nepheline–sodalite equilibrium at 600 °C, 1 bar, using a pure halite at T, P standard state.

(b) Calculate $\Delta_r G^\circ$ for the reaction using this pure halite at T, P standard state.

(c) Sodalite is 3nepheline·NaCl, or $Na_4Al_3Si_3O_{12}Cl$. The nepheline–sodalite reaction is a dissociation,

sodalite = 3 nepheline + NaCl

Using the pure phase at T and P standard state, sodalite and nepheline have unit activities, so the equilibrium constant for the reaction is $K = a_{NaCl}$ and for the reaction $\Delta_r\mu^\circ = -RT \ln a_{NaCl}$. The activity of NaCl can use a standard state of either the pure phase (halite) or the ideal one molal solution of NaCl in water. If $\Delta_f G^\circ_{NaCl(aq)}$ is $-393{,}133$ J mol^{-1} using the ideal one molal standard state, and $\Delta_f G^\circ_{NaCl(s)}$ using the halite standard state is $-384{,}138$ J mol^{-1}, both at 25 °C and 1 bar, what is the activity of NaCl at nepheline–sodalite equilibrium at 600 °C using the ideal one molal standard state? Here's a hint. In the expression $\Delta_r\mu^\circ = -RT \ln a_{NaCl}$, the trick is to change $\Delta_r\mu^{\circ,\text{old}}$ to $\Delta_r\mu^{\circ,\text{new}}$, so that $-RT \ln a^{\text{old}}_{NaCl}$ becomes $-RT \ln a^{\text{new}}_{NaCl}$. So first find $\Delta_r\mu^{\circ,\text{new}} - \Delta\mu^{\circ,\text{old}}$, and add it to $\Delta_r\mu^{\circ,\text{old}}$. Thus,

$$\Delta_r \mu^{\circ,\text{old}} = -RT \ln a_{\text{NaCl}}^{\text{old}}$$

plus $\Delta_r \mu^{\circ,\text{new}} - \Delta_r \mu^{\circ,\text{old}}$ gives

$$\Delta_r \mu^{\circ,\text{new}} = -RT \ln a_{\text{NaCl}}^{\text{new}}$$

We can assume that the *difference* between $\Delta_f G^\circ_{\text{NaCl(aq)}}$ and $\Delta_f G^\circ_{\text{NaCl(s)}}$ is the same at 600 °C as it is at 25 °C.

A8.3 A and B are two organic liquids which are completely miscible. The measurements in Table 8.1 show the partial pressures (P_A and P_B) of each in the vapor phase above various mixtures of A and B. Recall that 1 bar = 760 mm Hg.

Table 8.1. Partial Pressures.

x_B	P_B (mm Hg)	P_A (mm Hg)
0.0000	0.0	344.5
0.0588	9.2	323.2
0.1232	20.4	299.3
0.1853	31.9	275.4
0.2910	55.4	230.3
0.4232	88.9	174.3
0.5143	117.8	135.0
0.5812	139.9	108.5
0.6635	170.2	79.0
0.7997	224.4	37.5
0.9175	267.1	13.0
1.0000	293.1	0.0

(a) Calculate the ideal partial pressures of A and B, and the total ideal vapor pressure in each mixture.

(b) Plot vapor pressure vs. x_B, showing the actual and ideal partial pressures of A and B and the actual and ideal total vapor pressure for each mixture.

(c) Divide each partial pressure by the vapor pressure of the pure liquid. That is, calculate P_A/P_A° and P_B/P_B°. These will vary between 0 and 1.0.

(d) On a second graph, plot values of P_A/P_A° and P_B/P_B° vs. x_B, for both the actual and the ideal partial pressures. Draw the diagonal lines on this plot representing Raoult's law.

(e) Calculate the Raoultian activity coefficient for both A and B at each composition.

(f) Determine the Henryan activity coefficients for at least the first three x_B points. To do this you need the tangent to the Raoultian a_B curve at $x_B = 0$. This can be done by fitting a solution model to the curve and differentiating, or simply by drawing a best guess on the graph. The point here is not to get the right answer, but simply to do it one way or another so that in future you have no uncertainty when you see the symbols γ_R and γ_H.

(g) The relationship between the two kinds of activity coefficients, γ_R and γ_H is useful. Because

$$f_B = x_B \gamma_{R_B}$$

$$f_B = x_B K_H \gamma_{H_B}$$

then

$$\gamma_{H_B} = \gamma_{R_B}/K_H \qquad\qquad (8.23)$$

Show that this is the case from the data.

Note that because the vapor phase is almost an ideal gas, the partial pressures are fugacities, therefore $P_A/P_A^\circ = f_A/f_A^\circ$. And because the fugacity of a species or component is the same in all phases at equilibrium, then f_A/f_A° in the vapor is the same as f_A/f_A° in the liquid. This is a measure of the *activity* of A in the liquid, and of course similarly for B. Thus, in the liquid phase, $a_A = f_A/f_A^\circ = x_A \gamma_{R_A}$.

Measuring vapor pressures of components in solutions is a classical way of determining liquid component activities and activity coefficients.

9 The Equilibrium Constant

Reactions involving dissolved compounds are different in an important way from reactions involving only pure compounds, such as pure solids. To see why, consider two reactions, one between pure compounds and one between dissolved substances.

The first is a reaction in which all products and reactants are pure substances, the kind of reaction we have been considering up to now. It is

$$NaAlSiO_4(s) + 2\,SiO_2(s) = NaAlSi_3O_8(s) \tag{9.1}$$

The second is a reaction in which all products and reactants are dissolved in water and are capable of changing their concentration:

$$H_2CO_3(aq) = HCO_3^-(aq) + H^+(aq) \tag{9.2}$$

The temperature and pressure are normal, 298.15 K and 1 bar. As usual, we want to know which way each reaction will go. Reaction (9.1) presents no problem. We look up the values of $\Delta_f G^\circ$ for each compound, and calculate $\Delta_r G^\circ$:

$$\Delta_r G^\circ = \Delta_f G^\circ_{NaAlSi_3O_8} - \Delta_f G^\circ_{NaAlSiO_4} - 2\,\Delta_f G^\circ_{SiO_2}$$

$$= -3711.5 - (-1978.1) - 2(-856.64)$$

$$= -20.12\,kJ\,mol^{-1}$$

We see that the reaction as written is spontaneous; $NaAlSiO_4$ (nepheline) and SiO_2 (quartz) at 1 bar pressure should react together to form $NaAlSi_3O_8$ (albite). If the reaction does proceed (thermodynamics doesn't tell us whether it will or not, only that the energy gradient favors it), then nepheline and quartz get used up during the reaction. However, while being used up, they do not change their Gibbs energies. The reaction should actually proceed as long as any reactants are left. When either the nepheline or the quartz is used up completely, the reaction must stop. This reaction can be represented graphically as in Figure 9.1. Here we use bars to represent the magnitude of the combined Gibbs energy of the products and of the reactants. The difference in the height of the bars represents $\Delta_r G^\circ$, the driving force for the reaction. The middle bar represents an activation energy barrier that prevents the reaction from occurring. It is put there to form a link with the discussion in Chapter 2, but thermodynamics is unable to calculate the size of this barrier, or anything whatever about it. Nevertheless it is often there, and is the reason why one of the states is metastable. The point here is that the size of the bars does not change during the reaction,

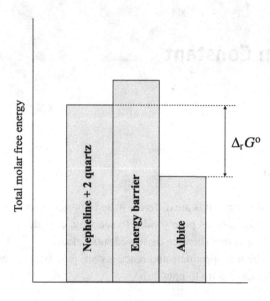

Figure 9.1 Molar Gibbs energies when all products and reactants are pure compounds. The Gibbs energy of reaction is given by $\Delta_r G°$ because all products and reactants are in their reference states (more accurately called standard states), and this does not change during the reaction until one of the reactants disappears.

if it proceeds, because none of the products or reactants changes in any way – only the amounts present change. The value of $\Delta_r G°$ never goes to zero.

Reaction (9.2) is different. We can start off the same way, by looking up the values of $\Delta_f G°$ for each compound:

$$\Delta_r G° = \Delta_f G°_{HCO_3^- (aq)} + \Delta_f G°_{H^+ (aq)} - \Delta_f G°_{H_2CO_3 (aq)}$$

$$= -586.77 + 0 - (-623.109)$$

$$= 36.339 \, \text{kJ mol}^{-1}$$

This is positive, and so the reaction goes spontaneously to the left. So far, so good. But as soon as the reaction starts, the concentrations of H^+ and HCO_3^- start to decrease, the concentration of H_2CO_3 starts to increase, and the Gibbs energies of all three change, as shown in Figure 9.2. All we can say from the tabulated data is that if all three aqueous species were present in their standard state concentrations, the reaction would start to go to the left. But suppose we are interested in some other concentrations? And what happens to the reaction after it has started? Because the solutes can change their concentrations and their Gibbs energies, the situation is quite different from the "all pure substances" situation. These problems are all handled easily by the *equilibrium constant*.

9.2 Reactions at Equilibrium

Chemical reactions can not only go one way or the other (our main problem), but also can stop going, for two reasons. Either one of the reactants is used up, or the reaction can reach an equilibrium state, with all products and reactants present in a balanced condition. The second possibility is the subject of this chapter – how much can we predict about this balanced state of equilibrium?

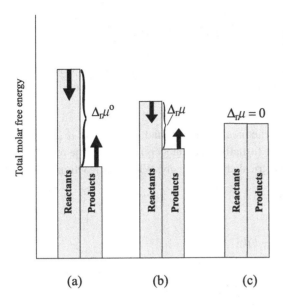

Figure 9.2 Molar Gibbs energies of reactants and products of a reaction between aqueous solutes. (a) A hypothetical starting condition, represented by the numbers in the tables of data (note superscript $°$ on $\Delta_r\mu°$, indicating standard conditions for all products and reactants). (b) Either the reaction in (a) after it has proceeded for some time, or a beginning state where the reactants and products are not all at 1 molal (no superscript $°$). The point is that, as shown by the arrows, the Gibbs energy of the reactants decreases and that of the products increases during the reaction. (c) Sooner or later, a state of equilibrium is reached, when the Gibbs energies of reactants and products are equal.

In Chapter 4 we defined the molar Gibbs energy, G, which always decreases in spontaneous reactions ($\Delta G < 0$). In Chapter 6, we used the fact that a reaction at equilibrium (e.g., calcite $\rightleftharpoons$ aragonite) does not go either way ($\Delta G = 0$) to calculate the T and P of equilibrium between phases. The expression $\Delta G = 0$ expresses a balance between the Gibbs energies of calcite and aragonite, that is, $G_{\text{calcite}} = G_{\text{aragonite}}$ (Section 6.2.1). If there is more than one reactant or product, the same relationship must hold (the G of reactants and products are equal), but each side is now a sum of G terms, and the G terms for solutes are properly written as μ rather than G. Of course, not all products and reactants need be solutes. For example, the reaction

$$\text{SiO}_2(s) + 2\text{H}_2\text{O} = \text{H}_4\text{SiO}_4(aq) \tag{9.3}$$

shows what happens when quartz dissolves in water. Molecules of SiO_2 dissolve and combine with water molecules to form the solute species H_4SiO_4. This dissolution process continues until the solution is saturated with silica, and then stops. The system is then at equilibrium, because

$$\mu_{\text{H}_4\text{SiO}_4} = \mu_{\text{SiO}_2} + 2\mu_{\text{H}_2\text{O}} \tag{9.4}$$

If we added some H_4SiO_4 to this solution it would then be supersaturated, $\mu_{\text{H}_4\text{SiO}_4}$ would be greater than its equilibrium value, and the reaction would tend to go to the left, precipitating quartz.[1]

---

[1] At the risk of becoming repetitious, we note that it is in our model reaction that quartz precipitates. In real life, something else might happen – nothing might precipitate, or some other SiO_2 phase such as silica gel might precipitate.

9.3   The Most Useful Equation in Thermodynamics

To find out what we can say about this balanced equilibrium state when several solutes and other phases are involved, let's consider a general chemical reaction

$$aA + bB = cC + dD \tag{9.5}$$

where A, B, C, and D are chemical formulae, and a, b, c, and d (called stoichiometric coefficients) are any numbers (usually small integers) that allow the reaction to be balanced in both composition and electrical charges, if any. When this reaction reaches equilibrium,

$$c\mu_C + d\mu_D = a\mu_A + b\mu_B$$

and

$$\Delta_r\mu = c\mu_C + d\mu_D - a\mu_A - b\mu_B$$
$$= 0 \tag{9.6}$$

By our definition of activity, Equation (7.43),

$$\mu_A = \mu_A^\circ + RT \ln a_A$$
$$\mu_B = \mu_B^\circ + RT \ln a_B$$
$$\mu_C = \mu_C^\circ + RT \ln a_C$$
$$\mu_D = \mu_D^\circ + RT \ln a_D$$

Substituting these expressions into (9.6), we get

$$\Delta_r\mu = c\mu_C + d\mu_D - a\mu_A - b\mu_B$$
$$= c(\mu_C^\circ + RT \ln a_C) + d(\mu_D^\circ + RT \ln a_D)$$
$$\quad - a(\mu_A^\circ + RT \ln a_A) - b(\mu_B^\circ + RT \ln a_B)$$
$$= (c\mu_C^\circ + d\mu_D^\circ - a\mu_A^\circ - b\mu_B^\circ) + RT \ln a_C^c + RT \ln a_D^d$$
$$\quad - RT \ln a_A^a - RT \ln a_B^b$$
$$= \Delta_r\mu^\circ + RT \ln \left(\frac{a_C^c a_D^d}{a_A^a a_B^b} \right)$$

There may be any number of reactants and products, and so to be completely general we can write

$$\Delta_r\mu = \Delta_r\mu^\circ + RT \ln \prod_i a_i^{\nu_i} \tag{9.7}$$

where i is an index that can refer to any product or reactant, ν_i refers to the stoichiometric coefficients of the products and reactants, with ν_i positive if i is a product, and negative if i is a reactant. $\prod_i$ (or $\prod_i$) is a symbol meaning "product of all i terms," which means that all the $a_i^{\nu_i}$ terms are to be multiplied together (much as $\sum_i a_i$ would mean that all a_i terms were to be added together). So in our case, the ν terms are c, d, $-a$, and $-b$, and

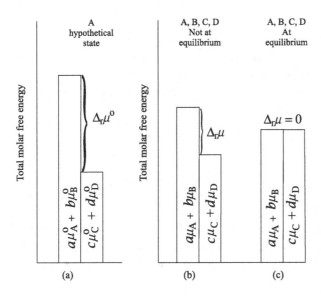

Figure 9.3 Possible Gibbs energy relationships in the reaction $a\mathrm{A} + b\mathrm{B} \rightarrow c\mathrm{C} + d\mathrm{D}$.

$$\prod_i a_i^{v_i} = a_{\mathrm{C}}^c\, a_{\mathrm{D}}^d\, a_{\mathrm{A}}^{-a}\, a_{\mathrm{B}}^{-b}$$

$$= \frac{a_{\mathrm{C}}^c\, a_{\mathrm{D}}^d}{a_{\mathrm{A}}^a\, a_{\mathrm{B}}^b}$$

In the general case, $\prod_i a_i^{v_i}$ is given the symbol Q, so (9.7) becomes

$$\Delta_{\mathrm{r}}\mu = \Delta_{\mathrm{r}}\mu^\circ + RT \ln Q \tag{9.8}$$

We must be perfectly clear as to what (9.8) means. In Figure 9.3 (a variation of Figure 9.2) are pictured the possible relationships between the Gibbs energies of the products and reactants in reaction (9.5).

First, the term $\Delta_{\mathrm{r}}\mu^\circ$ refers to the difference in Gibbs energies of products and reactants when each product and each reactant, whether solid, liquid, gas, or solute, is in its standard state. This means the pure phase for solids and liquids [e.g., most minerals, $H_2O(s)$, $H_2O(l)$, alcohol, etc.], pure ideal gases at 1 bar [e.g., $O_2(g)$, $H_2O(g)$, etc.], and dissolved substances [solutes, e.g., NaCl(aq), Na^+, etc.] in ideal solution at a concentration of 1 molal. Although we do have at times fairly pure solid phases in our real systems (minerals such as quartz and calcite are often quite pure), we rarely have pure liquids or gases, and we never have ideal solutions as concentrated as 1 molal. Therefore, $\Delta_{\mathrm{r}}\mu^\circ$ usually refers to quite a hypothetical situation. It is best not to try to picture what physical situation it might represent, but to think of it as just the difference in numbers that are obtained from tables.

$\Delta_{\mathrm{r}}\mu$, on the other hand, is the difference in Gibbs energy of reactants and products as they actually occur in the system you are considering, which may or may not have reached stable equilibrium.[2] The activities in the Q term (the concentrations, fugacities, mole fractions, etc. of the products and reactants) change during the reaction as it strives to reach equilibrium and at any particular moment result in a particular value of $\Delta_r\mu$. Thus $\Delta_r\mu°$ is a number obtained from tables that is independent of what is happening in the real system you are considering, but $\Delta_r\mu$ and Q are linked together – whatever activities (think *concentrations*) are in Q will result in a certain value of $\Delta_r\mu$.

If it makes more sense, you can write Equation (9.8) as

$$\Delta_r\mu - \Delta_r\mu° = RT \ln Q \tag{9.9}$$

which means that whatever terms are in Q control how different the chemical potentials ($\Delta_r\mu$) are from their standard tabulated values ($\Delta_r\mu°$). When all activities in Q are 1.0, then there is no difference, $\Delta_r\mu = \Delta_r\mu°$.

We are especially interested in the value of Q when our systems reach equilibrium, that is, when the product and reactant activities have adjusted themselves spontaneously such that $\Delta_r\mu = 0$. In this state, the $\prod_i a_i^{v_i}$ term is called K, instead of Q, and (9.8) becomes

$$0 = \Delta_r\mu° + RT \ln K$$

or

$$\Delta_r\mu° = -RT \ln K \tag{9.10}$$

Standard states usually refer to pure substances (except for the aqueous standard states) in which $\mu = G$, so this equation is often written

$$\boxed{\Delta_r G° = -RT \ln K} \tag{9.11}$$

This equation has been called, with some reason, the most useful in chemical thermodynamics, and it certainly merits the most careful attention. Most important is the fact that the activity product ratio (K) on the right-hand side is independent of variations in the system composition. Its value is controlled completely by a difference in standard (tabulated) state molar Gibbs energies ($\Delta_r G°$) and so is a function only of the temperature and pressure. It is a constant for a given system at a given temperature or temperature and pressure and is called the *equilibrium constant*. Its numerical value for a given system is not dependent on the system actually achieving equilibrium, or in fact even existing. Its value is fixed when the reacting substances are chosen. The left-hand side refers to a difference in Gibbs energies of a number of different physical and ideal states, which do not represent any real system or reaction. The right-hand side, on the other hand, refers to a single reaction that has reached equilibrium, or more exactly, to the activity product ratio that would be observed if the system had reached equilibrium.

---

[2] Strictly speaking, μ has meaning only in equilibrium states, so we cannot really consider the reaction which is not at equilibrium. What we really do is to consider the reaction as taking place in a series of metastable equilibrium states, as discussed more fully in connection with the extent-of-reaction variable in Chapter 13. At this stage, however, you may consider this a mere quibble, and think of reacting substances as having μ values if you wish.

The great usefulness of Equation (9.11) lies in the fact that knowledge of a few standard state Gibbs energies allows calculation of an indefinite number of equilibrium constants. Furthermore, these equilibrium constants are very useful pieces of information about any reaction. If K is very large, it tends to shows that a reaction will tend to go "to completion," that is, mostly products will be present at equilibrium, and if K is small, it tends to show that the reaction hardly goes at all before enough products are formed to stop it.[3] If you are a chemical engineer designing a process to produce some new chemical, it is obviously of great importance to know to what extent reactions should theoretically proceed. The equilibrium constant, of course, will never tell you whether reactants will actually react, or at what rate; there may be some reason for the reaction kinetics being very slow. It indicates the activity product ratio at equilibrium, not whether equilibrium is easily achievable.

Finally, <u>NEVER</u> write Equation (9.11) as

$$\Delta_r G = -RT \ln K$$

that is, omitting the superscript °, because doing so indicates a complete lack of understanding of the difference between $\Delta_r G$ and $\Delta_r G°$ and is just about grounds for failing any course in this subject.

Let's go over it once more. $\Delta_r G°$ (or $\Delta_r \mu°$) is the difference in molar Gibbs energy between products and reactants when they are all in their reference or standard states (pure solids and liquids, solutes at ideal 1 molal, gases at 1 bar), obtained directly from tabulated data such as Appendix B. Products and reactants are virtually never at equilibrium with each other under these conditions ($\Delta_r G°$ or $\Delta_r \mu°$ never becomes equal to zero). $\Delta_r G$ (or $\Delta_r \mu$) is the difference in free energy between products and reactants in the general case (when at least one of the products or reactants is *not* in its standard state) and becomes equal to zero when the reaction reaches equilibrium. $\Delta_r G$ cannot be used in place of $\Delta_r G°$ in (9.11) because this would mean, among other things, that every reaction at equilibrium ($\Delta_r G = 0$) would have an equilibrium constant of 1.0.

9.3.1   A First Example

Let's calculate the equilibrium constant for reaction (9.2),

$$H_2CO_3(aq) = HCO_3^-(aq) + H^+(aq)$$

First we write, as before,

$$\Delta_r G° = \Delta_f G°_{HCO_3^-} + \Delta_f G°_{H^+} - \Delta_f G°_{H_2CO_3}$$

---

[3] These are just generalizations which are not always true, because the activities of products and reactants at equilibrium can sometimes be very large or very small, so that the magnitude of K is not a sure guide. We emphasize this a bit more in the boxed statement on page 216.

Getting numbers from the tables, we find

$$\Delta_r G° = -586.77 + 0 - (-623.109)$$

$$= 36.339 \text{ kJ mol}^{-1}$$

$$= 36,399 \text{ J mol}^{-1}$$

The fact that this number is positive is not as significant as in our previous examples. In this case it means that *if* H_2CO_3, HCO_3^-, and CO_3^{2-} were all present in an ideal solution, and each had a concentration of 1 molal, the reaction would go to the left. This hypothetical situation is not of much interest. We want the value of K.

Inserting this result in Equation (9.11), we get

$$\Delta_r G° = -RT \ln K$$

$$36,339 = -(8.3145 \times 298.15)\ln K$$

so

$$\ln K = -36,339/(8.3145 \times 298.15)$$

$$= -14.659$$

or $$K = 4.30 \times 10^{-7}$$

$$= 10^{-6.37}$$

If you don't like dealing with natural logarithms, you can use the conversion factor $\log x = \ln x/2.30259$ (Appendix A). This gives

$$\log K = -36,339/(2.30259 \times 8.3145 \times 298.15)$$

$$= -6.37$$

directly.

This means that when these three aqueous species are at equilibrium,

$$\frac{a_{HCO_3^-} \cdot a_{H^+}}{a_{H_2CO_3}} = 10^{-6.37}$$

This is the answer to our question in Section 9.1 ("what happens to the reaction after it starts?"). The reaction continues until the ratio of the activities of the products and reactants equals the equilibrium constant, in this case $10^{-6.37}$. It doesn't matter what the starting activities were, and individual activities at equilibrium can be quite variable. In other words the values of $a_{H_2CO_3}$ and of $(a_{HCO_3^-} \cdot a_{H^+})$ are not determined, nor are the values of $a_{HCO_3^-}$ or a_{H^+} individually; only the ratio expressed by K is fixed. In specific cases, the values of these individual activities are determined by the bulk composition of the solution, and can be determined by *speciation* (Chapter 10). For now, we are content to determine K. In this case K is the ionization constant for carbonic acid, H_2CO_3. It is a very small number, meaning that carbonic acid is a weak acid.

9.4   Special Meanings for *K*

. . . .

Equilibrium constants are also sometimes equal to system properties of interest, such as vapor pressures, solubilities, phase compositions, and so on. This is because quite often it can be arranged that all activity terms drop out (are equal to 1.0) except the one of interest, which can then be converted to a pressure or composition.

9.4.1   *K* Equal to a Solubility

Quartz–Water Example

In our quartz–water example (Equation (9.3)), the equilibrium constant expression is

$$K = \frac{a_{H_4SiO_4}}{a_{SiO_2}\, a_{H_2O}^2} \tag{9.12}$$

At this point the expression is perfectly general, valid for any conditions, and *K* is calculable from Equation (9.11) if we know the Gibbs energies of the three species in their standard states. In this case we are dealing with pure quartz and water saturated with quartz. The quartz is in its standard state, and the water contains so little silica that it is almost pure.[4] By our definitions then (Equations (8.22)), $a_{SiO_2} = 1$ and $a_{H_2O} = 1$. Therefore

$$K = a_{H_4SiO_4}$$

$$= (m_{H_4SiO_4}\, \gamma_{H_4SiO_4})$$

$$= m_{H_4SiO_4} \quad \text{assuming } \gamma_{H_4SiO_4} = 1.0$$

This shows that, assuming $\gamma_{H_4SiO_4}$ is 1.0, which happens to be an excellent approximation in this case, we can calculate the concentration of silica ($m_{H_4SiO_4}$) in equilibrium with quartz, that is, the solubility of quartz.

Following our routine, we write for the reaction as written

$$\Delta_r G^\circ = \Delta_f G^\circ_{H_4SiO_4} - \Delta_f G^\circ_{SiO_2} - 2\,\Delta_f G^\circ_{H_2O} \tag{9.13}$$

Then, getting numbers from the tables,

$$\Delta_r G^\circ = -1307.7 - (-856.64) - 2(-237.129)$$

$$= 23.198 \text{ kJ mol}^{-1}$$

$$= 23{,}198 \text{ J mol}^{-1}$$

Then

$$\Delta_r G^\circ = -RT \ln K$$

$$23{,}198 = -(8.3145 \times 298.15)\ln K$$

---

[4] Dissolving minerals in water changes a_{H_2O} very little. So while strictly speaking a_{H_2O} is not 1.0 when saturated with some mineral, this assumption is usually quite good.

so

$$\log K = -23{,}198/(2.30259 \times 8.3145 \times 298.15)$$

$$= -4.064$$

Thus the molality of SiO_2 in a solution in equilibrium with quartz is about $10^{-4.064}$, or about 5.2 ppm.[5]

Doing It Backwards. So we see that Gibbs energies can sometimes be used to calculate a solubility. The same calculation also works in the other direction, that is, measuring a solubility can be used to calculate a value of $\Delta_r G^\circ$. In the quartz–water case, the reaction is particularly simple, in that, because the quartz and water are essentially pure phases, not only are a_{H_2O} and a_{SiO_2} equal to 1.0, but also their values of $\Delta_f G^\circ$ are known, as shown above. Therefore, a value of $\Delta_r G^\circ$ calculated from a solubility measurement can be used to calculate $\Delta_f G^\circ$ for aqueous silica. Thus if you measured the solubility of quartz to be 5.2 ppm at 25 °C, you could use (9.13) in the form

$$\Delta_f G^\circ_{H_4SiO_4} = \Delta_f G^\circ_{SiO_2} + 2\,\Delta_f G^\circ_{H_2O} + \Delta_r G^\circ \tag{9.14}$$

to calculate $\Delta_f G^\circ_{H_4SiO_4} = -1307.7\,\text{kJ mol}^{-1}$, and as a matter of fact that is usually how this quantity is determined.

A Strange Procedure. Note the strangeness of what we are doing here. On the left-hand side of $\Delta_r G^\circ = -RT \ln K$ (Equation (9.11)) we enter the standard Gibbs energies of the reactants and products, which in this case includes $\Delta_f G^\circ$ of H_4SiO_4 at a concentration of one molal (its concentration in its standard state) in a hypothetical ideal solution, and on the right-hand side we calculate its equilibrium concentration, only a few ppm. Remember what we said in deriving the equilibrium constant – the left-hand side consists of tabulated standard state data; it has nothing to do with real systems or with equilibrium. But from these data, equilibrium activity ratios and sometimes compositions can be calculated. Think about it.

9.4.2   *K* Equal to Fugacity of a Volatile Species

Hematite–Magnetite Example

The next example is the same in principle. Consider the reaction[6]

$$6\,Fe_2O_3(s) = 4\,Fe_3O_4(s) + O_2(g) \tag{9.15}$$

for which the equilibrium constant is

$$K = \frac{a^4_{Fe_3O_4}\, a_{O_2}}{a^6_{Fe_2O_3}}$$

---

[5] If the molality of H_4SiO_4 (aq) is x, then the molality of SiO_2 (aq) is also x, as there is 1 mole of SiO_2 in each.

[6] It is important in this reaction to note that we write oxygen as O_2 (g), that is, oxygen gas. There are also data for dissolved oxygen, written O_2 (aq) which are of course completely different. The same ambiguity does not exist for hematite and magnetite, but it is always a good idea to append the (g), (aq), (s), or (l) symbols for clarity.

9.4 Special Meanings for K

If the reaction involves pure hematite Fe_2O_3 and pure magnetite Fe_3O_4, then $x_{Fe_2O_3} = 1$ and $x_{Fe_3O_4} = 1$, so $a_{Fe_2O_3} = 1$ and $a_{Fe_3O_4} = 1$. Therefore

$$K = a_{O_2}$$
$$= f_{O_2}$$
$$= P_{O_2}\gamma_f$$

Assuming that the activity coefficient γ_f is 1.0, which is again in this case an excellent approximation, we can calculate the partial pressure of oxygen in a gas phase in equilibrium with the two minerals hematite and magnetite.

Following the routine, we write

$$\Delta_r G° = 4\,\Delta_f G°_{Fe_3O_4} + \Delta_f G°_{O_2} - 6\,\Delta_f G°_{Fe_2O_3}$$
$$= 4(-1015.4) + 0 - 6(-742.2)$$
$$= 391.6\,\text{kJ}\,\text{mol}^{-1}$$
$$= 391,600\,\text{J}\,\text{mol}^{-1}$$

Then

$$\Delta_r G° = -RT \ln K$$
$$391,600 = -(8.3145 \times 298.15)\ln K$$

and

$$\log K = -391,600/(2.30259 \times 8.3145 \times 298.15)$$
$$= -68.40$$

So the oxygen fugacity in equilibrium with hematite and magnetite at 25 °C and 1 bar is $10^{-68.40}$ bar. This is an incredibly small quantity, which would have absolutely no significance if it were simply a partial pressure, unconnected to thermodynamics. A partial pressure of this magnitude would be produced by one molecule of oxygen in a volume larger than that of a sphere with a diameter of the solar system (Section 11.11). However, it is in fact a parameter in the thermodynamic model, just as valid as any other part of the model. It can be used, for example, to calculate other parameters that might be more easily measurable. For example, the reaction

$$CH_4(g) + O_2(g) = CO_2(g) + 2\,H_2(g) \tag{9.16}$$

is one that you might be interested in if you were studying the bottom muds in Figure 2.1(c). The equilibrium constant for this is

$$\Delta_r G° = -394.359 + 2(0) - (-50.72) - 0$$
$$= -343.639\,\text{kJ}\,\text{mol}^{-1}$$
$$= -343,639\,\text{J}\,\text{mol}^{-1}$$

and

$$\log K = 343{,}639/(2.30259 \times 8.3145 \times 298.15)$$

$$= 60.203$$

which means that at equilibrium

$$\frac{f_{CO_2} \cdot f_{H_2}^2}{f_{CH_4} \cdot f_{O_2}} = 10^{60.203}$$

Now $10^{60.203}$ is just as ridiculous as $10^{-68.40}$ in a sense. But if we insert the value $f_{O_2} = 10^{-68.40}$ into this expression, we get

$$\frac{f_{CO_2}}{f_{CH_4}} f_{H_2}^2 = 10^{60.203} \cdot 10^{-68.40}$$

$$= 10^{-8.20}$$

which begins to look a little more reasonable. This tells you something about how the CO_2/CH_4 ratio varies with f_{H_2}. For example, you could say that according to the thermodynamic model, if f_{O_2} is controlled by hematite–magnetite, the CO_2 and CH_4 fugacities (partial pressures) are equal when f_{H_2} is $10^{-4.1}$ bar, and this might in fact be a measurable quantity in the muds.

The point is that by writing a few reactions and using thermodynamics, your thoughts about what might be happening in the bottom muds or any other environment take shape in a controlled fashion – controlled, that is, by the implied hypothesis of chemical equilibrium. Your system may not be at complete equilibrium, but your model is, because that is a good place to start. And the fact that one of your thermodynamic parameters, such as f_{O_2}, turns out to be impossibly small or large does not make it ridiculous; it just means you won't be able to measure it directly, and you might want to concentrate on other parameters to which your impossible one is connected by the model.

Muscovite Example

The hematite–magnetite example is just one of a great variety of geologically important reactions in which the fugacity of one species is numerically equal to an equilibrium constant. That one species can be O_2 as in the example, but it can also be another species, typically H_2O, CO_2, or H_2. For example, the assemblage muscovite plus quartz reacts at high temperatures to andalusite plus K-feldspar in the reaction

$$KAl_3Si_3O_{10}(OH)_2(s) + SiO_2(s) = Al_2SiO_5(s) + KAlSi_3O_8(s) + H_2O(g) \tag{9.17}$$

for which the equilibrium constant is

$$K = \frac{a_{Al_2SiO_5} \cdot a_{KAlSi_3O_8} \cdot a_{H_2O}}{a_{KAl_3Si_3O_{10}(OH)_2} \cdot a_{SiO_2}}$$

$$= a_{H_2O} \quad \text{(minerals are pure so } a = 1\text{)}$$

$$= \frac{f_{H_2O}}{f_{H_2O}^\circ}$$

$$= f_{H_2O} \quad \text{(ideal gas standard state at } P = 1 \text{ so } f^\circ = 1 \text{ bar)}$$

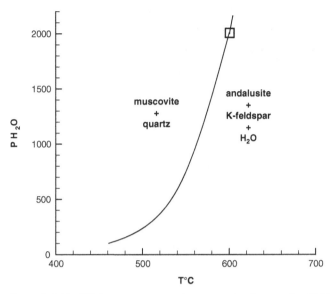

Figure 9.4 Equilibrium pressures and temperatures for reaction (9.17) as calculated by SUPCRT92. The square indicates the equilibrium state at 599.75 °C, 2000 bars.

so that as long as the minerals are all quite pure and are only very slightly soluble so that the water is also quite pure, K is numerically equal to the fugacity of pure water at the specified T and P. These conditions are met in experimental work (though perhaps not in nature) so that determination of the equilibrium T and P for the reaction allows determination of K, because the fugacity of pure water has been determined over a wide range of conditions.

The equilibrium diagram for this reaction calculated from data in Johnson *et al.* (1992) is shown in Figure 9.4. The water fugacity at 2000 bars, 599.75 °C (indicated by the square) is 1052 bars, so $K = 1052$ at this point. However, be sure to note the assumptions we have made in saying $K = 1052$. They are as follows.

1. The standard state for all minerals is the pure mineral at $T = 599.75$ °C and 2000 bars.
2. The standard state for water is ideal gaseous H_2O at $T = 599.75$ °C and a pressure of one bar.

These are the common choices in geochemical calculations, but of course others are possible. Every value of an equilibrium constant implies that standard state choices have been made.

9.5 *K* in Solid–Solid Reactions

It should be evident by now that the equilibrium constant is most useful in reactions between dissolved substances, those that change their activities during the reaction. Reactions of the other kind, between pure substances that do not change their activities during the reaction, like reaction (9.1), have no need of an equilibrium constant because in general they do not reach an equilibrium; they proceed until one of the reactants has disappeared. But what happens if you *do* calculate K for such a reaction – what does it mean? Let's do this for reaction (9.1) and see what happens:

$$\Delta_r G° = \Delta_f G°_{NaAlSi_3O_8} - 2\,\Delta_f G°_{SiO_2(s)} - \Delta_f G°_{NaAlSiO_4}$$

$$= -3711.5 - 2(-856.64) - (-1978.1)$$

$$= -20.12 \text{ kJ mol}^{-1}$$

$$= -20{,}120 \text{ J mol}^{-1}$$

Then

$$\log K = 20{,}120/(2.30259 \times 8.3145 \times 298.15)$$

$$= 3.52$$

and

$$K = 3349$$

As usual, this means that, *at equilibrium*,

$$\frac{a_{NaAlSi_3O_8}}{a_{NaAlSiO_4} a^2_{SiO_2}} = 10^{3.52}$$

If we in fact have pure nepheline, pure albite, and pure quartz involved in the reaction, then we come up with the same answer as before. The activities of $NaAlSi_3O_8$, $NaAlSiO_4$, and SiO_2 are 1.0 by our definitions, therefore we know that the ratio $a_{NaAlSi_3O_8}/(a_{NaAlSiO_4} a^2_{SiO_2})$ is fixed at 1.0 and can never be equal to 3349 – the three pure minerals can never reach equilibrium at 25 °C, 1 bar. But suppose the minerals are not pure – suppose they are solid solutions. Albite ($NaAlSi_3O_8$) forms a solid solution with anorthite ($CaAl_2Si_2O_8$) in the mineral plagioclase, and nepheline also usually occurs in a solid solution with kalsilite ($KAlSiO_4$), so the mole fractions and hence the activities of both $NaAlSi_3O_8$ and $NaAlSiO_4$ will generally be less than 1.0, even though there are pure minerals with these compositions. The activity is less than 1.0 when the minerals are *not* pure, but occur as components of solid solutions.

In this particular case, having $a_{NaAlSi_3O_8}$ less than 1.0 would not help to achieve equilibrium. Equilibrium could be achieved only by lowering $a_{NaAlSiO_4}$ or a_{SiO_2}. For example, in the presence of pure nepheline and pure albite, $a_{SiO_2(s)}$ would have to be 0.0173 to achieve equilibrium. This of course could not happen if quartz was present, but a_{SiO_2} might be controlled in some other way, such as by the amount of dissolved SiO_2 in a solution that is undersaturated with quartz. To calculate this SiO_2 concentration, the reaction would be written using $SiO_2(aq)$ rather than $SiO_2(s)$ (see page 209).

There is an important lesson here. When we write a chemical reaction, we look up a value of $\Delta_f G°$ for each chemical formula. The values of $\Delta_f G°$ are determined for those chemical species in very particular states – pure solids, ideal 1 molal solution, and so on. If our reaction is concerned with those species in those particular states, then the result is directly applicable to our problem – the value of $\Delta_r G$ is the same as the value of $\Delta_r G°$, and the reaction accordingly will go or not go. This case basically arises only when dealing with pure solids. When dealing with solutions (solid, liquid, or gaseous), $\Delta_r G°$ is only a starting point. The reacting species are never in their standard states and have values of Gibbs energy that add up to $\Delta_r G$, not $\Delta_r G°$. The chemical formulae in our reactions represent species in some kind of solution, and we deal with these solutions with our activity terms, which are basically concentrations (strictly speaking, concentrations corrected for the fact that they are not ideal solutions).

Reactions between solid phases such as (9.1) are in principle no different from any other kind of reaction, such as (9.2). The only difference is that there is in fact such a thing as relatively pure albite and quartz, and the like, to which the numbers in the tables apply directly, and we are sometimes interested in reactions between these pure compounds. In principle, however, each chemical formula in a chemical reaction, whether $Mg_2SiO_4(s)$ or $HCO_3^-(aq)$, can and usually does occur in a solution of some kind, with an activity controlled by its concentration.

Box 9.1 Calculation of Aqueous Silica

What is the silica content of a solution in equilibrium with plagioclase and nepheline solid solutions at 25 °C and 1 bar? The activities of the solid solution components are

$$a_{NaAlSiO_4} = 0.75$$

$$a_{NaAlSi_3O_8} = 0.5$$

The reaction is Equation (9.1) as before, *except* that we must use $SiO_2(aq)$ rather than $SiO_2(s)$. Thus we write

$$NaAlSiO_4(s) + 2\,SiO_2(aq) = NaAlSi_3O_8(s)$$

Getting data from Appendix B,

$$\Delta_r G° = \Delta_f G°_{NaAlSi_3O_8} - 2\,\Delta_f G°_{SiO_2°(aq)} - \Delta_f G°_{NaAlSiO_4}$$

$$= -3711.5 - 2(-833.411) - (-1978.1)$$

$$= -66.578\,\text{kJ mol}^{-1}$$

Then

$$\log K = -(-66,578)/(2.30259 \times 8.31451 \times 298.15)$$

$$= 11.664$$

$$K = 10^{11.664}$$

$$= \frac{a_{NaAlSi_3O_8}}{a_{NaAlSiO_4}\,a^2_{SiO_2(aq)}}$$

$$= 0.5/(0.75 \times a^2_{SiO_2(aq)})$$

$$a_{SiO_2(aq)} = [0.5/(10^{11.664} \times 0.75)]^{1/2}$$

$$= 10^{-5.92}$$

So, assuming $\gamma_{SiO2(aq)} = 1$, $m_{SiO2(aq)}$ in equilibrium with these two solid solutions is $10^{-5.92}$ molal, slightly less than would be in equilibrium with the pure minerals.

- This is quite a hypothetical situation. Solid solutions with these activities are common, but on the basis of experience, they would never equilibrate with aqueous silica at 25 °C. However, at higher temperatures, such equilibria are common.
- We used $SiO_2(aq)$ here, but H_4SiO_4 in Equation (9.12). We discuss this in Section 9.8.

9.6 Change of K with Temperature I

To get the effect of temperature on K, assuming as before (Section 6.4) that $\Delta_r H°$ and $\Delta_r S°$ are constants (not affected by temperature), we need only combine Equations (9.11) and (6.11),

$$\Delta_r G° = -RT \ln K$$
$$= \Delta_r H°_{298} - T \Delta_r S°_{298}$$

so

$$\ln K = \frac{-\Delta_r H°_{298}}{RT} + \frac{\Delta_r S°_{298}}{R} \tag{9.18}$$

or

$$\log K = \frac{-\Delta_r H°_{298}}{2.30259\,RT} + \frac{\Delta_r S°_{298}}{2.30259\,R} \tag{9.19}$$

As both $\Delta_r H°$ and $\Delta_r S°$ are assumed constant, this can be rewritten

$$\log K = a(1/T) + b$$

where a and b are constants, which is an equation in the form $y = ax + b$, meaning that $\log K$ is a linear function of $1/T$. An example of this is shown in Figure 9.5.

In Figure 9.5(a) are shown some solubility data for quartz, measured at a constant pressure of 1000 atm. As discussed in Section 9.4.1, these numbers can be interpreted as values of the equilibrium constant for the quartz dissolution reaction. The same data plotted as $\log m_{SiO_2(aq)}$ vs. $1/T$, where T is in kelvins, are shown in Figure 9.5(b). Obviously this shows a good linear correlation, indicating that $\Delta_r H°$ does not change greatly over the temperature range of 25 to 300 °C.

A plot of $\log K$ vs. $1/T$ can therefore be used to obtain an estimate of $\Delta_r H°$ for the reaction for which K is the equilibrium constant. According to the authors (Morey *et al.*, 1962), the slope of the line in Figure 9.5(b) (fitted by the method of least squares) is -1180 K, and so from (9.19)

$$\frac{\Delta_r H°}{2.30259\,R} = -(-1180)$$

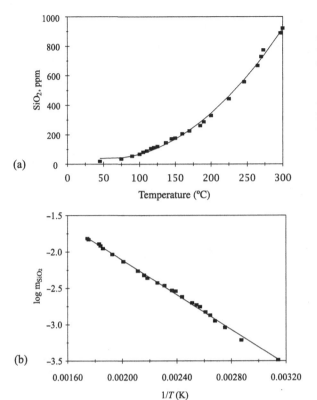

Figure 9.5 (a) The solubility of quartz in water as a function of temperature at a pressure of 1000 bar. (Morey *et al.* (From 1962).) (b) The same data converted to $\log m_{SiO_2(aq)}$ and plotted vs. the reciprocal of absolute temperature.

and

$$\Delta_r H° = 1180 \times 2.30259 \times 8.3145$$

$$= 22,590 \, \text{J mol}^{-1}$$

$$\approx 22.6 \, \text{kJ mol}^{-1}$$

However, although the data may appear to be quite linear, confirming a constant $\Delta_r H°$ and $\Delta_r S°$, even a gentle curvature can easily be obscured by small random experimental errors, and even a gentle curvature implies a significant change in slope and of $\Delta_r H°$. The assumption of constant $\Delta_r H°$ is not suitable for accurate work, but is often useful nonetheless.

9.6.1   Another Example

As an example of the effect of T on K, as well as some of the other points we have made, consider the reaction

$$CaCO_3(s) + SiO_2(s) = CaSiO_3(s) + CO_2(g) \tag{9.20}$$

This is an important reaction at high temperatures, when granites intrude limestones at depth in the Earth, but we will consider it at low temperatures and 1 bar pressure.

First, we get the equilibrium constant, as usual,

$$\Delta_r G^\circ = \Delta_f G^\circ_{CaSiO_3} + \Delta_f G^\circ_{CO_2} - \Delta_f G^\circ_{CaCO_3} - \Delta_f G^\circ_{SiO_2}$$

$$= -1549.66 + (-394.359) - (-1128.79) - (-856.64)$$

$$= 41.411 \text{ kJ mol}^{-1}$$

$$= 41,411 \text{ J mol}^{-1}$$

Then

$$\Delta_r G^\circ = -RT \ln K$$

$$41,411 = -(8.3145 \times 298.15) \ln K$$

and

$$\log K_{298} = -41,411/(2.30259 \times 8.3145 \times 298.15)$$

$$= -7.25$$

This means, as usual, that

$$\frac{a_{CaSiO_3} a_{CO_2}}{a_{CaCO_3} a_{SiO_2}} = 10^{-7.25}$$

and because all the solid phases are pure, their activities are all 1.0, and we write

$$a_{CO_2} = 10^{-7.25}$$

$$= f_{CO_2}$$

Of course, both $CaSiO_3$ (wollastonite) and $CaCO_3$ (calcite) often form solid solutions and in natural situations might have activities less than 1.0, as discussed above. However, we are interested here in the pure phases.

The calculated f_{CO_2} of $10^{-7.25}$ can be thought of as meaning that if calcite, wollastonite, and quartz were at equilibrium with a gas phase having a pressure of 1 bar at 25 °C, the partial pressure of CO_2 in that gas would be about $10^{-7.25}$ or 5.6×10^{-8} bar. As long as the three minerals remain pure and at equilibrium, the equilibrium constant will continue to be equal to f_{CO_2}, and so we can calculate the temperature at which the CO_2 pressure (fugacity) will reach 1 bar by calculating the change in K with T.

To do this, we will first get another expression for the effect of T on K that will be more convenient. From (9.18) you can see that the slope of the graph of $\ln K$ as a function of $1/T$ is $-\Delta_r H^\circ/R$, which is to say that at a temperature of 298 K,

$$\frac{d \ln K}{d(1/T)} = -\frac{\Delta_r H^\circ_{298}}{R}$$

Integrating this between 298 K and T, we get

$$\int_{298}^{T} d \ln K = -\frac{\Delta_r H^\circ_{298}}{R} \int_{298}^{T} d(1/T)$$

and so

$$\ln K_T - \ln K_{298} = -\frac{\Delta_r H^\circ_{298}}{R}\left(\frac{1}{T} - \frac{1}{298.15}\right)$$

or

$$\log K_T = \log K_{298} - \frac{\Delta_r H^\circ_{298}}{2.30259\,R}\left(\frac{1}{T} - \frac{1}{298.15}\right) \tag{9.21}$$

By substituting terms, you can easily show that these are equivalent to our previous Equations, (9.18) and (9.19). Remember, they are valid only for constant $\Delta_r H^\circ$ and $\Delta_r S^\circ$.

Now we need $\Delta_r H^\circ$ for reaction (9.20). This is

$$\Delta_r H^\circ = \Delta_f H^\circ_{CaSiO_3} + \Delta_f H^\circ_{CO_2} - \Delta_f H^\circ_{CaCO_3} - \Delta_f H^\circ_{SiO_2}$$

$$= -1634.94 + (-393.509) - (-1206.92) - (-910.94)$$

$$= 89.411 \text{ kJ mol}^{-1}$$

If we want to calculate the temperature T at which f_{CO_2} reaches 1 bar while in equilibrium with calcite, quartz, and wollastonite, then $K_T = 1$, $\log K_T = 0$, and, using our value of $\Delta_r H^\circ$, we get

$$0 = -7.25 - \frac{89,411}{2.30259 \times 8.3145}\left(\frac{1}{T} - \frac{1}{298.15}\right)$$

from which $T = 555$ K or about $282\,^\circ$C. The meaning of these calculations is illustrated in Figure 9.6.

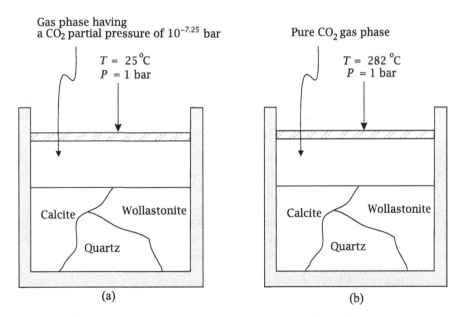

Figure 9.6 The fugacity of CO_2 in equilibrium with calcite, wollastonite, and quartz.

A Common Error

Remember that you <u>CANNOT</u> calculate $\ln K$ at T from

$$\Delta_r G^\circ_{298} = -RT \ln K \tag{9.11}$$

where $\Delta_r G^\circ$ comes from the normal tables, simply by changing T from 298.15 to some other value. In Equation (9.11) above, $\Delta_r G^\circ$ and K must refer to the same temperature. If you want K at some temperature other than 25 °C, *first* get $\Delta_r G^\circ$ at that new temperature from (6.11) or some other method, *then* get K from (9.11), using your new value of $\Delta_r G^\circ_T$ in place of $\Delta_r G^\circ_{298}$. Of course, this procedure has essentially been done for you in equations such as (9.19) and (9.21).

9.6.2   log K vs. $1/T$ Doesn't Always Work

In Section 9.6 we assumed a constant ΔH° and ΔS°, resulting in a simple linear relationship between $\log K$ and $1/T$. However, we also pointed out in Section 3.6.5 that

$$\frac{d\Delta H}{dT} = \Delta C_P \tag{3.37}$$

or

$$\frac{d\Delta_r H^\circ}{dT} = \Delta_r C^\circ_P \tag{9.22}$$

and

$$\frac{d\Delta_r S^\circ}{dT} = \frac{\Delta_r C^\circ_P}{T} \tag{9.23}$$

In other words, what we have really assumed is that $\Delta_r C^\circ_P$ is zero, or that the heat capacities of reactants and products are equal. Pure solids, liquids, and gases generally have C°_P values that increase monotonically (are constantly increasing) with T, so this assumption is reasonable over fairly small temperature intervals. But aqueous solutes have quite a different behavior, so reactions having both minerals and aqueous species have virtually no chance of having $\Delta_r C^\circ_P$ zero or constant over a range of temperatures. However, there are better ways of calculating equilibrium constants at elevated temperatures, e.g., Anderson *et al.* (1991).

9.6.3   Change of K with Pressure

The only commonly used relation between pressure and the equilibrium constant is the case where all reactants and products are solids. Assuming the solids are incompressible, applying relation (5.21) to each of the terms in $\Delta_r G^\circ$ results in

$$\frac{\partial \ln K}{\partial P} = -\frac{\partial \Delta_r G^\circ}{\partial P} \bigg/ RT$$

$$= -\Delta_r V^\circ / RT$$

so that

$$\ln K_{P_2} = \ln K_{P_1} - \frac{\Delta_r V^\circ}{RT}(P_2 - P_1)$$
(9.24)

9.7   The Amino Acid Example Again

Let's write Equation (5.11) one more time:

$$C_8H_{16}N_2O_3(aq) + H_2O(l) = C_6H_{13}NO_2(aq) + C_2H_5NO_2(aq)$$
(9.25)

or

leucylglycine + water = leucine + glycine
(9.26)

and

$$\Delta_r G^\circ = \Delta_f G^\circ_{leucine} + \Delta_f G^\circ_{glycine} - \Delta_f G^\circ_{leucylglycine} - \Delta_f G^\circ_{water}$$
$$= -13{,}903 \, J\,mol^{-1}$$

We now know that our calculation of this $\Delta_r G^\circ$ (Section 5.5.2), the reaction in which a peptide bond between two amino acids is broken, was only a beginning. The value of $-13{,}903 \, J\,mol^{-1}$ means that if all reactants and products had unit activity (leucine, glycine, and leucylglycine had concentrations of 1 molal, and water was pure), the reaction would start to go to the right; leucylglycine would start to break down to leucine and glycine. But we note again the fundamental difference between this reaction between dissolved compounds, and reaction (9.1) between solid compounds. Repeating (9.1) here,

$$NaAlSiO_4(s) + 2\,SiO_2(s) = NaAlSi_3O_8(s)$$
(9.27)

Then

$$\Delta_r G^\circ = \Delta_f G^\circ_{NaAlSi_3O_8} - \Delta_f G^\circ_{NaAlSiO_4} - 2\,\Delta_f G^\circ_{SiO_2}$$
$$= -20.12 \, kJ\,mol^{-1}$$

The value of $\Delta_r G^\circ$ of $-20{,}120 \, J\,mol^{-1}$ means that reaction (9.27) will also go to the right. But this reaction will continue to go (strictly, it *should* continue to go, according to our model) until either $NaAlSiO_4(s)$ or $SiO_2(s)$ is used up. Thus $NaAlSiO_4(s)$ and $SiO_2(s)$ *are not stable together* – one of them must disappear.

This is not the case with leucylglycine. We cannot say that leucylglycine is not stable in water – what happens to it depends entirely on its concentration and on the concentrations of other things in solution such as leucine and glycine. The unit activities are only a starting point, and a very unrealistic one at that. The next step is to calculate the equilibrium constant for (9.25):

$$\Delta_r G^\circ = -RT \ln K$$

$$-13{,}903 = -(8.3145 \times 298.15)\ln K$$

and

$$\log K_{298} = 13{,}903/(2.30259 \times 8.3145 \times 298.15)$$

$$= 2.436$$

Thus

$$\frac{a_{\text{leucine}}\, a_{\text{glycine}}}{a_{\text{leucylglycine}}\, a_{\text{water}}} = 10^{2.436}$$

The activity (mole fraction) of water in biochemical systems is usually close to 1.0, so we see that, although leucylglycine is not "unstable" in water, its concentration at equilibrium must be quite a bit less than that of its constituent amino acids. For example, if leucine and glycine had concentrations of say $10^{-3}\,m$ (activities of 10^{-3}), the equilibrium activity of leucylglycine would be $10^{-8.436}$ (concentration $10^{-8.436}\,m$). So with concentrations of 10^{-3}, 10^{-3}, and $10^{-8.436}$, leucine, glycine, and leucylglycine would not react at all, but would be at equilibrium. In fact, with a concentration of leucylglycine of less than $10^{-8.436}$, the reaction as written would go to the left – leucylglycine would form from the two amino acids. So remember this – unless the reaction consists only of pure phases,

> *you cannot reliably tell which way the reaction will go by looking at $\Delta_r G^\circ$.*

You can always tell which way the reaction will go by looking at $\Delta_r G$.

Look at Equation (9.8) one more time. When leucine, glycine, leucylglycine, and water all have unit activities, (9.8) becomes

$$\Delta_r \mu = \Delta_r \mu^\circ + RT \ln Q$$

$$-13{,}903 = -13{,}903 + RT \ln\left(\frac{1 \times 1}{1 \times 1}\right)$$

In other words, $\Delta_r \mu$ is the same as $\Delta_r \mu^\circ$; the driving force for the reaction can be obtained directly from the tables, as for solid–solid reactions. When products and reactants have reached equilibrium,

$$\Delta_r \mu = \Delta_r \mu^\circ + RT \ln K$$

$$0 = -13{,}903 + RT \ln\left(\frac{10^{-3} \times 10^{-3}}{10^{-8.436} \times 1}\right)$$

Now the $\ln K$ term exactly balances the $\Delta_r \mu^\circ$ term, and the driving force for the reaction is zero. If $a_{\text{leucylglycine}} < 10^{-8.436}$, the driving force ($\Delta_r \mu$) becomes positive.

9.7.1   Peptides Favored at Higher Temperatures

To round out our discussion of this reaction, let's calculate the effect of temperature on the equilibrium constant in reaction (8.18). From Appendix B we find the data in Table 9.1.

Table 9.1. Some data from Appendix B.

Substance	Formula	$\Delta_f H°$ (J mol^{-1})	$S°$ (J mol^{-1} K^{-1})
Leucine	$C_6H_{13}NO_2$(aq)	−632,077	215.48
Glycine	$C_2H_5NO_2$(aq)	−513,988	158.32
Leucylglycine	$C_8H_{16}N_2O_3$(aq)	−847,929	299.16
Water	H_2O(l)	−285,830	69.91

$$\Delta_r H° = \Delta_f H°_{leucine} + \Delta_f H°_{glycine} - \Delta_f H°_{leucylglyine} - \Delta_f H°_{water}$$
$$= -632,077 - 513,988 - (-847,929) - (-285,830)$$
$$= -12,306 \text{ J mol}^{-1}$$

These aqueous species are not ionized, so perhaps our assumption of constant $\Delta_r C_P°$ will not be too bad over small temperature intervals. Suppose we wanted the value of K at 100 °C. Equation (9.21) then becomes

$$\log K_T = \log K_{298} - \frac{\Delta_r H°_{298}}{2.30259\,R}\left(\frac{1}{T} - \frac{1}{298.15}\right)$$
$$= 2.436 - \frac{-12,306}{2.30259 \times 8.31451}\left(\frac{1}{373.15} - \frac{1}{298.15}\right)$$
$$= 2.00$$

Alternatively, by calculating $\Delta_r S°$, you could use Equation (6.11) first, then Equation (9.11). Thus

$$\Delta_r S° = S°_{leucine} + S°_{glycine} - S°_{leucylglyine} - S°_{water}$$
$$= 215.48 + 158.32 - 299.16 - 69.91$$
$$= 4.73 \text{ J mol}^{-1}$$

Then

$$\Delta_r G°_{373} = \Delta_r H°_{298} - T\,\Delta_r S°_{298}$$
$$= -12,306 - 373.15 \times 4.73$$
$$= -14,071 \text{ J mol}^{-1}$$

from which

$$\log K_{373} = \frac{-\Delta_r G°_{373}}{2.30259\,RT}$$
$$= -\frac{-14,071}{2.30259 \times 8.31451 \times 373.15}$$
$$= 1.97$$

There will often be a small discrepancy in log K calculated in different ways, as here (2.00 vs. 1.97), because of slight inconsistencies in the data. In other words, to get answers that are exactly the same no matter which way the calculation is done, the data in the tables for each compound must satisfy the relation

$$\Delta_f G° = \Delta_f H° - 298.15 \times \Delta_f S°$$

Because enthalpy, entropy, and Gibbs energy data come from different experiments, using a variety of methods, this relation is often not satisfied exactly in the tabulated data.

The interesting aspect of this calculation of K is that according to the data, leucylglycine (and perhaps all peptide bonds in proteins) becomes *more stable* as temperature increases. Thus for the same concentrations of leucine and glycine ($10^{-3}m$) as before, we find the leucylglycine concentration is $10^{-8.0}$ m at 100 °C, compared to $10^{-8.436}$ m at 25 °C. That is, its concentration is more than doubled. This result is quite interesting to those scientists trying to figure out how life could have begun in the early days of the Earth, 3.5 billion years ago. The fact that increasing temperatures do not impair but in fact aid the bonding of simple amino acids, the building blocks of life, has led to thoughts that perhaps life began when the oceans were at higher temperatures, or in particular locations (volcanic environments) where heat was available.

This result is typical of the value of thermodynamics. It does not and cannot tell you how life began, but it can tell you which processes are possible and which impossible, and what the effects of changing the constraints on your system will be. This guides the development of scientific ideas in an essential way and provides a universally agreed-upon bedrock from which to start. However, it is up to you to think of the processes to ask thermodynamics about, and this is the creative part of science.

9.8 Some Conventions Regarding Components

There are two ways in which the way chemical formulae are used may prove confusing.

1. Aqueous species are used in both hydrated and non-hydrated forms. For example, dissolved silica is written as $SiO_2(aq)$ or as $SiO_2 \cdot 2H_2O$ (or H_4SiO_4, which is the same thing).
2. Formulae can be presented in various multiples. For example forsterite may be listed as Mg_2SiO_4 or as $MgSi_{0.5}O_2$.

These are quite simple relationships, but they can cause quite a bit of confusion.

9.8.1 Hydrated vs. Non-Hydrated Species

H_4SiO_4 Example

One way to write the dissolution reaction for quartz is

$$SiO_2(s) + 2H_2O = H_4SiO_4(aq) \tag{9.28}$$

Another way to write the same reaction is

$$SiO_2(s) = SiO_2(aq) \tag{9.29}$$

The only difference between these two ways of writing a reaction for the dissolution of quartz is that in (9.28) we have assumed that the dissolved silica is in the form of a molecule containing one SiO_2 attached to two H_2O molecules, whereas in (9.29) we have made no assumption as to the form of the dissolved silica. So what do we actually know about dissolved silica? What we know, besides the concentrations under various conditions, is as follows.

1. Under most conditions, the aqueous silica molecule has only one Si (i.e., it is *monomeric*, not *polymeric*).
2. Except in very basic solutions it is uncharged, electrically neutral.

We might as well then write a formula for this species that is as simple as possible, while observing these two facts, and $SiO_2(aq)$ does this. Therefore, $SiO_2(aq)$ does *not* refer to a species of dissolved silica which is not attached to any H_2O or other molecules; it refers to the silica that exists as a monomeric uncharged species of whatever nature in solution. It might be attached to two H_2O, or six H_2O, or be a mixture of several such species; it doesn't matter.

The other common formula, $H_4SiO_4(aq)$, originates historically in the belief that Si in water must be tetrahedrally coordinated by oxygens, as it is in crystals. That may well be true, but there may be other oxygens in the form of H_2Os also attracted to the Si. The exact nature of the complexes of Si and many other elements of interest is a continuing research topic. The important point from the thermodynamic point of view is that what we *call* the dissolved silica, whether $H_4SiO_4(aq)$ or $SiO_2(aq)$, doesn't matter, because as long as we derive the properties of each in a consistent manner, each will give the right answer in calculations.

In what way are the properties of these species different? Because these two formulae refer to the same physical substance, dissolved silica, their concentrations and activities are identical. But because they are related by the equation

$$SiO_2(aq) + 2\,H_2O(l) = H_4SiO_4(aq)$$

their standard state properties such as $\Delta_f G^\circ$ and $\Delta_f H^\circ$ must be different by exactly twice the corresponding property of $H_2O(l)$. Thus

$$\Delta_f G^\circ_{H_4SiO_4(aq)} = \Delta_f G^\circ_{SiO_2(aq)} + 2\,\Delta_f G^\circ_{H_2O}(l)$$

The Gibbs energy of formation of H_4SiO_4 is *defined* as the sum of the Gibbs energies of $SiO_2(aq)$ and (twice that of) $H_2O(l)$. In other words, the relationship between $H_4SiO_4(aq)$ and $SiO_2(aq)$ is strictly a formal one. They are derived from the same experimental data and will yield the same results in calculations. Similar relationships exist for other hydrated species, such as H_2CO_3 and $Al(OH)_3$.

H₂CO₃ Example

The same relationship also holds for other species. For example, when CO_2 gas dissolves in water, it hydrolyzes (reacts with water) to a very small extent, forming some H_2CO_3 molecules in solution. It is rather difficult to determine the exact amount of H_2CO_3, and this problem is avoided by simply calling the total amount of carbon dioxide in solution either $CO_2(aq)$ or $H_2CO_3(aq)$, exactly as the dissolved silica is called $SiO_2(aq)$ or $H_4SiO_4(aq)$. Then, for the same reason as before, we find that

$$\Delta_f G°_{H_2CO_3(aq)} = \Delta_f G°_{CO_2(aq)} + \Delta_f G°_{H_2O(l)}$$

Again, this relationship is strictly formal, although in this case it can be more confusing, because there is in fact some literature on the subject of how much dissolved CO_2 actually hydrolyzes to the species H_2CO_3 and how much remains as CO_2 molecules. In other words, H_2CO_3 is sometimes used as a species, and sometimes in the conventional sense we are discussing. The thermodynamic properties of H_2CO_3 in these two senses will of course be completely different. In this book we use the conventional sense for $H_2CO_3(aq)$. Another way of looking at this is to see H_2CO_3 as an alternative component, rather than as an aqueous species.

9.8.2 What Is a Mole of Olivine?

Another way in which the choice of formula can differ is that some choices can be multiples of other choices. This is most often seen in choosing solute species in solids, because there are no "real" species, just a crystal structure that is a solid solution. For example, the mineral olivine is a solid solution of two components, forsterite (Mg_2SiO_4) and fayalite (Fe_2SiO_4). The solution is represented by $(Mg, Fe)_2SiO_4$, because the Mg and Fe atoms share the same positions in the crystal structure.

But what reason do we have to choose Mg_2SiO_4 and Fe_2SiO_4 as our components? The formula simply shows us the stoichiometry of the components – the ratios or relative amounts of the elements. Why not $MgSi_{0.5}O_2$, or $Mg_4Si_2O_8$? The same question could also arise in discussing aqueous species, except in that case we often have experimental evidence about the nature of the species in solution. That kind of evidence does not exist for the three-dimensional crystal structures of solid solutions – we are free to choose any component that is stoichiometrically correct. Does it make any difference? Yes.

In Section 7.2 we noted that some confusion might arise in the definition of mole fractions, but here the choice makes no difference. That is,

$$\frac{n_{FeSi_{0.5}O_2}}{n_{FeSi_{0.5}O_2} + n_{MgSi_{0.5}O_2}} = \frac{n_{Fe_2SiO_4}}{n_{Fe_2SiO_4} + n_{Mg_2SiO_4}}$$

So that is not the problem. The problem is that the choice of formula, the mole of substance, affects the energy content per mole and hence the activity.

Suppose our system consists of a certain mass (a crystal) of pure forsterite, Mg_2SiO_4. The Gibbs energy of the system is a finite, unknown quantity, which depends on the mass of the crystal. A crystal with twice the mass has a **G** twice as large. But the *molar G* does not vary with the size of the crystal. The molar G is defined as $G = \mathbf{G}/n$ (Section 2.4.1), where n is the number of moles in the crystal. The point is, the number of moles of *what*? Obviously the number of moles of Mg_2SiO_4 in the crystal will be exactly half the number of moles of $MgSi_{0.5}O_2$ in the crystal, because Mg_2SiO_4 contains twice the number of atoms that $MgSi_{0.5}O_2$ does. Therefore, $G_{Mg_2SiO_4} = 2\,G_{MgSi_{0.5}O_2}$. Or, if you prefer, you can say that $G_{Mg_2SiO_4} = 2\,G_{MgSi_{0.5}O_2}$ simply because it contains twice the mass and, therefore, twice the energy of whatever kind.

This difference in the Gibbs energy of the mole is translated into a difference in activities. Because $G_{Mg_2SiO_4} = 2\,G_{MgSi_{0.5}O_2}$ and $G^\circ_{Mg_2SiO_4} = 2\,G^\circ_{MgSi_{0.5}O_2}$, we have

$$G_{Mg_2SiO_4} - G^\circ_{Mg_2SiO_4} = 2\,(G_{MgSi_{0.5}O_2} - G^\circ_{MgSi_{0.5}O_2})$$

and

$$RT \ln a_{Mg_2SiO_4} = 2\,RT \ln a_{MgSi_{0.5}O_2}$$

and therefore

$$a_{Mg_2SiO_4} = a^2_{MgSi_{0.5}O_2}$$

The problem this poses can be seen in considering Raoult's law, which we said was $a_i = x_i$. But if $a_i = a^2_{0.5i}$ we have a problem. Because x_i is independent of how we write the formula for i, we see that a_i and $a_{0.5i}$ cannot both be equal to x_i, even if Raoult's law is followed exactly. If $a_{0.5i}$ vs. x_i is a straight line, then a_i vs. x_i will describe a parabola.

This is a well-known problem, and generally the formula for components is chosen such that the simple statement of Raoult's law is followed as closely as possible. Again, this relationship between activities is entirely formal and tells us nothing about forsterite or olivine. However, it is important to remember that choosing a formula for your components has consequences for activities.

The problem is more difficult in other systems. How does one choose components in a complex silicate melt, for example? In a melt there are no stoichiometric restrictions to be observed, but the formal relationship between the activities of various component choices that we have discussed remains true. So if you measure the activity of some component in a melt, and determine the deviations of these activities from Raoult's law by calculating activity coefficients, the question is, what part of these activity coefficients represents non-ideal behavior, and what part represents a poor choice of components? Generally speaking, extremely large or extremely small activity coefficients mean that the component involved has been badly chosen, which is to say that it does not come very close to representing the "real" situation in the system. In these situations, thermodynamics provides no help whatsoever. It points out the consequences of choices relative to each other, and from there on the investigator is on her own. In other words, the choice of components, as much as the choice of system to investigate, is a part of the "art of doing science," that part which relies on skill and intuition, and can never be taught.

9.9 Summary

This chapter contains a sudden increase in the amount of practical, usable material. If you ever have occasion to use thermodynamics in a practical situation, it will very likely involve the use of the equilibrium constant.

The molar Gibbs energy of a dissolved substance changes with the concentration of the substance. The activity is a dimensionless concentration-like term that is used to give the Gibbs energy in a particular state, in terms of its difference from its value in some reference or standard state (Equation (7.43)). When a reaction has reached equilibrium, the activities of the various products and reactants can have a variety of values individually, but their ratio, as expressed in the equilibrium constant K, has a fixed and calculable value.

The equilibrium constant is calculated from numbers (Gibbs energies) taken from tables of standard data (derived experimentally, as discussed in Chapter 5). These standard data give the term $\Delta_r \mu^\circ$ or $\Delta_r G^\circ$, which is a constant for a given T and P. It has nothing to do with whether or not your system or reaction has reached equilibrium ($\Delta_r \mu = 0$). However, it can be used to calculate K, which gives the ratio of product and reactant activities your reaction will have if it ever reaches equilibrium.

The superscript $^\circ$ therefore has considerable significance. It should not be omitted or inserted carelessly in your calculations.

Exercises

E9.1 Hydrogen sulfide is one of the three most important volatiles in the Earth's crust together with carbon dioxide and water. It plays an important role in many crustal processes such as sulfide ore formation. The equilibrium constant for the reaction $H_2S = H^+ + HS^-$, that is, the first ionization constant of $H_2S(aq)$, as measured by Suleimenov and Seward (1997) is shown in Table 9.2. From these data the authors calculated ΔG°, ΔH°, and ΔC_P° for the ionization reaction. The values for 25 °C are shown in Table 9.3.

Table 9.2. The first ionization constant of aqueous hydrogen sulfide, from Suleimenov and Seward (1997), Table 2.

T (°C)	25	50	100	150	200	250	300	350
$\log K$	−6.99	−6.68	−6.49	−6.49	−6.73	−7.19	−7.89	−8.89

(a) Calculate the $\log K$ values of the first ionization constant of $H_2S(aq)$ at the temperatures in Table 9.2 using Equation (9.21), often called the van 't Hoff equation. Use (i) the $\Delta_r H^\circ$ in Table 9.3 and (ii) $\Delta_r H^\circ$ from data in Appendix B.

(b) Plot these $\log K$ values and those in Table 9.2 vs. T on the same graph, and also vs. $1/T$. When using the $1/T$ axis it is helpful to reverse it, to have $1/T$ values decreasing to the right, so that T values are increasing.

Table 9.3. Some thermodynamic properties for the first ionization reaction of aqueous hydrogen sulfide.

$\Delta_r G°$ (J mol^{-1})	$\Delta_r H°$ (J mol^{-1})	$\Delta_r C_P°$ (J mol^{-1} K^{-1})
39,666	22,733	−265.3

(c) Check that the $\Delta G°$ value in Table 9.2 gives the same value of log K as in Table 9.2 (see Equation (9.11)).

(d) Can you use the $C_P°$ value in Table 9.3 to improve the estimation of the high-T ionization constants?

After looking at these results, you will probably be more careful about accepting results from complex computer programs without some knowledge of what they are doing.

E9.2 In a recent large-scale experiment, the amount of $CO_2(g)$ produced from the oxidation of organic material in soils seemed to be far less than expected. Then someone suggested that perhaps cement blocks, which contain $Ca(OH)_2$ (portlandite) and were enclosing the experiment, were soaking up $CO_2(g)$ in the form of calcite.

(a) Show that this is indeed possible from the reaction

$$Ca(OH)_2 + CO_2(g) = CaCO_3(s) + H_2O$$

(b) What would be the equilibrium partial pressure of $CO_2(g)$ in a room containing both portlandite and calcite?

(c) Show that in the presence of liquid water, lime (CaO(s)) would not be expected in the cement blocks.

(d) What would be the partial pressure of water at equilibrium with lime (CaO) and portlandite?

(e) What is the equilibrium constant of the reaction in part (a) at 150 °C? What is the temperature at which the reaction direction would be reversed, i.e., the temperature above which calcite could not form from portlandite and $CO_2(g)$ at atmospheric pressure?

E9.3 Write an equation for the reaction of methane and oxygen to give carbon dioxide and liquid water. Calculate the equilibrium constant. Given the partial pressures of $CO_2(g)$ and oxygen (Chapter 7, Exercise E3), what is the equilibrium amount of methane in the atmosphere? Compare your answer with the amount in the table. Why the discrepancy?

E9.4 The standard Gibbs energy change for the reaction

$$2Cu(s) + \tfrac{1}{2}O_2(g) = Cu_2O(s)$$

as a function of temperature is sometimes given as

$$\Delta_r G° \text{ J mol}^{-1} = -168,600 + 75.729\, T \text{ (K)}$$

What are the $\Delta_r H°$ and $\Delta_r S°$ of this reaction? (You can tell this directly from the equation given, or you can work it out from the tables, or both.) What is the f_{O_2} in equilibrium with Cu and Cu_2O at 600 °C?

E9.5 Calculate the vapor pressure of water at 25 and 100 °C. The reaction is simply

$$H_2O(l) = H_2O(g)$$

E9.6 Write a reaction for the dehydration of diaspore to form corundum and gaseous water (water vapor). Calculate the equilibrium constant. If the fugacity of water in the atmosphere is controlled by evaporation as in Exercise E9.5, which way will the reaction go?

E9.7 Calculate the solubility of amorphous silica (i.e., the concentration of H_4SiO_4) in a solution having a water activity of 0.9.

E9.8 Calculate the vapor pressure (i.e., the f_{S_2}) of orthorhombic sulfur at 50 °C.

E9.9 Calculate the fugacity of oxygen in equilibrium with solid Ca and solid CaO (lime). Why is pure solid Ca never found in the Earth's crust?

E9.10 Suppose that meteoric water that is percolating down through a bauxite deposit comes to equilibrium with gibbsite and kaolinite. What would be the silica concentration in the water?

E9.11 At another bauxite deposit close to the one in the last question, the groundwater in the bauxite is also in contact with granite boulders containing quartz and feldspars. The weathering of these boulders results in a silica, $SiO_2(aq)$, concentration in the groundwater of 100 ppm. Would you expect to find kaolinite or gibbsite in the bauxite? Or is there another aluminum oxide or hydroxide you would expect?

E9.12 When carbonate rocks are metamorphosed, dolomite, $(CaMg(CO_3)_2)$ and quartz (SiO_2) often react to form diopside, $(CaMg(SiO_3)_2)$ and carbon dioxide. Write a balanced reaction for this process, and show whether this reaction should proceed at 25 °C or not. If the three minerals were at equilibrium together at 25 °C, what would be the fugacity of CO_2? At what temperature would f_{CO_2} become equal to 1 bar?

E9.13 Calculate the solubility of H_2S in water if $f_{H_2S} = 1$ bar, at 25 and 100 °C.

E9.14 (a) Calculate the concentration of silica in the ocean, assuming that it is controlled by the solubility of radiolaria shells, which are made of amorphous silica.

   (b) The SiO_2 concentration in seawater is actually about 7 ppm. Use this number to calculate the Gibbs energy of formation of $H_4SiO_4(aq)$ and of $SiO_2(aq)$ from the elements, assuming equilibrium with amorphous silica. Compare the results with the values in Appendix B.

E9.15 (a) Calculate the first and second ionization constants of aqueous hydrogen sulfide, $H_2S(aq)$. Combine these to get the equilibrium constant for the reaction

$$H_2S(aq) = 2H^+ + S^{2-}$$

   (b) Calculate this equilibrium constant directly. It should be identical.

   (c) Use the solubility product of PbS and this equilibrium constant to calculate the lead ion content of a solution saturated with hydrogen sulfide at 25 °C, 1 bar $(m_{H_2S(aq)} = 0.1)$ and a pH of 4.0. Assume all activity coefficients are 1.0.

E9.16 Calculate the ionization constant of acetic acid at 25 and $100\,^{\circ}C$.

E9.17 Calculate the proportions (mole fractions) of an equilibrium mixture of NO_2 and N_2O_4 gases at $25\,^{\circ}C$, 1 bar.

E9.18 A sample of seawater has a bicarbonate concentration of $0.09\,m$ ($\approx a_{HCO_3^-}$), a pH of 8.0, and a calcium ion activity ($a_{Ca^{2+}}$) of 10^{-6}. Will calcite precipitate? What activity of Mg^{2+} would be required to precipitate dolomite?

E9.19 Calculate the calcium ion concentration in a solution having $a_{H_2CO_3(aq)} = 0.1$ at a pH of 5.0, and in equilibrium with (a) calcite; (b) aragonite. What should happen if this solution were in contact with both minerals? Is this consistent with their values of $\Delta_f G^{\circ}$?

E9.20 There are two main reactions in the oxidation of pyrite at the Earth's surface. One of course involves oxygen,

$$FeS_2 + \tfrac{7}{2}O_2 + H_2O = Fe^{2+} + 2\,SO_4^{2-} + 2\,H^+ \tag{1}$$

The other involves ferric iron as oxidant,

$$FeS_2 + 14\,Fe^{3+} + 8\,H_2O = 15\,Fe^{2+} + 2\,SO_4^{2-} + 16\,H^+ \tag{2}$$

The ferric iron in this reaction is produced from the oxidation of ferrous iron by oxygen,

$$Fe^{2+} + H^+ + \tfrac{1}{4}O_2 = Fe^{3+} + \tfrac{1}{2}H_2O \tag{3}$$

All three reactions, along with countless others, proceed simultaneously in the mine waste environment. Calculate the equilibrium constant for each of these reactions using an aqueous standard state for oxygen, $O_2(aq)$. What do you conclude about the equilibrium level of molecular oxygen (O_2) in groundwater in contact with pyrite and dissolved ferrous and ferric iron?

E9.21 (a) The fugacity of S_2 in equilibrium with pyrite and pyrrhotite at $602\,^{\circ}C$, 1 bar is $10^{-1.95}$ bar. The pyrrhotite in this equilibrium is $Fe_{0.92}S$, which can be considered to be a solid solution composition in the system $FeS-S_2$. The activity of FeS in this pyrrhotite is 0.46 based on a standard state of pure stoichiometric FeS at the same P and T. The pyrite is pure stoichiometric FeS_2. Calculate $\Delta_r G^{\circ}$ for the reaction in which pyrite breaks down to form pyrrhotite and S_2 gas at this P, T.

(b) Is the pyrrhotite involved in this $\Delta_r G^{\circ}$ term FeS or $Fe_{0.92}S$?

(c) The fugacity of sulfur in equilibrium with iron and stoichiometric FeS (troilite) at $602\,^{\circ}C$, 1 atm is $10^{-12.5}$ atm. Calculate $\Delta_r G^{\circ}$ for this reaction and combine it with the previous result to get the standard Gibbs energy of formation of pyrite from its elements at $602\,^{\circ}C$, 1 atm.

E9.22 A wollastonite-bearing contact metamorphic zone is observed adjacent to a granite which has intruded a quartz-bearing limestone horizon. Heat-flow calculations indicate that the maximum temperature achieved at a given distance from the contact is given by

$$T = 760 - 7.66d + 0.0396d^2$$

where T is the temperature in °C, and d is the distance in feet. Stratigraphic considerations put the pressure at the time of intrusion at 2000 bars.

(a) If the contact zone is 50 feet wide, what was the f_{CO_2} in the pore fluid of the limestone?

(b) If you assume that the pore fluid was a H_2O-CO_2 solution, what do you need to know to calculate its composition?

E9.23 Barium sulfide (BaS) never occurs in natural systems. Why might this be so? The maximum possible sulfide ion ($a_{S^{2-}}$) activity near the Earth's surface is about 0.1.

E9.24 The equilibrium constant for the reaction $CO_2(g) + H_2O(l) = H_2CO_3(aq)$ is $10^{-1.468}$. Calculate the activity of the carbonate ion ($a_{CO_3^{2-}}$) in a solution if $f_{CO_2(g)} = 0.1$ bars and the pH is 5.0.

E9.25 Calculate the zinc content ($a_{Zn^{2+}}$) of a solution having a carbonate ion activity of $10^{-9.16}$ in equilibrium with smithsonite ($ZnCO_3$).

E9.26 If a solution having a zinc ion concentration of 0.173 m at a pH of 5.0 was in contact with quartz, would willemite, $Zn_2SiO_4(s)$, precipitate? A relevant reaction would be

$$2\,Zn^{2+} + SiO_2(quartz) + 2\,H_2O(l) = Zn_2SiO_4(s) + 4\,H^+$$

E9.27 Show why pure water has a pH of 7 by calculating the (self-)ionization constant of $H_2O(l)$. Calculate the fugacity of $I(g)$ and the fugacity of $I_2(g)$ in equilibrium with solid iodine ($I_2(s)$).

E9.28 Calculate the oxygen fugacity of water in equilibrium with hydrogen gas having a pressure of 1 bar. Calculate the hydrogen fugacity in equilibrium with water in equilibrium with our atmosphere, which has a mole fraction of $O_2(g)$ of 0.21.

Additional Problems

The formation of metallic ore deposits from hot aqueous solutions circulating in the Earth's crust some millions of years ago is a subject of considerable interest to geologists trying to find those deposits. Obviously, if we understood the chemical and physical processes involved in their formation we would have a much better chance of finding them. Deciphering those processes involves detailed examination of existing deposits plus development of conceptual models of these processes, constrained both by the field observations and by the laws of chemistry and physics. Good field observations and thermodynamics are both essential parts of developing these models.

Say we have an ore deposit with millions of tons of lead in the form of galena in some carbonate rocks. There is good evidence that the PbS has precipitated from a hot ($\approx 100\,°C$) saline solution (a "hydrothermal solution") that has traveled through a sandstone aquifer before depositing the PbS. Many questions arise. Where did the lead come from, and why did it precipitate there? You figure that the solutions must have carried *at least* 10^{-5} molal lead, otherwise it would take forever to precipitate millions of tons of PbS, so presumably there must have been the same amount of sulfide, S^{2-}. Or was there?

A9.1 The first question that occurs is, what is the solubility of PbS? Calculate the solubility product,

$$PbS = Pb^{2+} + S^{2-}$$

If $m_{Pb^{2+}}$ is 10^{-5}, what is $m_{S^{2-}}$? We assume concentration and activity are the same.

A9.2 So then the question is maybe the sulfide is not mostly S^{2-} but mostly H_2S or HS^-. To think about this calculate the second ionization constant of $H_2S(aq)$, at $25\,°C$. The reaction is

$$HS^- = H^+ + S^{2-}$$

From the first and second ionization constants, what do you conclude is the dominant form of reduced sulfur (i.e., H_2S, HS^-, or S^{2-}) in water at a pH ≤ 7 at low temperatures. In geochemistry, "low temperatures" does not mean cryogenic temperatures. It means much lower than igneous and metamorphic temperatures; in the case of carbonate-hosted lead–zinc deposits, not more than $150\,°C$ or so.

A9.3 If the ore solution had a near neutral or slightly acid pH ,

$$PbS + 2H^+ = Pb^{2+} + H_2S(aq)$$

would be more relevant than the solubility product. Calculate this $\log K$ value at $25\,°C$. Use this $\log K$ to plot values of $m_{Pb^{2+}}$ vs. pH at $25\,°C$. Use a constant m_{H_2S} of 0.01 molal. Water at this temperature is saturated with H_2S at about 0.1 molal.

A9.4 But using the lead ion Pb^{2+} to represent galena solubility in natural solutions ignores the fact that Pb^{2+} combines with other components of such solutions, greatly increasing the total lead content. It's not that the calculated Pb^{2+} concentration is incorrect, it's just that it represents only a tiny fraction of the lead in a saline solution, or brine. In brines with lots of Cl^- and other species, the lead content can be increased by a factor of 1000 or more. Multiply your Pb^{2+} values by 1000 and plot these on the same graph.

A9.5 Another way to increase the lead content is to reduce the H_2S content. Maybe it is not 0.01 molal, but 10^{-3} or even 10^{-5}. Who knows? But lowering H_2S from 10^{-2} to 10^{-5} has the same effect as multiplying the lead content by 1000, so you already know the result of doing this.

A9.6 To be a little more precise, in a saline solution, much of the lead will be carried as $PbCl_4^{2-}$, and the relation of this to the lead ion is

$$Pb^{2+} + 4\,Cl^- = PbCl_4^{2-}$$

$\Delta_f G°$ for aqueous $PbCl_4^{2-}$ is $-557{,}560\,J\,mol^{-1}$ at $25\,°C$. If the solution is 3 molal in NaCl ($a_{Cl^-} \approx 3$), and the activity of Pb^{2+} is 10^{-5}, what is $a_{PbCl_4^{2-}}$? There are many other species which contribute to the total lead content. In the next chapter you will see how we could determine the pH of such solutions.

A9.7 Experiments (Barrett and Anderson, 1982) show that an equation for the lead content of a solution 1 molal in NaCl saturated with H_2S at $80\,°C$ as a function of pH is

$$\log m_{Pb} = -2.051\,pH - 2.098 \tag{9.30}$$

where m_{Pb} is the total lead molality, including all the lead attached to chloride ions. The concentration of H_2S in these solutions is 0.02 molal so there is no shortage of reduced sulfur. Plot this on the same diagram.

A9.8 You might conclude from these calculations that "low-temperature" hydrothermal solutions can have reasonably high metal contents, but only if sulfide is very low. How then to precipitate metal sulfides? One idea is that saline solutions carrying both lead ions (Pb^{2+}) and sulfate ions (SO_4^{2-}), circulating in the crust of the Earth, encounter a carbonate unit containing methane, $CH_4(g)$, produced by the heating of organic material in the rock. The reaction could be written

$$Pb^{2+} + SO_4^{2-} + CH_4(g) = PbS(s) \downarrow + CO_2(g) + 2H_2O(l)$$

(a) Calculate the equilibrium constant for this reaction at 25 °C.

(b) Consider the solution as it enters the methane-bearing rock. If the fugacity of both methane and CO_2 is 1 bar, the activity of SO_4^{2-} is 0.0018 (Table 10.2), and the lead ion concentration is 10^{-5} molal, will galena precipitate? Assume that activity coefficients are 1.0 and that the water has unit activity, despite the chloride content.

A9.9 Then it occurs to you that with so much lead and sulfate in solution, maybe anglesite ($PbSO_4$) should precipitate. Why do we get PbS ores and not anglesite ores? Calculate $\log K$ for the reactions

$$PbSO_4 = Pb^{2+} + SO_4^{2-}$$
$$PbSO_4 + H_2S(aq) = PbS + SO_4^{2-} + 2H^+$$

(a) What is the solubility of anglesite? For example, if the activity of SO_4^{2-} is 0.0018 (Table 10.2), what lead ion activity (concentration) would be required to precipitate anglesite?

(b) For coexisting galena and anglesite, plot the SO_4^{2-}/H_2S ratio as a function of pH for pH < 7.

A9.10 Clearly the pH is an important parameter. What limits can we put on it? Hydrothermal solutions will react with silicates and carbonates, changing and perhaps controlling the pH . Calculate $\log K$ for the reaction

$$CaCO_3(s) + 2H^+ = Ca^{2+} + CO_2(g) + H_2O$$

Assume the Ca^{2+} activity is constant at its value in seawater (one of the hypotheses for the origin of these solutions) 0.0025 (Table 10.2) and plot the CO_2 pressure required for equilibrium with calcite at pH values < 7.

A9.11 Keep in mind the following.

- Carrying the lead to the site of deposition and deposition (precipitation) are two different aspects of the origin of hydrothermal lead deposits.

- There is a limit to the acidity a natural solution can have. It will react with minerals to reduce its acidity (e.g., Additional Problem A9.10; also Equation (10.10)).

- Similarly, there is a relationship between how deep the ore is and a realistic CO_2 pressure. It cannot be greater than a few hundred bars if the ore deposition was not very deep (Additional Problem A9.10).

- Lead concentrations of less than about 10^{-5} molal would require unrealistic lengths of time to deposit enough lead to make a commercial ore deposit.

Write a few lines on how these reactions and their equilibrium constants clarify (or not) the formation of a hydrothermal galena ore deposit. What other questions occur that are *not* answered by these calculations?

A9.12 Is cinnabar or metacinnabar the stable phase under ordinary conditions? If the pressure is raised, does the other phase become stable, or not? Other conditions being the same, at what temperature do cinnabar and metacinnabar become at equilibrium with each other? Sketch a P–T diagram showing their relationship.

A9.13 Native sulfur (orthorhombic) and the stable form of solid HgO are placed together in air. Will cinnabar form? What is the equilibrium f_{O_2} for these three solids to coexist?

A9.14 Show that the $\Delta_f G°$ for sphalerite (ZnS), which is $-201.29 \, \text{kJ mol}^{-1}$, can be calculated from $\Delta_f G° = \Delta_f H° - T \Delta_f S°$, using other data from the tables.

A9.15 If you knew that sphalerite is more stable than wurtzite under ordinary conditions, and that its entropy was greater than that of wurtzite, what would the P–T phase diagram for ZnS look like? Would you expect wurtzite to be more or less dense than sphalerite? Why?

A9.16 The principal ore of zinc is sphalerite, and the ores are usually formed by precipitation from aqueous solutions in the Earth's crust. If the solution contains 10^{-5} molal $H_2S(aq)$ and the pH is 4.0, what is the maximum concentration of Zn^{2+} the solution can have? Make a sketch of $\log a_{Zn^{2+}}$ vs. pH if $m_{H_2S(aq)}$ is fixed at 10^{-5}.

A9.17 Sphalerite is not stable at the Earth's surface; it tends to oxidize to other minerals. The fugacity of $CO_2(g)$ in the atmosphere is $10^{-3.5}$ bars. Would you expect zincite (ZnO) or smithsonite ($ZnCO_3$) to be the stable form of zinc at the Earth's surface? Would you change your answer if you consider that groundwater, which tends to oxidize the zinc ores, generally has a higher $CO_2(g)$ fugacity (about 10^{-2} bars, due to the oxidation of organic matter) than does the atmosphere?

A9.18 With a $CO_2(g)$ fugacity of $10^{-3.5}$ bars, what is the equilibrium $CO_2(aq)$ content of the groundwater at 25 °C? At 50 °C?

A9.19 Calculate the solubility product of smithsonite. Using the groundwater composition in Additional Problem A9.18, and assuming that $m_{H_2CO_3(aq)}$ is the same as $m_{CO_2(aq)}$, above what pH will smithsonite precipitate from a groundwater containing 10^{-6} molal Zn^{2+} at 25 °C?

A9.20 At a pH of 8, what concentration of $SiO_2(aq)$ would cause willemite, $Zn_2SiO_4(s)$, to precipitate, given the above zinc concentration? If you have both the silica at 10^{-6} molal and the CO_2 fugacity at $10^{-3.5}$ bars in the groundwater with the Zn^{2+}, would you expect smithsonite or willemite?

Table 9.4. Extra data for gold.

Formula	Form	Molecular weight (g mol^{-1})	$\Delta_f H^\circ$ (kJ mol^{-1})	$\Delta_f G^\circ$ (kJ mol^{-1})	S° (J mol^{-1} K^{-1})	C_P° (J mol^{-1} K^{-1})
Au	s	196.9670	0	0	47.40	25.418
AuCl	s	232.4200	-34.727	-15.062	92.885	
AuCl$_3$	s	303.3260	-45.187	-115.06	148.114	
Au$^+$	aq	196.9670	199.075	163.176	102.508	
Au^{3+}	aq	196.9670	405.555	433.462	-242.254	
AuCl$_4^-$	aq	338.7790	-322.2	-235.14	266.9	

A9.21 Calculate the solubility products of auric chloride (AuCl$_3$) and aurous chloride (AuCl). What would be the gold concentration in a solution in equilibrium with AuCl and having $m_{Cl^-} = 0.5$, $\gamma_{Cl^-} = 0.8$? Some extra data are given in Table 9.4.

10 Rock–Water Systems

Real Problems

We have now completed our survey of the thermodynamic principles required to model natural systems. It only remains to gain practice in formulating problems involving natural systems in thermodynamic terms. Quite often, that is the hardest part. Once the problem has been set up in terms of relevant reactions and components, the equations can be solved by anyone who has absorbed the previous chapters. However, choosing the appropriate components and setting up the relevant balanced reactions only comes from experience. In this chapter we explore a few situations that have been investigated by thermodynamic methods.

Is the Sea Saturated with Calcium Carbonate?

If you have ever been to Florida or the Bahamas, you may be aware that there are vast areas adjacent to the coasts where the sea bottom at shallow levels is a white mud, which turns out to be made of almost pure aragonite. Carbonate muds extend well out to the deep sea as well; in fact, a fairly large proportion of the sea bottom is composed of calcium carbonate. There are also countless calcitic atolls and reefs throughout the tropical zones of the world. Given this amount of contact between the sea and calcium carbonate, both calcite and aragonite, plus the fact that there are vigorous oceanic currents stirring things up constantly, plus the fact that things have not changed drastically for millions of years, you would think that there would be little doubt that the system consisting of the oceans plus their bottom sediments must have reached equilibrium by now. If these were the only factors involved, perhaps they would have, but the situation is quite a bit more complicated. Why would anyone want to know? Reactions involving carbonate in the oceans are fundamental to an understanding of the global CO_2 cycle, which in turn is linked to climate change and other things we would like to understand.

How Do You Tell If a Solution Is Saturated?

To explore this problem further, we must first find out how to determine whether a solution is saturated, undersaturated, or supersaturated with a given mineral or compound. One answer would be to just observe the solution in contact with the mineral. If the mineral dissolves, the solution is undersaturated. If the mineral grows in size, the solution is supersaturated. If nothing happens, the solution is saturated – it is at equilibrium.

This method and variations of it are used, but it is very difficult for a number of reasons. We would like to be able to predict the state of saturation for a sample of water without performing difficult experiments on it. We would like simply to determine the chemical composition of the solution and calculate theoretically the state of saturation.

In other words, we want a thermodynamic answer. Having just spent several chapters developing a method for determining which way a reaction will go, we should be able to put it to use here. The reaction could be written

$$\text{solid mineral} = \text{dissolved mineral} \tag{10.1}$$

If this reaction goes to the right, the solution is undersaturated. If it goes to the left, the solution is supersaturated. All we need to do is to determine the molar Gibbs energy of the dissolved mineral and compare it with the molar Gibbs energy of the pure mineral, and the question is answered.

10.2.1   Solubility Products

But dissolution reactions that result in uncharged solutes such as $H_4SiO_4(aq)$ and $H_2CO_3(aq)$ are unusual. Most solutes are ionized to some extent, that is, they break up into charged particles, called ions. In other words, we write the dissolution reaction not as in (10.1), but as

$$\text{solid mineral} = \text{aqueous ions} \tag{10.2}$$

For example, calcium carbonate (calcite or aragonite), when it dissolves, breaks up into calcium and carbonate ions (Figure 10.1), written as

$$CaCO_3(s) = Ca^{2+} + CO_3^{2-} \tag{10.3}$$

The equilibrium constant for reaction (10.3) can be found in our routine way:

$$\Delta_r G^\circ = \Delta_f G^\circ_{Ca^{2+}} + \Delta_f G^\circ_{CO_3^{2-}} - \Delta_f G^\circ_{CaCO_3(s)}$$

$$= -553.58 + (-527.81) - (-1128.79)$$

$$= 47.40 \, \text{kJ} \, \text{mol}^{-1}$$

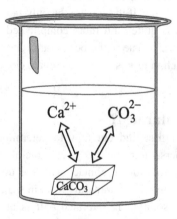

Figure 10.1 When calcite dissolves, the solute consists of electrically charged ions.

for calcite, or

$$\Delta_r G^\circ = -553.58 + (-527.81) - (-1127.75)$$

$$= 46.36 \, \text{kJ mol}^{-1}$$

for aragonite. This gives

$$\Delta_r G^\circ = -RT \ln K$$

$$47{,}400 = -(2.30259 \times 8.3145 \times 298.15)\log K$$

$$\log K = -8.304$$

for calcite, or $\log K = -8.122$ for aragonite.
   This equilibrium constant is

$$K = \frac{a_{Ca^{2+}} \, a_{CO_3^{2-}}}{a_{CaCO_3(s)}}$$

Now if, as in Figure 10.1, we are dealing with the solubility of pure calcium carbonate its activity is 1.0, so the equilibrium constant becomes

$$K_{sp} = a_{Ca^{2+}} \, a_{CO_3^{2-}}$$

and is called a *solubility product* constant, or just a solubility product. Whereas in the case of quartz solubility we found the equilibrium constant to be equal to the solubility itself (Section 9.4.1), here we find the equilibrium constant to be equal to a product of two ion activities. Therefore, "the solubility of calcite" has a somewhat ambiguous meaning. If it refers to the concentration (activity) of calcium in solution, this obviously depends on how much carbonate ion is in solution, and vice versa. It is the *combination* of calcium ion and carbonate ion activities that determines whether calcite is over- or undersaturated in a solution. For example, suppose we have determined that the activity of CO_3^{2-} in a solution (say the one in Figure 10.1) is 10^{-5}. A solution having $a_{Ca^{2+}} \, a_{CO_3^{2-}} = 10^{-8.304}$ will be in equilibrium with calcite, so in this case the equilibrium activity of the calcium ion is $a_{Ca^{2+}} = 10^{-3.304}$. In a solution with $a_{CO_3^{2-}} = 10^{-6}$, the equilibrium value of $a_{Ca^{2+}}$ is $10^{-2.304}$, and so on.
   Having now learned about the solubility product, please do not tack the sp subscript on to every equilibrium constant you calculate. There are equilibrium constants for many types of reactions. The solubility product is an equilibrium constant for a reaction having a solid mineral or compound on the left side and its constituent ions on the right.

10.2.2   IAP, K_{sp}, Ω, and SI

Of course natural solutions, such as seawater, are not necessarily at equilibrium. In Figure 10.2 we see a river carrying dissolved material, including calcium and carbonate ions, entering the sea. Carbonate ions are already there, because the sea is in contact with the atmosphere, which contains carbon dioxide, and when CO_2 dissolves it produces carbonate and bicarbonate ions. Because calcium and carbonate are being added, there may be a tendency for them to increase beyond the equilibrium value, and for calcite to

Table 10.1. Relations between IAP, K_{sp}, and SI.

IAP, K_{sp}	Ω	SI $\left(= \log\left(\frac{IAP}{K_{sp}}\right)\right)$	Result
IAP < K_{sp}	< 1	Negative	Mineral dissolves
IAP > K_{sp}	> 1	Positive	Mineral precipitates
IAP = K_{sp}	1	0	Equilibrium

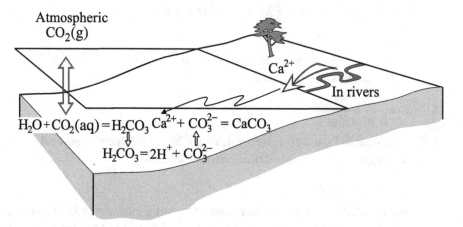

Figure 10.2 Calcite may precipitate in natural bodies of water, but the calcium and the carbonate may come from different sources.

precipitate as a result. The product of the calcium and carbonate ion activities *which are actually present in a solution*, regardless of any theory, is called the *ion activity product* (IAP) for that solution. It follows that when IAP > K_{sp}, calcite will precipitate, and when IAP < K_{sp} calcite will dissolve. The IAP/K_{sp} ratio is called Ω, and the logarithm of the ratio is called the *saturation index* (SI), so that when SI > 0 calcite precipitates, and when SI < 0 calcite dissolves (Table 10.1).

These relationships should be fairly intuitive, but if you have difficulty convincing yourself, consider the following equation:

$$\Delta_r\mu = \Delta_r\mu^\circ + RT\ln Q \tag{9.8}$$

In the case we are considering – calcite and its ions, reaction (10.3) – when $Q = K_{sp}$, $\Delta_r\mu = 0$, Therefore, if $Q = IAP > K_{sp}$, $\Delta_r\mu > 0$, and reaction (10.3) will go to the left (calcite precipitates). And if $Q = IAP < K_{sp}$, $\Delta_r\mu < 0$, reaction (10.3) will go to the right (calcite dissolves).

10.3   Determining the IAP – Speciation

Now to answer our question as to whether the sea is saturated with calcite, we need only determine its IAP and compare it with K_{sp} for calcite. This is easier said than done. Apart from some specialized electrochemical techniques, it is generally not possible to analyze a solution for the concentration of a specific ion such as Ca^{2+} or CO_3^{2-}. Analyses are made for the *total* calcium, or the *total* carbonate in the solution. There are many kinds of ions and uncharged molecular species such as ion-pairs in a solution, so the total calcium concentration would consist of the sum of the concentrations of all the ionic and molecular species containing Ca, and similarly for carbonate. Thus

$$m_{Ca,total} = m_{Ca^{2+}} + m_{CaHCO_3^+} + m_{CaSO_4^\circ} + \cdots$$

$$m_{CO_3,total} = m_{CO_3^{2-}} + m_{HCO_3^-} + m_{H_2CO_3} + \cdots$$

Here, $CaSO_4^\circ$ represents an electrically neutral species resulting from the joining of a Ca^{2+} ion and a SO_4^{2-} ion in solution. There are several more species in each of these summations in real seawater which we won't mention. But, however many there are, the concentrations and activities of all of them can be calculated if for every species there is a known equilibrium constant relating it to other species and/or minerals, and if there is a suitable equation for calculating the activity coefficients (γ_H) of each of the species.

What it amounts to is the well-known fact that if in a set of equations you have the same number of equations as you have variables, it is possible to solve for every variable. Given the total concentrations of the various constituents of seawater (not only calcium and carbonate, but however many you are interested in), plus an equilibrium constant for each, plus the knowledge that the total number of positive charges must equal the total number of negative charges, it is always possible to achieve this goal. When more than a few ions are involved, the procedure can be carried out only on a computer, and many programs are now available for doing this.

Minimizing the Gibbs Energy

It is also worth pointing out that the process of speciation is mathematically equivalent to finding the system configuration which minimizes the Gibbs energy. Speciation determines the *equilibrium* concentrations and activities of species in all possible chemical reactions in the solution. All possible reactions have achieved a minimum Gibbs energy, or G for every reaction is less than than it would be in any other molecular configuration. When all possible reactions achieve their minimum Gibbs energy, then so does the system; **G** has its minimum possible value. Examples are Figure 13.3 for the reaction $N_2O_4 = 2NO_2$ and Figure 1.3 in the van 't Hoff Equilibrium Box chapter in the online resources.

10.3.1   Speciation Using a Calculator

Let's start with a simple problem we can do "by hand." What are the species activities in a dilute solution of acetic acid? To calculate these activities, you need to know

- what ions there are in solution, and
- the equilibrium constants for the formation of these ions, at the temperature and pressure you are considering.

This is a simple case, in which there are only four ions: H^+, OH^-, HAc, and Ac^-, where we let HAc stand for $CH_3COOH(aq)$ and Ac^- for CH_3COO^-. There are two equilibrium constants,

$$a_{H^+} a_{OH^-} = a$$

$$\frac{a_{H^+} a_{Ac^-}}{a_{HAc}} = b$$

one charge balance,

$$m_{H^+} = m_{OH^-} + m_{Ac^-}$$

and one mass balance,

$$m_{HAc} + m_{Ac^-} = c$$

The equilibrium constants require activities, but we start with the assumption that all activity coefficients are 1.0. With four equations and four unknowns, a solution is possible. Some manipulation of the equations then results in

$$m_{H^+}^3 + b\, m_{H^+}^2 - (a + bc)\, m_{H^+} + ab = 0 \qquad (10.4)$$

This is what the equation looks like with all activity coefficients equal to 1.0. The real version includes several activity coefficients, which of course are generally non-zero. This cubic equation has three roots, but only one is reasonable. With $a = 10^{-14}$, $b = 10^{-4.76}$, and $c = 0.1$, the results are

$$m_{H^+} = 10^{-2.88}$$

$$m_{OH^-} = 10^{-11.12}$$

$$m_{HAc} = 0.0987$$

$$m_{Ac^-} = 0.0013$$

To include activity coefficients, you need to calculate an ionic strength from the results of this calculation, calculate an activity coefficient for each ion from an appropriate expression, solve the exact form of Equation (10.4) again, and iterate until all the answers don't change. You can readily imagine that with a few more species this procedure would become impossible without a computer. In fact with more species there is no closed-form, exact solution like Equation 10.4, and the equations must be solved by an iterative method. This procedure is called *speciation* – calculating the activity and concentration of every known ion and complex or ion-pair in a solution of a given bulk composition and at a given T and P. It is basically a problem in numerical analysis, and has not much to do with thermodynamics.

10.3.2   Speciation Using a Computer

There are now hundreds of programs designed to compute such species in complex, multicomponent systems. Thorough reviews of methods and programs include Van Zeggeren and Storey (1970) and Smith and Missen (1982).

Program PHREEQC

A good program which is freely available at the USGS web site is PHREEQC. This program is capable of many modeling operations besides speciation, but let's see what it does with the same problem. To tell the program what to do, you prepare an input file (in PHREEQC for Windows this can be done in one of the program windows):

```
DATABASE        llnl.dat
TITLE           Acetic acid 0.1 molal
SOLUTION 1
        units           mol/kgw
        temp            25
        pH              7       charge
        Acetate         0.1
END
```

The command charge tells the program to change the concentration of H^+ until the charges balance. After nine iterations, an output file is produced which includes these lines:

Species	Molality	Activity	Log Molality	Log Activity	Log Gamma
H+	1.344e-003	1.292e-003	-2.872	-2.889	-0.017
OH-	7.756e-012	7.442e-012	-11.110	-11.128	-0.018
H2O	5.553e+001	9.983e-001	-0.001	-0.001	0.000
Acetate	1.000e-001				
HAcetate	9.866e-002	9.866e-002	-1.006	-1.006	0.000
Acetate-	1.344e-003	1.290e-003	-2.872	-2.889	-0.018
H(0)	2.650e-017				
H2	1.325e-017	1.325e-017	-16.878	-16.878	0.000
O(0)	0.000e+000				
O2	0.000e+000	0.000e+000	-58.442	-58.442	0.000

The program has found these results:

$$m_{H^+} = 10^{-2.872}$$

$$m_{OH^-} = 10^{-11.110}$$

$$m_{HAc} = 0.09866$$

$$m_{Ac^-} = 0.001344$$

These are very close to the results from Equation (10.4) because in this case the activity coefficients are in fact close to 1.0. PHREEQC finds them to be

$$\gamma_{H^+} = 0.962$$

$$\gamma_{OH^-} = 0.959$$

$$\gamma_{HAc} = 1.0 \quad \text{(the value assumed for all uncharged species)}$$

$$\gamma_{Ac^-} = 0.959$$

10.3.3   The Species Activity Diagram for Carbonate Species

Having determined the activities and concentrations of all (or most) of the species in a solution, it is helpful to be able to see how these species activities vary as a function of other important variables like pH or $\log f_{O_2}$. These are species activity or species distribution diagrams. One of the more informative of these is the species–pH diagram for aqueous CO_2. There are three carbonate species, H_2CO_3, HCO_3^-, and CO_3^{2-}, which would result from dissolving carbon dioxide gas in water.[1] We have already calculated the equilibrium constant for one of the relevant ionic equilibria (Section 9.3.1):

$$H_2CO_3(aq) = HCO_3^- + H^+; \quad K = 10^{-6.37} \tag{10.5}$$

Another is

$$HCO_3^- = CO_3^{2-} + H^+; \quad K = 10^{-10.33} \tag{10.6}$$

And let's say that our total carbonate is $0.10\,m$, so that

$$m_{CO_3,total} = m_{CO_3^{2-}} + m_{HCO_3^-} + m_{H_2CO_3} = 0.10\,m \tag{10.7}$$

These three equations contain four variables, $m_{CO_3^{2-}}$, $m_{HCO_3^-}$, $m_{H_2CO_3}$, and m_{H^+}, so we need another equation, which is the charge balance,

$$m_{H^+} = m_{HCO_3^-} + 2\,m_{CO_3^{2-}} \tag{10.8}$$

An additional four unknowns are the activity coefficients of each of the four species for which we could write four more equations, making a total of eight equations and eight unknowns. This is a routine speciation problem. Program PHREEQC gives

$$a_{H^+} = 10^{-3.683}$$

$$a_{H_2CO_3} = 0.0998$$

$$a_{HCO_3^-} = 2.072 \times 10^{-4}$$

$$a_{CO_3^{2-}} = 4.435 \times 10^{-11}$$

Much more interesting than this single example, however, would be to see how the activities or concentrations of H_2CO_3, HCO_3^-, and CO_3^{2-} vary as a function of pH. Natural solutions contain many components in addition to CO_2 and water, so the pH can be quite different from the one we have just calculated. To do this, we simply choose specific pH values from 0 to 14 and solve for the activities of the three carbonate species. The result is shown in Figure 10.3. This diagram makes it easy to see which species is dominant (has the largest concentration) at any given pH. For example, in seawater, with a pH of about 8.1, carbonate is present almost entirely as the bicarbonate species.

An interesting feature of this diagram is the fact that the intersection of the lines representing $a_{H_2CO_3}$ and $a_{HCO_3^-}$ occurs at a pH of 6.37, which is the pK value of the first

---

[1] We use H_2CO_3 in the conventional sense discussed in Section 9.8.1.

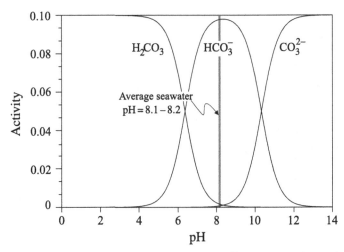

Figure 10.3 Activities of H_2CO_3(aq), and bicarbonate and carbonate ions as a function of pH, for a total concentration of $0.10\,m$.

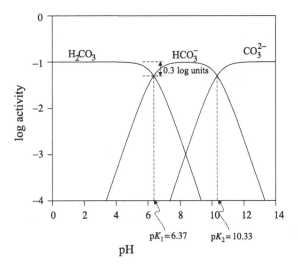

Figure 10.4 Same as Figure 10.3, but with a log activity scale. The pH values of the crossover points are the pK values of carbonic acid ionization.

ionization constant of H_2CO_3, and similarly for the intersection of the HCO_3^- and CO_3^{2-} lines, as shown in Figure 10.4. The reason for this is easy to see when you look at the equilibrium constant expressions, for example

$$\frac{a_{HCO_3^-}\, a_{H^+}}{a_{H_2CO_3}} = 10^{-6.37}$$

At the crossover or intersection point, $a_{HCO_3^-} = a_{H_2CO_3}$, so that

$$a_{H^+} = 10^{-6.37}$$

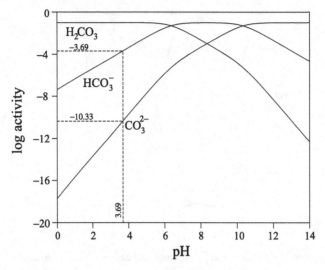

Figure 10.5 Same as Figure 10.4, but with an expanded activity axis. At a pH of 3.69, the HCO_3^- and CO_3^{2-} activities are easily calculated.

which means that the intersection of the lines representing $a_{H_2CO_3}$ and $a_{HCO_3^-}$ occurs at a pH of 6.37.

Another interesting fact is that the crossover points occur at an activity of 0.05 (Figure 10.3), because, at each of the two crossover points, the activity of the third species (the one not involved in the crossover) is negligibly small, so that the two "crossing species" make up virtually the total activity or concentration, and therefore they both have a value of one-half the total concentration. Because $\log \frac{1}{2} = -0.30$, this means that on the log scale the crossovers occur 0.3 log units below the plateau representing the total concentration, as shown in Figure 10.4. A third interesting fact is that the slopes of the lines representing the activities of the species are either $+1$ or -1, just below their intersections (although they may change to $+2$ and -2 farther down, as shown in Figure 10.5). The combination of these three properties of these diagrams makes it very easy to rapidly sketch such a diagram, given some pK values.

Species distribution diagrams are quite useful in seeing (and remembering) the relationships between species in dissociation reactions. For example, in Figure 10.5 you can see that $a_{CO_3^{2-}}$ goes to some very low values in acid solutions, a fact that explains (when you look at the solubility product) why carbonate minerals have such high solubilities in acid solutions. You also see that the activities of dissolved species never go to zero, at least in the model. In reality, of course, very low activities may mean that that species does not exist in the system.

As another example, Figure 10.6 shows the distribution of phosphate species at the same total concentration of $0.10\,m$. Such diagrams are clearly useful for any solute that can exist in a variety of species, differing only by the number of hydrogen ions (protons) they have.

Table 10.2. Properties of the major ions in near-surface seawater.

Major ions	Concentration (m)	Amount occurring as free ions (%)	Activity coefficient γ_H	Activity a
Na^+	0.475	99	0.76	0.357
Mg^{2+}	0.054	87	0.36	0.017
Ca^{2+}	0.010	91	0.28	0.0025
K^+	0.010	99	0.64	0.0063
Cl^-	0.56	100	0.64	0.36
SO_4^{2-}	0.028	54	0.12	0.0018
HCO_3^-	0.0024	69	0.68	0.0011
CO_3^{2-}	0.0003	9	0.20	0.0000054

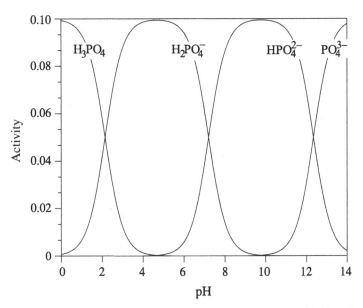

Figure 10.6 The activity distribution diagram for phosphoric acid. The pK values are 2.15, 7.21, and 12.34.

10.4 Combining the IAP and the K_{sp}

So *speciation* allows us to determine the concentrations and activities of all known species in any chemically analyzed solution. Applied to seawater, it allows us to determine the IAP of calcite and of many other minerals. This is now done routinely on oceanographic research vessels, and a certain amount of variability is found in the composition of seawater from various locations. However, the composition of *average* seawater is quite well known and is not greatly different from that proposed in the classic work of Garrels and Thompson (1962), who first applied this method, from which we get the data in Table 10.2.

Table 10.3. IAP, Ω, and SI for calcium carbonate in average seawater.

	IAP	K_{sp}	Ω	SI
Seawater	13.5×10^{-9}			
Calcite		4.97×10^{-9}	2.72	0.43
Aragonite		7.55×10^{-9}	1.79	0.25

From these data, we find

$$\text{IAP} = a_{Ca^{2+}} \, a_{CO_3^{2-}}$$

$$= 0.0025 \times 0.0000054$$

$$= 1.35 \times 10^{-8}$$

$$= 10^{-7.870}$$

and we found (Section 10.2.1) that for calcite $K_{sp} = 10^{-8.304}$, and for aragonite $K_{sp} = 10^{-8.122}$. Therefore, average seawater is slightly supersaturated with calcium carbonate (Table 10.3).

This is a reasonably interesting result, as far as it goes. It means that marine organisms should have no difficulty in precipitating their carbonate shells, and once precipitated, they should not redissolve. This is consistent with the vast amounts of aragonitic mud on the Florida and Bahamas coastlines. These muds are made up almost exclusively of the shells of microscopic marine organisms, which sink to the bottom when the organisms die and do not redissolve.

10.4.1 What Part of the Sea Is Saturated with $CaCO_3$?

So far, thermodynamics and observations fit together fairly well. However, we know from oceanographic surveys that although vast areas of the sea are underlain by these carbonate muds, especially in water depths less than 3 to 4 km, the deepest parts of the ocean basins (4 to 6 km depth) have little or no carbonate in the bottom muds. Down to a variable depth, but usually between 3 and 4 km, the bottom muds are close to 100% calcium carbonate. Then, within a relatively short increase in depth, the percentage of carbonate in the muds drops off rapidly, becoming zero or close to zero at another variable depth, but usually 4 to 5 km. The depth at which the rapid increase in carbonate dissolution begins is called the Lysocline, and the depth below which there is little or no carbonate is called the carbonate compensation depth (CCD). Carbonate-secreting organisms are active at the surface virtually everywhere, and their carbonate shells are settling down through the water column everywhere, not just in shallow water. But while they coat the ocean floor at shallow depths, they never reach depths greater than 5 km or so – they dissolve completely at these depths. So obviously the oceans are saturated with $CaCO_3$ at and near the surface but undersaturated at great depths.

The exact explanation for this is one of the many continuing problems of chemical oceanography, but from our point of view it illustrates two things.

- The oceans, like most natural systems, are not at chemical, thermal, or mechanical equilibrium, but in spite of this, our equilibrium thermodynamic model is quite useful if we know how to apply it. We have shown how it is useful at the ocean surface; it is also useful at great depth. But obviously it would not be useful to apply it to the oceans as a whole system – they are too far from equilibrium.

- The explanation for the CCD and its variations is complex, involving the kinetics of carbonate dissolution, variations in ocean chemistry, temperature, and pressure, worldwide circulation patterns, and other factors. Equilibrium chemical thermodynamics does not suffice for an understanding of this natural system, but it is invariably the starting point for all other types of investigation. You must have an understanding of the equilibrium state before you can understand the departures from this state.

10.5   Mineral Stability Diagrams

The problem of calcite saturation in the sea is only one of a large number of problems in oceanography, geology, soil chemistry, and many other areas of science that involve solid $\rightleftharpoons$ fluid reactions, and a variety of diagrams have been used to illustrate the thermodynamic relationships involved. Humans find two-dimensional diagrams much easier to understand than multidimensional systems of equations.

10.5.1   The Reaction of Feldspar with Water

One of the most common minerals on Earth is feldspar, and its reaction with water to form other minerals such as clays is of great interest in several fields. In soils, feldspars (solid solutions of $NaAlSi_3O_8$, $KAlSi_3O_8$, and $CaAl_2Si_2O_8$) react to form clay minerals, helping to control soil acidity. In petroleum reservoirs, the same reaction forming clay minerals can have serious effects on the rock permeability and oil recovery. K-feldspars often have appreciable amounts of Pb substituting for K in their structures, and reaction of K-feldspar with formation fluids is thought to release this Pb to the fluid, which may then go on to form a lead ore-body elsewhere on its travels through the Earth's crust. During metamorphism, when rocks containing feldspars are subjected to high temperatures and pressures deep in the crust, feldspars participate in a variety of reactions with fluids and with other minerals, all of which are of interest to geologists studying the history of the Earth.

The question for us is, how do we apply thermodynamics to these reactions? The first thing we must do is write a reaction that seems interesting. This is one of the most difficult steps, but one that is rarely discussed because there are no rules to guide us. There is a very large number of reactions that could be written involving K-feldspar, depending on what other things are in the system, but only a few of these are useful.

Only experience and scientific insight can distinguish between what might turn out to be useful and what will not. The reactions that appear in texts have, of course, proven to be useful.

If you were to write a reaction between K-feldspar and kaolinite with no previous experience or prejudices, but just putting them on opposite sides of an equal sign and adding other compounds to balance, you would likely wind up with something like this:

$$KAlSi_3O_8(s) + \tfrac{3}{2}H_2O(l) = \tfrac{1}{2}Al_2Si_2O_5(OH)_4(s) + 2\,SiO_2 + KOH \qquad (10.9)$$

We have not specified whether SiO_2 is quartz or $SiO_2(aq)$ – you have your choice. If you choose quartz, and quartz is present in the system, its activity will be 1.0; if you choose $SiO_2(aq)$, its activity will be (a dimensionless number equal to) the molality of silica in solution. Pure KOH is a solid, but it ionizes completely in solution, and you again have a choice of the (s) form or the (aq) form of data. If we choose to use quartz, KOH(aq), and the maximum microcline form of $KAlSi_3O_8$ (or K-feldspar), the equilibrium constant turns out to be $10^{-7.88}$, and, because all the products and reactants except KOH have $a = 1$, at least if the water is reasonably pure, this turns out to be the molality of KOH in our system. What system? We have not been very specific about what our system consists of, other than that it contains K-feldspar, kaolinite, and water, at 25 °C, 1 bar. We might be thinking about a soil with groundwater.

But, you say, groundwater contains organic solutes, CO_2, and lots of other things. Why are they not in my system? Well, you can put them in if you wish, in balanced reactions, but you are under no obligation to do so. Your system is a model system, a simplified version of the real thing, and what goes in is under your control. Here we are choosing to look at the predicted KOH concentration. It may be that we have calculated something that is not useful, but that is our fault, not the fault of thermodynamics. Or we may have overlooked some factor which will invalidate our result, but that remains to be seen. We will see examples of such pitfalls shortly.

Knowing the KOH concentration that would equilibrate with K-feldspar and kaolinite is interesting, as far as it goes, but it is not the best way to look at this system. With a little insight, we can see that by subtracting OH^- from each side, and changing from quartz to $SiO_2(aq)$, we get

$$KAlSi_3O_8(s) + \tfrac{1}{2}H_2O(l) + H^+ = \tfrac{1}{2}Al_2Si_2O_5(OH)_4(s) + 2\,SiO_2(aq) + K^+ \qquad (10.10)$$

In this reaction, we have three potentially measurable things – the silica concentration ($SiO_2(aq)$), the K^+ concentration, and the pH ($-\log a_{H^+}$). Given any two, we could predict the third, using the thermodynamic model. Or we could construct a three-dimensional diagram using these three parameters as axes. That's a bit too ambitious for us, so we will combine the a_{H^+} and a_{K^+} parameters into a ratio and plot this against $a_{SiO_2}(aq)$. This is now a standard procedure.

The way it is done is worth remembering, because it is used with various kinds of reactions. Write the equilibrium constant, take the logarithm of both sides, put your y-axis parameter on the left, the x-axis parameter on the right, and combine the rest of the terms. The equilibrium constant is

10.5 Mineral Stability Diagrams

$$K = \frac{a_{Al_2Si_2O_5(OH)_4}^{1/2} a_{SiO_2(aq)}^2 a_{K^+}}{a_{KAlSi_3O_8} a_{H_2O(l)}^{1/2} a_{H^+}} \tag{10.11}$$

If kaolinite, K-feldspar, and water are reasonably pure, their activities are 1, and

$$\log K = \log\left(\frac{a_{K^+}}{a_{H^+}}\right) + 2 \log a_{SiO_2(aq)}$$

or

$$\log\left(\frac{a_{K^+}}{a_{H^+}}\right) = -2 \log a_{SiO_2(aq)} + \log K \tag{10.12}$$

or

$$y = ax + b$$

where y is $\log(a_{K^+}/a_{H^+})$, x is $\log a_{SiO_2(aq)}$, $a = -2$, and $b = \log K$. Thus Equation (10.12) is the equation of a straight line having a slope of -2, if we plot $\log(a_{K^+}/a_{H^+})$ against $\log a_{SiO_2(aq)}$. To get K, we apply our routine method,

$$\Delta_r G° = \tfrac{1}{2} \Delta_f G°_{Al_2Si_2O_5(OH)_4}^{kaolinite} + 2 \Delta_f G°_{SiO_2(aq)} + \Delta_f G°_{K^+}$$
$$- \Delta_f G°_{KAlSi_3O_8}^{microcline} - \tfrac{1}{2} \Delta_f G°_{H_2O(l)} - \Delta_f G°_{H^+}$$

Getting numbers from the tables,

$$\Delta_r G° = \tfrac{1}{2}(-3799.7) + 2(-833.411) + (-283.27)$$
$$- (-3742.9) - \tfrac{1}{2}(-237.129)$$
$$= 11.523 \, kJ \, mol^{-1}$$
$$= 11,523 \, J \, mol^{-1}$$

Then

$$\Delta_r G° = -RT \ln K$$
$$11,523 = -(8.3145 \times 298.15)\ln K_{298}$$

so

$$\log K_{298} = -11,523/(2.30259 \times 8.3145 \times 298.15)$$
$$= -2.019$$

Therefore,

$$\log\left(\frac{a_{K^+}}{a_{H^+}}\right) = -2 \log a_{SiO_2(aq)} - 2.019 \tag{10.13}$$

This is a line on a plot of $\log(a_{K^+}/a_{H^+})$ vs. $\log a_{SiO_2(aq)}$ having a slope of -2 and a y-intercept (the value of y when $x = 0$) of -2.019, as shown in Figure 10.7. The line we have just calculated is the thick line.

We now have a line on a graph. What does it mean? The meaning is implicit in the methods we used to get the line. We put the activities of the minerals and water equal

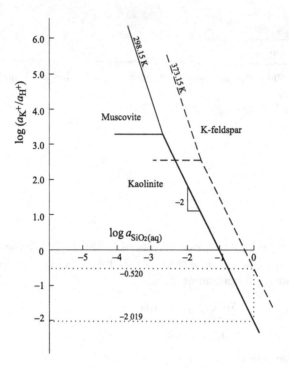

Figure 10.7 A plot of $\log(a_{K^+}/a_{H^+})$ vs. $\log a_{SiO_2(aq)}$ at 298.15 K and 373.15 K.

to 1, and we used the equilibrium constant. That means that the line is the locus of solution conditions (a_{K^+}/a_{H^+}) and $a_{SiO_2(aq)}$ for which the pure minerals and water are in equilibrium with each other. For any values of (a_{K^+}/a_{H^+}) and $a_{SiO_2(aq)}$ that do not lie on the line, our solution cannot be in equilibrium with both minerals, although it might be in equilibrium with one or the other. Applying Le Chatelier's principle to reaction (10.10), we see that increasing $a_{SiO_2(aq)}$ or increasing (a_{K^+}/a_{H^+}) favors the formation of K-feldspar, so a field of K-feldspar lies to the right and above our line, and a field of kaolinite lies to the left and below.

The next problem is that, having chosen a fairly complex system like this, there are more possible reactions than the one we have chosen. K-feldspar can react to form not only kaolinite, but also muscovite, and kaolinite can also react to form muscovite. These reactions are

$$\tfrac{3}{2}KAlSi_3O_8(s) + H^+ = \tfrac{1}{2}KAl_3Si_3O_{10}(OH)_2(s) + 3\,SiO_2(aq) + K^+ \tag{10.14}$$

and

$$KAl_3Si_3O_{10}(OH)_2(s) + \tfrac{3}{2}H_2O + H^+ = \tfrac{3}{2}Al_2Si_2O_5(OH)_4(s) + K^+ \tag{10.15}$$

Using the same methods as before, we find that reaction (10.14) has a slope of -3 on our graph, and an intercept ($\log K$) of -4.668, and reaction (10.15) has a slope of 0 (it is independent of a_{SiO_2}) and an intercept of 3.281. It can be shown (using the phase rule) that these three lines must intersect at a point, so another way to draw them is to calculate the point of intersection, which is $\log a_{SiO_2(aq)} = -2.650$, $\log(a_{K^+}/a_{H^+}) = 3.281$, and draw lines with slopes 0, -2, and -3 through this point. We now have a kind of phase diagram, showing which minerals are stable, not as a function of T and P, but of the composition of a solution in equilibrium with the minerals.

Effect of Temperature

There are other mineral phases to be added to our diagram, but first let's look at the effect of temperature. If we want the same diagram for a temperature of $100\,^{\circ}$C, we must calculate K at this temperature. To do this, we can use (9.19) or (9.21). We'll use both in the following calculations. For reaction (10.10), $\Delta_r H^{\circ} = 43{,}437\,\mathrm{J\,mol}^{-1}$ and $\Delta_r S^{\circ} = 106.449\,\mathrm{J\,mol}^{-1}/K$. Therefore

$$\Delta_r G^{\circ}_{373} = \Delta_r H^{\circ}_{298} - T\,\Delta_r S^{\circ}_{298}$$

$$= 43{,}437 - 373.15 \times 106.449$$

$$= 3715.556\,\mathrm{J\,mol}^{-1}$$

and

$$\log K = -3715.556/(2.30259 \times 8.31451 \times 373.15)$$

$$= -0.520$$

$$= \log\left(\frac{a_{K^+}}{a_{H^+}}\right) + 2\log a_{SiO_2(aq)}$$

giving

$$\log\left(\frac{a_{K^+}}{a_{H^+}}\right) = -2\log a_{SiO_2(aq)} - 0.520$$

as the equation of the K-feldspar–kaolinite boundary at $100\,^{\circ}$C.

For reaction (10.14), $\Delta_r H^{\circ} = 74{,}473\,\mathrm{J\,mol}^{-1}$ and $\log K_{298} = -4.668$, and so

$$\log K_{373} = \log K_{298} - \frac{\Delta_r H^{\circ}_{298}}{2.30259\,R}\left(\frac{1}{T} - \frac{1}{298.15}\right)$$

$$= -4.668 - \frac{74473}{2.30259 \times 8.31451}\left(\frac{1}{373.15} - \frac{1}{298.15}\right)$$

$$= -2.046$$

For reaction (10.15), $\Delta_r H^{\circ} = -18{,}635\,\mathrm{J\,mol}^{-1}$ and $\Delta_r S^{\circ} = -1.165\,\mathrm{J\,mol}^{-1}/K$, and so

$$\log K_{373} = \frac{-\Delta_r H^{\circ}_{298}}{2.30259\,RT} + \frac{\Delta_r S^{\circ}_{298}}{2.30259\,R}$$

$$= \frac{18{,}635}{2.30259 \times 8.31451 \times 373.15} + \frac{-1.165}{2.30259 \times 8.31451}$$

$$= 2.548$$

These three lines intersect at $\log(a_{K^+}/a_{H^+}) = 2.55$, $\log a_{SiO_2(aq)} = -1.53$, as shown in Figure 10.7.

We can add two final reactions to our diagram. These relate kaolinite and muscovite to gibbsite:

$$Al_2Si_2O_5(OH)_4(s) + H_2O(l) = 2\,Al(OH)_3(s) + 2\,SiO_2(aq) \tag{10.16}$$

and

$$KAl_3Si_3O_{10}(OH)_2(s) + 3\,H_2O(l) + H^+ = 3\,Al(OH)_3(s) + 3\,SiO_2(aq) + K^+ \tag{10.17}$$

These two reactions also meet at a point with the muscovite–kaolinite reaction, as shown in Figure 10.8.

The Problem of Metastable Phases

One of the pitfalls in choosing data and drawing diagrams, such as we have done, is that we may not have considered all the possible reactions in our system. That is, there may be phases that are more stable than the ones we have chosen – we may have chosen metastable phases. This is illustrated in our system by the fact that there is another aluminosilicate phase, pyrophyllite, which has more silica in it than does kaolinite and so is stable at higher values of $a_{SiO_2(aq)}$. Considering the reactions

$$KAlSi_3O_8(s) + H^+ = \tfrac{1}{2}\,Al_2Si_4O_{10}(OH)_2(s) + SiO_2(aq) + K^+ \tag{10.18}$$

and

$$\tfrac{1}{2}Al_2Si_4O_{10}(OH)_2 + \tfrac{1}{2}H_2O(l) = \tfrac{1}{2}Al_2Si_2O_5(OH)_4(s) + SiO_2(aq) \tag{10.19}$$

results in the two boundaries shown in Figure 10.8, which completely enclose the K-feldspar–kaolinite boundary. This means that at 100 °C, and according to data from Johnson et al. (1992), K-feldspar and kaolinite are not stable together in the presence of water. They should react to form pyrophyllite. In nature, of course, they may not. K-feldspar may well react directly to form kaolinite, in spite of thermodynamics, but it would be a metastable reaction; that is, a reaction involving metastable phases.

One other thing. The solubility of quartz, calculated by the method in Section 9.4.1, is $10^{-3.078}\,m$ at 100 °C. This is shown in Figure 10.8 as a vertical line. A solution in equilibrium with quartz at 100 °C must lie on this line. Solutions to the right of it are supersaturated, and solutions to the left are undersaturated, with quartz. Therefore, if your system contains quartz, as most soils do, then not only is the assemblage K-feldspar + kaolinite metastable with respect to pyrophyllite, but also pyrophyllite is itself metastable. According to our diagram, pyrophyllite can only exist in an aqueous solution if that solution is supersaturated with silica. If the silica were to precipitate as quartz, pyrophyllite should break down to form kaolinite, releasing silica to the solution, by reaction (10.19). Again, nature may not do this. It is quite common for natural solutions, especially those at or near the Earth's surface, to be supersaturated with silica, even in the presence of quartz, which is one reason why we use $\log a_{SiO_2(aq)}$ as a variable in our diagrams.

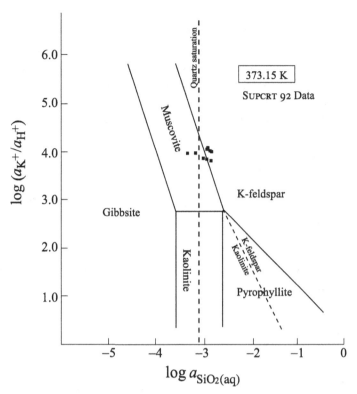

Figure 10.8 A plot of $\log(a_{K^+}/a_{H^+})$ vs. $\log a_{SiO_2(aq)}$ at 373.15 K, using data from Johnson *et al.* (1992), showing the stability field of pyrophyllite and the quartz saturation line. Each black square represents a brine from the Kettleman North Dome oil field (Merino, 1975).

10.6 Summary

In this chapter we have seen how a knowledge of equilibrium constants and activities can be used to construct diagrams that relate fluid compositions to the minerals in equilibrium with those fluids. Real fluid compositions (activities) can then be plotted on the diagrams to see how well the hypothesis of equilibrium holds, or, assuming equilibrium, to predict what phases are coexisting with fluids, and what will happen if we change the fluid composition. There are many other applications of equilibrium thermodynamics, to be sure, but if you understand equilibrium constants and activities thoroughly, you will have no problem in understanding most other applications.

In the case of the Kettleman North Dome fluids, the mineralogy of the rocks in contact with the fluids is known. You might say, well, if we know what minerals are there, why go to all this trouble to show that the fluid compositions reflect this? Would you not expect them to reflect the mineralogy of their host rocks? For one thing, we didn't know whether the composition of oil field brines or any other kind of natural fluid was controlled by host rocks or not, until this kind of test had been done. For all we knew, perhaps disequilibrium reigned supreme. Furthermore, we're still learning.

Why is it important? This question gets at the whole idea of using science to try to understand natural processes. Until you have a quantitative model that can simulate or account for natural data, such as shown in Figure 10.8, you cannot hope to change or control the situation to your benefit. For example, the petroleum geologists at Kettleman North Dome might want to inject something to change the fluid characteristics. Without a thermodynamic model, they would have no way of predicting what would happen. The equilibrium model has its limitations, but it is a good place to start.

Exercises

E10.1 Calculate the solubility product of gibbsite. What would be the trivalent aluminum concentration of a solution in equilibrium with gibbsite at a pH of 6.0?

E10.2 (a) What is an appropriate diagram to use to consider the relationship between water compositions and rocks containing nepheline, wollastonite, and grossularite? What is the slope of the grossularite $\rightarrow$ nepheline + wollastonite boundary on this diagram?

$$2NaAlSiO_4(s) + 2CaSiO_3(s) + Ca^{2+} = SiO_2(aq) + Ca_3Al_2Si_3O_{12}(s) + 2Na^+$$

$$(10.20)$$

Grossularite is $Ca_3Al_2Si_3O_{12}$, nepheline is $NaAlSiO_4$, and wollastonite is $CaSiO_3$.

(b) What is buffered by coexisting grossularite + nepheline + wollastonite?

E10.3 (a) Calculate the equilibrium constants for the following equations:

$$Al(OH)_3(gibbsite) + 3H^+ = Al^{3+} + 3H_2O$$

$$Al(OH)_3(gibbsite) + 2H^+ = Al(OH)^{2+} + 2H_2O$$

$$Al(OH)_3(gibbsite) + H^+ = Al(OH)_2^+ + H_2O$$

$$Al(OH)_3(gibbsite) = Al(OH)_3^\circ(aq)$$

$$Al(OH)_3(gibbsite) + H_2O = Al(OH)_4^- + H^+$$

(b) Derive an equation suitable for plotting each reaction on a graph having the log of the activity of dissolved Al species and pH as axes.

E10.4 Consider a hot-spring solution in contact with a rock containing microcline, muscovite, kaolinite, quartz, and other minerals (an altered granite) at 100 °C. You have no analysis of this water, but you have your $log(a_{K^+}/a_{H^+})$ vs. $log\,a_{SiO_2(aq)}$ diagram. What does the diagram tell you about the composition of the solution,

(a) if quartz does not equilibrate with the solution, and

(b) if quartz equilibrates with the solution?

(c) If this hot-spring solution had a pH of 5.0 and was observed or believed to be altering the microcline in the granite to kaolinite, what could you say about the activity and the concentration (in ppm) of K^+ in the solution? Remember that one of the minerals in all granites is quartz.

E10.5 Gibbsite is a principal mineral in bauxite, which is mined as a source of aluminum. Would you expect to find quartz in bauxite? Would you expect to find kaolinite? Why?

E10.6 (a) How would a change in redox conditions affect the reactions in Exercise E10.5?

(b) What would be the potassium ion activity in groundwater in a soil containing microcline, kaolinite, and quartz, at a pH of 5.5, assuming equilibrium? See Equation (10.10).

(c) If you measured the a_{K+} in such a groundwater and found a different result, what reason would you suspect?

E10.7 Calculate and plot the muscovite–gibbsite and kaolinite–gibbsite boundaries (Equations (10.16) and (10.17)) at 298.15 K and 373.15 K in Figure 10.8. You will know you have the right answers when these two curves meet the muscovite–kaolinite boundary at a single point.

E10.8 According to the data in Appendix B, diaspore is more stable than gibbsite in water. Use your knowledge of Gibbs energy relationships to *predict* how substituting diaspore for gibbsite in Figure 10.8 will affect the position of the boundaries with muscovite and kaolinite. Calculate the boundaries to confirm your prediction.

E10.9 Consider the reaction between kaolinite and pyrophyllite. Which one would be expected to occur in a rock containing quartz? What might be the explanation if all three minerals were found in the same rock?

E10.10 If a rock containing quartz and kaolinite was in contact with water, what would be the silica concentration in the water, assuming equilibrium? How would a change in redox conditions affect this result?

E10.11 If a rock containing kaolinite and pyrophyllite (but no quartz) was in contact with water, what would be the silica concentration of the water, assuming equilibrium with both minerals?

E10.12 What would be the silica concentration of the water in the previous question if the temperature was not 25 °C but 75 °C? You can do this three different ways. If you do this you will get three very slightly different answers. Why?

Additional Problems

A10.1 As a part of your research into the metabolic processes in starfish, it becomes of interest to know the dominant species of arsenic in seawater. Get an answer to this by first calculating the three ionization constants of arsenic acid, H_3AsO_4, and then constructing a species distribution diagram ($\log a$ vs. pH). Use a total As activity of 10^{-6}. Knowing that the pH of seawater is 8.1, use the diagram to determine the dominant form of As. What assumptions have you made? The species distribution diagram need only be a sketch (not to scale, with important parameters noted). Relevant data are given in Table 10.4.

Table 10.4. Data for Additional Problem A10.1.

Formula	Form	Molecular Weight	$\Delta_f H°$ (kJ mol^{-1})	$\Delta_f G°$ (kJ mol^{-1})	$S°$ (J mol^{-1} K^{-1})	$C_P°$ (J mol^{-1} K^{-1})
Arsenic						
H_3AsO_4	aq	141.9432	−902.5	−766.0	184.0	—
$H_2AsO_4^-$	aq	140.9352	−909.56	−753.17	117.0	—
$HAsO_4^{2-}$	aq	139.9272	−906.34	−714.60	−1.7	—
AsO_4^{3-}	aq	138.9192	−888.14	−648.41	−162.8	—
As	s	74.9216	0	0	35.1	24.64
As_2	g	149.8432	222.2	171.9	239.4	35.003
As_4	g	299.6864	143.9	92.4	314.0	—
As_4O_6	Octahedral	395.6828	−1313.94	−1152.43	214.2	—
As_4O_6	Monoclinic	395.6828	−1309.6	−1153.93	234.0	—
As_2O_5	s	229.8402	−924.87	−782.3	105.4	116.52
AsH_3	g	77.9456	66.44	68.93	222.78	38.07
As_2S_3	Orpiment	246.0352	−169.0	−168.6	163.6	116.3
Sulfur						
S	Orthorhombic	32.0640	0	0	31.80	22.64

(a) There are two crystalline forms of As_4O_6(s). Which is more stable at 25 °C, 1 bar? If you assume that the phase boundary between the two forms has a positive slope, what can you conclude about the relative densities (or relative values of $V°$) of the two forms? At what temperature are the two forms at equilibrium? Sketch a possible phase diagram.

(b) Calculate the fugacities (vapor pressures) of As_2(g) and As_4(g) in equilibrium with solid metallic arsenic. Calculate the equilibrium constant for the reaction

$$2\,As_2(g) = As_4(g)$$

and show that it is consistent with the two calculated fugacities.

(c) Calculate the entropy of formation ($\Delta_f S°$) of As_2S_3. Use this with $\Delta_f H°$ of As_2S_3 to calculate $\Delta_f G°_{As_2S_3}$, and compare your answer with the tabulated value.

(d) Will metallic arsenic oxidize in air? What values would f_{O_2} have to have to prevent such oxidation?

A10.2 Table 10.5 shows the $\log K$ values at 25 °C of some common sodium aluminum silicate reactions, in which $NaAlSiO_4$ is nepheline, $NaAlSi_3O_8$ is albite, and $NaAl_3Si_3O_{10}(OH)_2$ is paragonite.

Table 10.5. Some sodium aluminum silicate reactions.

Reaction	$\log K$
$NaAlSiO_4 + 2\,SiO_2(aq) = NaAlSi_3O_8$	11.0362
$\frac{1}{2}NaAl_3Si_3O_{10}(OH)_2 + 3\,H^+ = \frac{1}{2}NaAlSiO_4 + SiO_2(aq) + 2\,H_2O(l) + Al^{3+}$	2.1708
$\frac{1}{2}NaAl_3Si_3O_{10}(OH)_2 + 3\,H^+ = \frac{1}{2}NaAlSi_3O_8 + 2\,H_2O(l) + Al^{3+}$	7.6889

(a) Use these equilibrium constants to sketch or describe a diagram showing the stability fields of the three minerals as a function of $\log(a_{Al^{3+}}/a_{H^+}^3)$ and $\log a_{SiO_2(aq)}$. The boundaries of the three phase fields intersect at a point. Calculate the value of the two parameters at this point.

(b) After sketching this diagram, you become aware that there is yet another sodium aluminum silicate mineral, namely analcime ($NaAlSi_2O_6 \cdot H_2O$), which has been neglected. Without knowing its thermodynamic properties, sketch a revised diagram, showing how it might fit in. Remember that the formulae for nepheline, analcime, and albite can be written with the same amounts of Na and Al but increasing amounts of SiO_2. Note too that, for the purposes of this question, you can disregard the H_2O in the analcime formula.

(c) Calculate and plot the quartz solubility line and the amorphous silica solubility line.

A10.3 This problem has been adapted and simplified from a case study in Chapter 6 of Zhu and Anderson (2002).

Figure 10.9 shows an example of a typical acid mine drainage (AMD) setting, in which water draining through a pile of rock containing sulfide minerals becomes extremely acidic and iron-rich. This drainage water then enters an underlying aquifer and contaminates it for some distance downstream. Four monitoring wells are used to obtain samples of the groundwater. If this drainage water appeared on the surface it might look like the photo on the front cover.

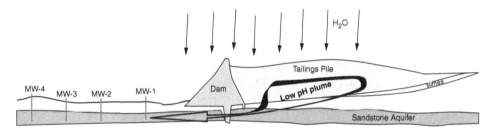

Figure 10.9 Cross-section through a mine tailings dump which is generating an acid plume in the underlying sandstone.

The aquifer is a sandstone, mostly quartz with a small percentage of other minerals, such as calcite ($CaCO_3$) and K-feldspar ($KAlSi_3O_8$). The acidic water tends to dissolve the calcite, and to alter the K-feldspar to kaolinite. It also precipitates ferric hydroxide, $Fe(OH)_3$, in the sandstone, as the iron oxidizes. No samples of the aquifer sandstone itself at this site have been examined, but it is suspected that the mineralogy of the aquifer changes in a systematic way downstream.

Analyses of solutions from the four wells are shown in Table 10.6. The compositions change slowly over space and time, but nevertheless are assumed to represent small volumes of local equilibrium, an assumption that must always be kept in mind.

Table 10.6. Speciation by program REACT (The Geochemist's Workbench, 2015) of water samples from the four monitoring wells. Units are molality. Sulfate has been adjusted for charge balance. The species are printed as they normally appear on computer output. The charge is indicated by the number of + or − signs; e.g., Fe(OH)2+ has a charge of +1. Adapted from a case study in Zhu and Anderson (2002).

Species	MW-1	MW-2	MW-3	MW-4
pH	3.8	4.5	6.5	6.7
Na+	0.1042	0.09159	0.02116	0.01994
SO4−	0.09304	0.09217	0.01644	0.01387
MgSO4	0.02076	0.02064	0.003045	0.00141
Mg++	0.02022	0.02038	0.006776	0.003224
Al(SO4)2−	0.01891	0.001718	1.403×10^{-8}	
Cl−	0.01522	0.01106	0.01024	0.00755
AlSO4+	0.01355	0.001266	3.178×10^{-8}	
NaSO4−	0.009255	0.008018	0.0006026	0.000522
Al+++	0.005244	0.0004949	2.308×10^{-8}	
CaSO4	0.003889	0.004748	0.005081	0.003967
Ca++	0.003754	0.004644	0.01002	0.007916
FeOH++	0.003411	0.002153	7.460×10^{-8}	
Fe(OH)2+	0.001742	0.004922	2.460×10^{-5}	6.390×10^{-6}
K+	0.001365	0.001594	0.000442	0.0002955
Mn++	0.0006491	0.0006886	4.223×10^{-6}	6.005×10^{-6}
MnSO4	0.0005561	0.0005624	1.711×10^{-6}	2.347×10^{-6}
SiO2(aq)	0.0005513	0.0004912	0.0001613	0.0001231
Fe+++	0.0003474	5.278×10^{-5}	6.000×10^{-6}	6.00×10^{-8}
HSO4−	0.0003293	6.077×10^{-5}	1.730×10^{-7}	
O2(aq)	0.0002736	0.0002899	0.0003161	0.0003309
KSO4−	0.0001691	0.0001963	1.831×10^{-5}	1.137×10^{-5}
MgCl+	0.0001607	0.0001246	5.416×10^{-5}	2.099×10^{-5}
CO3−	1.00×10^{-8}	1.00×10^{-7}	2.918×10^{-6}	2.350×10^{-6}

(a) Ionic strength is defined as

$$I = \frac{1}{2} \sum_i m_i z_i^2$$

where I is the ionic strength, and the $\sum_i$ term is the sum over all species in the solution of the product of the species molality m_i and the charge on that species squared, z_i^2. To avoid some tedious calculations, the ionic strength of the sample solutions is shown in Table 10.7.

Table 10.7. Ionic strength values of the AMD samples.

Sample	MW1	MW2	MW3	MW4
Ionic Strength I	0.3498	0.3028	0.0828	0.0642

(b) Calculate the activity coefficients of Ca^{2+}, Fe^{3+}, and CO_3^{2-}, using the Davies equation,

$$\log \gamma_{H_i} = \frac{-A z_i^2 \sqrt{I}}{1 + \sqrt{I}} + 0.2 A z_i^2 I$$

where $A = 0.5091$.

(c) Calculate the solubility product of ferric hydoxide, $Fe(OH)_3(s)$.[2]

(d) Calculate the saturation index of calcite, quartz, and of ferric hydroxide in each of the four wells.[3]

(e) Plot the saturation index values.

(f) Comment on each SI profile. Is it supersaturated? Is it undersaturated? It is changing because of ... ?

A10.4 Suppose you have dissolved some fluorite (CaF_2) in water at room temperature and pressure, so that the number of moles of F in solution is exactly twice the number of moles of Ca in solution, and the total fluorine molality (moles of F per kg water) is 10^{-4}. You have adjusted the pH to 7.0, using something that does not react in any way with calcium or fluorine. You want to know how close the solution is to saturation with fluorite.

To do this you need to know not the total amounts of Ca and F, but the activities of the Ca^{2+} and F^- ions. Then you need to know the solubility product of fluorite, and calculate the saturation index.

To calculate the activities of the ions, you need to know

- what ions there are in solution (besides Ca^{2+} and F^-), and
- the equilibrium constants for the formation of these ions, at the temperature and pressure you are considering.

The most important species in this case (not all the species known, just the most abundant ones) are CaF^+ and HF, and the reactions for which you need to know the equilibrium constants are

$$CaF^+ = Ca^{2+} + F^- \tag{10.21}$$

$$HF = H^+ + F^- \tag{10.22}$$

In addition you know that, because these are the only important ions,

$$m_{Ca^{2+}} + m_{CaF^+} = 5 \times 10^{-5}$$

$$m_{F^-} + m_{HF} = 10^{-4}$$

---

[2] Don't forget that because you know the pH , and you know that $a_{H^+} \cdot a_{OH^-} = 10^{-14}$, you can calculate the a_{OH^-} of each sample.

[3] Because quartz does not have a solubility product, its SI is simply the log of the $SiO_2(aq)$ molality of the fluid divided by the equilibrium molality of $SiO_2(aq)$ at the same T.

(a) Appendix B does not have data for species CaF^+, so you cannot calculate $\log K$ for reaction (10.21). But, from other sources, this $\log K$ is 0.6817 (Source A) and 0.94 (Source B). Calculate two values for the $\Delta_f G^\circ$ of CaF^+.

(b) Calculate the equilibrium constant at 25 °C for reaction (10.22) from data in Appendix B.

(c) Calculate the solubility product of fluorite from data in Appendix B. Source A says that $\log K_{sp} = -10.6$ and Source B says that $\log K_{sp} = -10.037$.

(d) Table 10.8 shows some results of speciating this solution. Calculate the IAP ($= a_{Ca^{2+}} \cdot a_{F^-}^2$) or $\log(IAP)$ from each source, assuming that the activity coefficients are 1.0.

Table 10.8. Species concentrations from three different sources. Activities are very slightly different.

Species	By hand	Source A	Source B
Ca^{2+}	4.9956×10^{-5}	4.998×10^{-5}	4.996×10^{-5}
F^-	9.9989×10^{-5}	9.996×10^{-5}	9.995×10^{-5}
HF	1.5133×10^{-8}	1.386×10^{-8}	1.382×10^{-8}
CaF^+	4.3504×10^{-8}	2.400×10^{-8}	4.111×10^{-8}

(e) Using these data, calculate three different values of the fluorite saturation index, $SI = \log(IAP/K_{sp})$.

If you feel ambitious, do the "hand calculation." You can also do the same problem in PHREEQC or PHREEQC for Windows, which is much easier. You can even compare results using other databases; several come with both PHREEQC and PHREEQC for Windows, and are the sources for the $\log K$ values given above. Source A is llnl.dat and Source B is phreeqc.dat. Generally speaking, every program and every choice of database will give a different answer, so, in a sense, no one knows the "right" answer.

11 Redox Reactions

11.1 Introduction

Normal seawater contains about 2660 ppm ($0.028\,m$) of sulfur in the form of sulfate (SO_4^{2-}). Sulfur in this form has a valence of $+6$, meaning that it has six fewer electrons per atom than has native sulfur, which exists (though not in the ocean) as a yellow crystalline solid. In some parts of the ocean, however, sulfur exists in the form of dissolved hydrogen sulfide, $H_2S(aq)$. Sulfur in this form has a valence of -2, meaning that it has two extra electrons compared with native sulfur, and in this form it is a deadly poison. Those parts of the ocean containing this electron-rich form of sulfur contain no living organisms other than a few kinds of bacteria. Obviously, the number of electrons that each sulfur atom has is not a question of interest only to atomic physicists. Changing sulfate-sulfur to H_2S-sulfur or vice versa involves transferring electrons from one to the other, and this electron transfer is the basic element of redox (reduction–oxidation) reactions.

Many naturally occurring elements in addition to sulfur show similar variations in their number of electrons, with similarly large differences in their chemical properties. It would be difficult to overemphasize the importance to us of these variations in valence, or numbers of electrons per atom. Biochemistry, for example, is in large part a study of redox reactions. Because natural environments show great variability in their redox state, we need to develop some kind of measurement, an index, which will be useful in characterizing these redox states, much as we use pH as a measurement or index to characterize the acidity of various states, or temperature as a measurement or an index of the hotness of states. In this chapter we develop two such indexes of redox state.

11.2 Electron Transfer Reactions

You may not have noticed it, but we have considered two kinds of reactions in previous chapters. In some, such as (9.3),

$$SiO_2(s) + 2H_2O = H_4SiO_4(aq) \tag{9.3}$$

all elements on the right side have the same number of electrons that they have on the left side – there is no change in valence of any element. In others, such as (9.16),

$$CH_4(g) + O_2(g) = CO_2(g) + 2H_2(g) \tag{9.16}$$

there is such a change. For example, the carbon in CH_4 is C^{4-}, and the carbon in CO_2 is C^{4+}. Each carbon atom in methane that changes to a carbon atom in carbon dioxide must get rid of eight electrons – it is *oxidized*. Where do the electrons go? Obviously, they go to the other actors in the reaction. Oxygen in O_2 has a valence of zero (O^0), while in carbon dioxide it is -2 (O^{2-}), so, in changing from O_2 to CO_2, two oxygens gain four electrons. The other four electrons go to hydrogen, which has a valence of $+1$ (H^+) in methane and zero in hydrogen gas. Both oxygen and hydrogen are *reduced*, if the reaction goes from left to right as written. Similarly in reaction (9.15),

$$6\,Fe_2O_3(s) = 4\,Fe_3O_4(s) + O_2(g) \tag{9.15}$$

we see that all of the iron atoms in Fe_2O_3 are ferric iron (Fe^{3+}), while one out of three iron atoms in Fe_3O_4 is ferrous iron (Fe^{2+}). The iron is partially reduced, while some oxygen in Fe_2O_3 is oxidized to $O_2(g)$ – there is a transfer of electrons from iron to oxygen, or from oxygen to iron, depending on which way the reaction goes. Without such electron transfers, these and many other reactions, including many necessary to life processes, could not proceed.

11.3   The Role of Oxygen

Both of our examples involve oxygen, which is the most common *oxidizing agent* in natural systems. In the presence of oxygen, many elements are oxidized (lose electrons, gain in valence), while oxygen is reduced. You need only think of rusty nails, green staining on copper objects, and burning logs to realize the truth of this. The process of oxidation obviously takes its name from the fact that oxygen is the premier oxidizing agent, but it is actually defined in terms of electron loss, or increase in valence. In other words, the electrons need not come from or go to oxygen; many redox reactions take place without oxygen.

Consider, for example, what happens when you put a piece of iron in a solution of copper sulfate (Figure 11.1). After a while you see the characteristic color of metallic

Figure 11.1 An iron nail in a solution of copper sulfate.

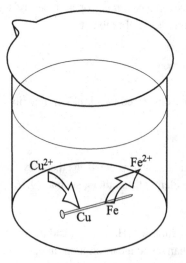

copper forming on the surface of the iron, and the iron gradually crumbles and eventually disappears. Metallic copper precipitates, and iron dissolves. The reaction is essentially

$$Cu^{2+} + Fe \rightarrow Cu + Fe^{2+} \tag{11.1}$$

We need not include the sulfate, because it is not involved in this process – being negatively charged, the SO_4^{2-} ions provide an overall charge balance in the solution. In this example, copper is reduced and iron is oxidized, without the aid of oxygen. (Of course, if we wait long enough, and our solution is open to the atmosphere, oxygen dissolved in the solution will eventually oxidize the copper and the ferrous iron.)

What does thermodynamics tell us about this reaction? Looking up the data from Appendix B, we find

$$\Delta_r G^\circ = \Delta_f G^\circ_{Cu} + \Delta_f G^\circ_{Fe^{2+}} - \Delta_f G^\circ_{Cu^{2+}} - \Delta_f G^\circ_{Fe}$$

$$= 0 + (-78.90) - 65.49 - 0$$

$$= -144.390 \, kJ \, mol^{-1}$$

$$= -144,390 \, J \, mol^{-1}$$

This means that with both metallic copper and iron present, and cupric and ferrous ions present at 1 molal concentration (and acting ideally), the reaction would proceed spontaneously, as observed. But more interestingly,

$$\Delta_r G^\circ = -RT \ln K$$

$$-144,390 = -(8.3145 \times 298.15)\ln K$$

and

$$\log K = 144,390/(2.30259 \times 8.3145 \times 298.15)$$

$$= 25.295$$

Thus

$$K = \frac{a_{Fe^{2+}}}{a_{Cu^{2+}}}$$

$$= 10^{25.295}$$

This means that to reach equilibrium, the activity ($\approx$ concentration) of Fe^{2+} would have to be enormously greater than that of Cu^{2+}, so the reaction will always proceed as written, and cannot be made to go the other way (at least, not by simply adjusting the ion concentrations).

11.4 A Simple Electrolytic Cell

Obviously in this case, and in most natural circumstances, the electron transfer takes place on a molecular level, and we have no control over it. However, what if we could separate the iron and copper, and have the electrons travel through a wire from one to the other? Well, why not? In Figure 11.2, the beaker has been divided into two parts. In one we

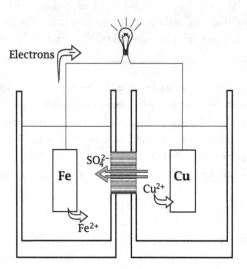

Figure 11.2 An electrolytic cell. Iron dissolves on one side, and copper precipitates on the other. A liquid junction called a salt bridge allows sulfate to migrate between the solutions. The cell reaction is identical to the reaction in Figure 11.1, Equation (11.1).

have a solution of ferrous ions and sulfate ions in contact with a piece of iron (an iron electrode), and in the other we have a solution of cupric ions and sulfate ions in contact with a piece of copper (a copper electrode). Considered separately (and in the absence of oxygen, which would oxidize both electrodes), nothing at all happens. But if we connect a wire between the electrodes, a current begins to flow, because reaction (11.1) wants to occur, and now it can. Iron dissolves, forming more Fe^{2+} in solution, and the electrons, instead of attaching themselves to some immediately adjacent copper ions, must travel through the wire before being able to do that. That is, when they get to the copper electrode, they jump onto some immediately adjacent Cu^{2+} ions, causing copper to precipitate. If the two cells are completely separate, a positive charge would soon build up in the iron solution and a negative charge in the copper solution, stopping the reaction; so we have to provide some kind of connection (a *liquid junction*), which allows sulfate ions to migrate from one solution to the other.

By separating the two parts of the redox reaction, we have caused a current to flow through a wire. We have a simple battery, which we could use to power a light bulb, or do other useful things. In many applications of such cells, however, the objective is to obtain thermodynamic data, not to generate electricity, so the cell is operated in a balanced (equilibrium) condition in which no current is allowed to flow, and no changes in the cell compositions take place.

11.4.1 The Cell Voltage

An electrical circuit delivering direct current such as we have described obviously must have a voltage difference between the two electrodes. It is in fact another kind of potential (see Sections 1.4 and 4.3), this time measured in volts. If we leave the wire connected to both electrodes, the cell will continue to operate until all of the iron has dissolved, or until there are no more copper ions to react. That is, the battery will run down, and whatever voltage we had at the beginning decreases to zero during the experiment. The voltage of the cell at any particular point is measured by attaching a voltmeter (which has an extremely high resistance) or a potentiometer (which imposes a voltage in the circuit equal and opposite to the cell voltage) across the electrodes, instead of a piece of wire. Either way, the current flow is stopped, and the voltage measured under equilibrium (no current flow in either direction) conditions. This equilibrium state is, of course, different from the equilibrium state reached when the cell "runs down" and can react no longer. It is a higher-energy state that is prevented from reaching a lower-energy state by some constraint, in this case an opposing voltage or a high resistance. It is in fact another example of a metastable equilibrium state.

What determines the magnitude of this cell voltage, when it is not zero? Intuitively, we would suspect that it depends a lot on what metals we use to make the electrodes, and if we thought a bit more, we might think that the concentrations of the ions in solution would have an effect too. This is exactly the case, and we must develop an equation relating the voltage to the activities of the reactants and products in the cell reaction.

11.4.2   Half-Cell Reactions

An oxidation reaction cannot take place without an accompanying reduction reaction – the electrons have to go somewhere – but it is convenient to nonetheless split cell reactions into two complementary "half-cell" reactions. In our copper–iron case, these half-cells are

$$\text{Oxidation:} \quad \text{Fe(s)} = \text{Fe}^{2+} + 2\,\text{e} \tag{11.2}$$

and

$$\text{Reduction:} \quad \text{Cu}^{2+} + 2\,\text{e} = \text{Cu(s)} \tag{11.3}$$

where e is an electron, and the cell reaction is the sum of these, Equation (11.1).

We also imagine that each half-cell reaction has a half-cell voltage associated with it, and that the cell voltage is the sum of the two half-cell voltages. If we had these half-cell potentials, or voltages, we could tabulate them and mix and match electrodes to calculate the potential of any cell we wanted, much as we tabulate Gibbs energies of compounds so as to be able to calculate the $\Delta_r G°$ of any reaction. Of course, half cell voltages cannot be measured, just as the G of compounds cannot be measured, but we can get around this, just as we did with $\Delta_f G°$.

With Gibbs energies, we tabulate the difference between the G of a compound and the sum of the Gs of its constituent elements. The elements and their Gibbs energies always cancel out in balanced reactions. With electrodes, we measure and tabulate the difference of every electrode against a standard electrode, and the potential of this standard electrode cancels out in balanced cell reactions. The standard electrode chosen is the *standard hydrogen electrode* (SHE, Figure 11.3). So if we measure both the copper electrode and the iron electrode and many others in separate experiments against the SHE, we will then be able to calculate the potential of any cell from these tabulated values. Although we can tabulate potentials for every kind of electrode, we wouldn't want to tabulate potentials for every conceivable concentration of product and reactant ions; that's very inefficient. It would be better to tabulate the potentials of each electrode for some standard conditions and to have an equation that could give the potential of the electrode (and of cells constructed from electrodes) at any particular concentrations we are interested in; see Equation (11.14).

11.4.3 Standard Electrode Potentials

A cell for measuring the potential of the copper–SHE cell is shown in Figure 11.3. The hydrogen electrode is a device for using the reaction

$$2H^+ + 2e \rightleftharpoons H_2(g) \tag{11.4}$$

as an electrode. The idea of using a gas as an electrode seems rather bizarre at first. It is accomplished by bubbling hydrogen gas over a specially treated piece of platinum (platinum coated with fine-grained carbon). The platinum serves simply as a source or sink of electrons; the reaction between hydrogen gas and hydrogen ions takes place at the surface of the platinum and is catalyzed by the carbon.

In order that the hydrogen electrode always give the same potential, the activities of both the hydrogen gas and the hydrogen ion must always be the same. These have been standardized at $f_{H_2(g)} = 1$ bar and $a_{H^+} = 1$, that is, a gas pressure of one bar and an acid concentration of about 1 molal. A hydrogen electrode operating under these conditions, the standard hydrogen electrode, is assigned a half-cell potential of zero volts by convention.

The half-cell reactions for the cell in Figure 11.3 are

$$\text{Oxidation:} \quad H_2(g) = 2H^+ + 2e \tag{11.5}$$

and

$$\text{Reduction:} \quad Cu^{2+} + 2e = Cu(s) \tag{11.6}$$

and the sum of these is the cell reaction

$$Cu^{2+} + H_2(g) = Cu(s) + 2H^+ \tag{11.7}$$

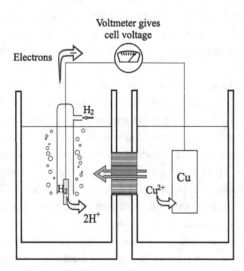

Figure 11.3 An electrolytic cell for measuring the potential of the copper electrode against the standard hydrogen electrode. If the activities of all ions and elements are 1.0, the cell voltage is the standard cell voltage, $\mathcal{E}°$. The direction of electron flow is indicated for reduction at the copper electrode. When a measurement is made, no current flows.

As we mentioned, the cell voltage depends on the activities of all the ions and compounds in the cell reaction; in this case it depends not only on the hydrogen gas pressure and a_{H^+}, but also on a_{Cu} and $a_{Cu^{2+}}$. "Standard conditions" are defined as $a = 1$ for all products and reactants in the cell reaction, so if the hydrogen electrode is operating under SHE conditions ($a_{H_2(g)} = 1; f_{H_2(g)} = 1$ bar), the copper electrode is pure Cu ($a_{Cu(s)} = 1$), and the cupric ion concentration and activity coefficient are adjusted to give $a_{Cu^{2+}} = 1$, the cell voltage will be the *standard* cell voltage, $\mathcal{E}°$.

11.5   The Nernst Equation

11.5.1   Work Done by Cells

In lighting the light bulb or running a small motor, our cell in Figure 11.2 is doing work. We have already seen, in Chapter 4, how much work can be done by a chemical reaction, but you may have forgotten this because we have put so much emphasis on reactions that do *no* work, other than the minimum necessary $P \Delta V$ work. But here we have a chemical reaction that is certainly doing $P \Delta V$ work (the Fe^{2+} and Cu in the cell reaction will have a slightly different molar volume from that of the Fe and Cu^{2+}, so some work is done against the atmospheric pressure on the solution), but in addition to this, it is doing work in lighting the bulb. According to Section 4.9, the maximum amount of work we can get from our cell reaction is given by the $\Delta_r G$ of that reaction, and when $\Delta_r G$ decreases to zero, we reach stable equilibrium and can get no more work from the cell.

The electrical work w required to move a charge of $\mathcal{F}$ coulombs through a potential difference $\mathcal{E}$ volts is

$$-w = \mathcal{F}\mathcal{E}$$

(joules = coulombs × volts)

where $\mathcal{F}$ is the charge per mole of electrons, so if n is the number of electrons appearing in the reaction as written, there are $n\mathcal{F}$ coulombs of charge, and the work is

$$-w = n\mathcal{F}\mathcal{E} \tag{11.8}$$

Because some convention must be adopted in order for us to know whether the voltage $\mathcal{E}$ is positive or negative, you may see Equation (11.8), as well as (11.9) and (11.10), below, written without the minus sign in some references. The conventions we have adopted (Section 11.6.1) require the minus sign.

This electrical work is by definition (Chapter 4) the ΔG associated with the process, as long as the electrical work is the only non-$P \Delta V$ work done. Therefore, for any process in which $n\mathcal{F}$ coulombs are moved through a potential difference $\mathcal{E}$,

$$\Delta G = -n\mathcal{F}\mathcal{E} \tag{11.9}$$

or

$$\Delta G° = -n\mathcal{F}\mathcal{E}° \tag{11.10}$$

for standard state conditions. As applied to electrochemical cells, these equations are more properly $\Delta\mu = -n\mathcal{F}\mathcal{E}$ and $\Delta\mu° = -n\mathcal{F}\mathcal{E}°$ because many of the individual free energy terms refer to constituents in solution and hence are partial molar terms. These equations connect electrochemistry to the world of thermodynamics. They allow us to calculate the voltage that will be observed in any cell for which we know the cell reaction and the $\Delta_r G$ or the $\Delta_r G°$.

11.5.2   Relation between Cell Activities and Voltage

Consider the general cell reaction

$$b\text{B} + c\text{C} = d\text{D} + e\text{E} \tag{11.11}$$

Let's say that this reaction reaches equilibrium with an external measuring system, giving cell voltage $\mathcal{E}$. If operated under standard conditions, it would give cell voltage $\mathcal{E}°$, and the corresponding Gibbs energies of reaction are

$$\Delta_r\mu = -n\mathcal{F}\mathcal{E} \tag{11.12}$$

and

$$\Delta_r\mu° = -n\mathcal{F}\mathcal{E}° \tag{11.13}$$

From Equation (9.8) we have

$$\Delta_r\mu = \Delta_r\mu° + RT\ln Q \tag{9.8}$$

where

$$Q = \prod_i a_i^{v_i}$$

$$= \frac{a_E^e a_D^d}{a_B^b a_C^c}$$

Recall from Chapter 9 that this activity term is referred to as Q rather than K because it refers to a metastable equilibrium. Substitution of Equations (11.12) and (11.13) gives

$$\mathcal{E} = \mathcal{E}° - \frac{RT}{n\mathcal{F}}\ln Q$$

or

$$\mathcal{E} = \mathcal{E}° - 2.30259\frac{RT}{n\mathcal{F}}\log Q \tag{11.14}$$

This is the Nernst equation, after the physical chemist W. Nernst, who derived it at the end of the nineteenth century. As above, n is the number of electrons transferred in the cell reaction (two in reaction (11.7)), $\mathcal{F}$ the Faraday of charge, R the gas constant, and T the temperature (in kelvins). The constant 2.30259 is used to convert from natural to base 10 logs. At 25 °C the quantity 2.30259 $RT/\mathcal{F}$ has the value 0.05916, which is called the Nernst slope. The importance of (11.14) is that it allows calculation of the potentials of cells having non-standard state concentrations (i.e., real cells) from tabulated values of standard half-cell values or tabulated standard Gibbs energies.

Equation (11.11) could also be considered to represent a half-cell reaction, except that the electron is not shown. So evidently we could use the Nernst equation to calculate half-cell potentials if we knew what value to assign to the chemical potential of an electron. It turns out, of course, that because the electrons always cancel out in balanced reactions, we could assign any value we like to the electron Gibbs energy and it would make no difference to our calculated cell potentials. The easiest value to assign is zero, and that is what is done. Therefore, the Nernst equation is used to calculate both half-cell and cell potentials.

11.6 Some Necessary Conventions

In our discussion so far, we have skipped lightly over some points which are not important to a general understanding, but which if neglected will result in our getting the wrong answers in calculations. For example, we said (Section 11.4.2) that the cell voltage is the *sum* of the two half-cell voltages. Actually, it is a little more complicated. Because half-cell and cell reactions may be written forwards or backwards, and a voltage by itself is not obviously positive or negative, there has to be a set of rules to keep things straight. Unfortunately, there is more than one set of rules. We present here the rules set out by the IUPAC (International Union of Pure and Applied Chemistry), which are followed by most people today. Be warned, however, that several geochemical sources use a different set of rules (e.g., Garrels and Christ, 1965).

11.6.1 The IUPAC Rules

Some of these points have already been discussed. We include all the rules here for completeness.

1. Cell reactions are written such that the left-hand electrode supplies electrons to the outer circuit (i.e., oxidation takes place), and the right-hand electrode accepts electrons from the outer circuit (i.e., reduction takes place).

2. The cell potential is given by

$$\mathcal{E} = \mathcal{E}_{\text{right electrode}} - \mathcal{E}_{\text{left electrode}}$$

that is,

$$\mathcal{E} = \mathcal{E}_{\text{reduction electrode}} - \mathcal{E}_{\text{oxidation electrode}}$$

3. The cell potential is related to the Gibbs energy by

$$\Delta_r G = -n \mathcal{F} \mathcal{E}$$

4. The electrode potential of a half-cell is equal in magnitude and sign to the potential of a cell formed with the electrode on the right and the standard hydrogen electrode ($\mathcal{E}° = 0$) on the left.

5. Standard half-cell reactions are tabulated and calculated as reductions; for example,

$$Zn^{2+} + 2\,e = Zn(s) \qquad \mathcal{E}° = -0.763\,V$$

However, the half-cell potential is a sign-invariant quantity, that is,

$$Zn(s) = Zn^{2+} + 2\,e \qquad \mathcal{E}° = -0.763\,V$$

6. For the reaction

$$bB + cC = dD + eE$$

the Nernst expression is

$$\mathcal{E} = \mathcal{E}° - \frac{RT}{n\mathcal{F}} \ln\left(\frac{a_E^e a_D^d}{a_B^b a_C^c}\right)$$

7. In view of item 5, the Nernst expression for a half-cell is given by

$$\mathcal{E} = \mathcal{E}° - \frac{RT}{n\mathcal{F}} \ln\left(\frac{\text{reduced form}}{\text{oxidized form}}\right)$$

11.6.2   Examples

Let's calculate the potential of the cell in Figure 11.2. By convention 5, both half-cells are written and calculated as reductions, no matter what is happening in the real cell. Thus, for the copper half-cell,

$$Cu^{2+} + 2e = Cu(s)$$

and

$$\begin{aligned}
\Delta_r G° &= \Delta_f G°_{Cu(s)} - \Delta_f G°_{Cu^{2+}} \\
&= 0 - 65.49\,kJ\,mol^{-1} \\
&= -65,490\,J\,mol^{-1} \\
&= -n\mathcal{F}\mathcal{E}°
\end{aligned}$$

Because two electrons are involved, $n = 2$, so

$$\begin{aligned}
\mathcal{E}°_{Cu\ half.cell} &= -\frac{-65,490}{2 \times 96,485} \\
&= 0.339\,V
\end{aligned}$$

For the iron half-cell,

$$Fe^{2+} + 2e = Fe(s)$$

and

$$\Delta_r G^\circ = \Delta_f G^\circ_{Fe(s)} - \Delta_f G^\circ_{Fe^{2+}}$$
$$= 0 - (-78.90) \, \text{kJ mol}^{-1}$$
$$= 78,900 \, \text{J mol}^{-1}$$
$$= -n\mathcal{F}\mathcal{E}^\circ$$

Then

$$\mathcal{E}^\circ_{Fe \, half.cell} = -\frac{78,900}{2 \times 96,485}$$
$$= -0.409 \, \text{V}$$

For the complete cell,

$$\mathcal{E}^\circ_{Cu-Fe \, cell} = \mathcal{E}^\circ_{reduction \, half.cell} - \mathcal{E}^\circ_{oxidation \, half.cell}$$
$$= \mathcal{E}^\circ_{Cu \, half.cell} - \mathcal{E}^\circ_{Fe \, half.cell}$$
$$= 0.339 - (-0.409)$$
$$= 0.748 \, \text{V}$$

Note that if we wrote the cell backwards,

$$Cu(s) + Fe^{2+} = Cu^{2+} + Fe(s)$$

both half-cell reactions would still be written and calculated as reductions, but the cell voltage would now be

$$\mathcal{E}^\circ_{Cu-Fe \, cell} = \mathcal{E}^\circ_{reduction \, half.cell} - \mathcal{E}^\circ_{oxidation \, half.cell}$$
$$= \mathcal{E}^\circ_{Fe \, half.cell} - \mathcal{E}^\circ_{Cu \, half.cell}$$
$$= -0.409 - (0.339)$$
$$= -0.748 \, \text{V}$$

Thus the signs of both $\Delta_r G^\circ$ and $\mathcal{E}^\circ$ of the complete cell reaction depend on how the cell is written, but the signs of the half-cell reactions do not.

These are the standard potentials. Suppose the cell is operating under non-standard (real) conditions. Let's say $a_{Cu^{2+}}$ is not 1 but 0.1, and $a_{Fe^{2+}}$ is 1.5. Using the Nernst expression for the complete cell, reaction (11.1), at 25 °C,

$$\mathcal{E} = \mathcal{E}^\circ - 2.30259 \frac{RT}{n\mathcal{F}} \log Q$$
$$= \mathcal{E}^\circ - \frac{0.05916}{n} \log \left(\frac{a_{Fe^{2+}}}{a_{Cu^{2+}}} \right)$$
$$= 0.748 - \frac{0.05916}{2} \log \left(\frac{1.5}{0.1} \right)$$
$$= 0.748 - 0.0348$$
$$= 0.713 \, \text{V}$$

So the real cell with these concentrations would have a potential of 0.713 V, rather than 0.748 V.

The Nernst expression can also be used for the half-cells. Thus, for the Cu half-cell,

$$\mathcal{E} = \mathcal{E}^\circ - 2.30259 \frac{RT}{n\mathcal{F}} \log Q$$

$$= \mathcal{E}^\circ - \frac{0.05916}{n} \log \left(\frac{a_{Cu}}{a_{Cu^{2+}}} \right)$$

$$= 0.339 - \frac{0.05916}{2} \log \left(\frac{1}{0.1} \right)$$

$$= 0.339 - 0.0296$$

$$= 0.309 \text{ V}$$

and for the Fe half-cell,

$$\mathcal{E} = \mathcal{E}^\circ - 2.30259 \frac{RT}{n\mathcal{F}} \log Q$$

$$= \mathcal{E}^\circ - \frac{0.05916}{n} \log \left(\frac{a_{Fe}}{a_{Fe^{2+}}} \right)$$

$$= -0.409 - \frac{0.05916}{2} \log \left(\frac{1}{1.5} \right)$$

$$= -0.409 + 0.0052$$

$$= -0.404 \text{ V}$$

The cell potential is then

$$\mathcal{E}_{Cu-Fe \text{ cell}} = \mathcal{E}_{\text{reduction half.cell}} - \mathcal{E}_{\text{oxidation half.cell}}$$

$$= \mathcal{E}_{Cu \text{ half.cell}} - \mathcal{E}_{Fe \text{ half.cell}}$$

$$= 0.309 - (-0.404)$$

$$= 0.713 \text{ V}$$

as before.

There certainly are other ways to do these calculations and have them come out right. However, all conventions have their good and bad points, and the IUPAC conventions are the most commonly used.

11.7   Measuring Activities

In geochemistry, possibly the greatest interest in galvanic cells is in understanding the concept of *Eh* and its use in determining redox conditions in natural environments (Section 11.8.1). In chemistry, it is in determining the activities of solutes.

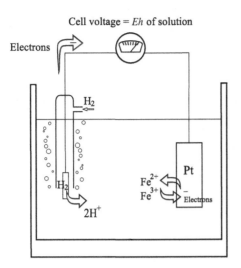

Cell voltage = Eh of solution

Electrons

H$_2$

H$_2$

2H$^+$

Fe^{2+}
Fe^{3+}

Pt

Electrons

Figure 11.4 How to measure the *Eh* of a solution containing both ferrous and ferric ions. The direction of electron flow is indicated for reduction at the platinum surface. When a measurement is made, no electrons are flowing.

11.7.1 $\mathcal{E}°$ as a Source of $\Delta_f G°$

Electrochemical cells are of course of great practical importance in the form of batteries and fuel cells. In thermodynamics, the relationship between the cell potential and the Gibbs energy is often used the other way around. That is, cell potentials are one of the most accurate and useful sources of information about Gibbs energies of reactions and dissolved substances. Not all Gibbs energy data come from calorimetry. Most data for both $\Delta_f G°_{Fe^{2+}}$, $\Delta_f G°_{Cu^{2+}}$, and other ionic species come from the measurement of $\mathcal{E}°$ in cells as shown in Figure 11.3.

Actually the cells in Figures 11.1 and 11.3 are complicated by a factor we have not mentioned. These are cells having a junction between two solutions with different compositions, and this results in an additional source of emf, called a junction potential. It arises because the two solutions must inevitably diffuse into one another to some extent, however small, and ionic concentration gradients are established, which create a potential difference. This is a major topic in electrochemistry, but need not be treated in detail here. Suffice it to say that in the case of Figure 11.3, the liquid junction is not necessary. The copper electrode could be placed in the same solution as the hydrogen electrode, as in Figure 11.4, creating a cell without a liquid junction, simplifying the thermodynamic interpretation.

The Activity of HCl

As an example of the determination of the activity of a solute, we consider HCl. A cell much like that in Figure 11.4 but having a standard calomel or silver chloride reference electrode (Section 11.8.3) in place of the platinum electrode, and having a solution of HCl rather than a solution of ferrous and ferric ions, can be abbreviated as

$Pt|H_2(g)|HCl(aq)|AgCl|Ag$

where | indicates a phase boundary. With the half-cell reactions

$$\frac{1}{2}H_2(g) = H^+ + e$$

and

$$AgCl + e = Ag(s) + Cl^-$$

the cell reaction is

$$\frac{1}{2}H_2(g) + AgCl = Ag + H^+ + Cl^-$$

or

$$\frac{1}{2}H_2(g) + AgCl = Ag + HCl$$

for which the Nernst equation is

$$\mathcal{E} = \mathcal{E}° - 2.30259\frac{RT}{n\mathcal{F}}\log Q$$

$$= \mathcal{E}° - 2.30259\frac{RT}{n\mathcal{F}}\log\left(\frac{a_{Ag}a_{HCl}}{f_{H_2}^{1/2}a_{AgCl}}\right) \tag{11.14}$$

Because the activity of AgCl and that of Ag are fixed at 1.0, the only variables are f_{H_2} and the concentration of HCl, so at a fixed standard state f_{H_2} of 1.0 bars, the cell voltage is a direct measure of the activity of HCl, as long as $\mathcal{E}°$ is known. Determination of $\mathcal{E}°$ requires an extrapolation to infinite dilution, and was the subject of a great deal of research involving the behavior of charged solutes in dilute solutions. For our purposes, it is sufficient to see how cell voltages can be related to solute activities.

11.8 Measuring Redox Conditions

So far we have considered only cells that we might construct ourselves, in the laboratory. For the scientist interested in natural environments, this is background information. What we really want to know is how to characterize natural environments as being either reducing or oxidizing on some numerical scale. Natural environments don't normally have electrodes sticking out of them, so what is the connection?

In the absence of electrodes and voltages, the redox state of a solution is characterized by the relative concentrations of reduced and oxidized ionic or molecular species in the solution. Thus in a solution containing Fe ions, the solution is relatively reduced if there are more Fe^{2+} ions than Fe^{3+} ions, and vice versa. In a solution containing carbon species, the solution is relatively reduced if there are more CH_4 molecules than CO_2 molecules. In a solution containing sulfur, the solution is relatively reduced if there are more H_2S molecules than SO_4^{2-} ions, and vice versa. And so on. For those elements that have more than one valence state in natural environments, the two (or more) states will be present in various ratios, depending on whether the environment is reducing or oxidizing. In a solution containing all of these, the $a_{Fe^{2+}}/a_{Fe^{3+}}$, a_{CH_4}/a_{CO_2}, and $a_{H_2S}/a_{SO_4^{2-}}$ ratios will all be different, but each will be controlled by the same factors (T, P, and the bulk composition of the solution), so each one should give us the same index of redox conditions, if equilibrium prevails.

But what is this index? We will consider two commonly used ones, Eh and f_{O_2}.

11.8.1   Redox Potential, Eh

Suppose you have a sample solution that contains both ferrous and ferric ions. The ferrous/ferric ratio is a measure of how reduced/oxidized the sample solution is. For any change in this ratio, some reaction involving electron transfer must take place. What we must do is insert an electrode that will supply/absorb these electrons. In other words, we need an electrode that responds to the ferrous/ferric ratio, that is, one that has a half-cell potential that varies with this ratio. We could insert this electrode into the solution and connect it to a SHE, and the measured cell potential would depend on the ferrous/ferric ratio in the solution. Finding such an electrode is easier than you might think.

Note that in the Cu and Fe electrodes we have considered, the "reduced form" in both cases is the metal, Cu or Fe. The "oxidized form" is an ion in solution, Cu^{2+} or Fe^{2+}, and the electrode, being made of the metal, is a necessary part of the half-cell. However, in the SHE, both the reduced form (H_2) and the oxidized form (H^+) are in the solution (one as a gas phase); neither is part of the electrode. The platinum electrode itself is nothing but a source or sink for electrons. We are now considering another case where both the reduced form (Fe^{2+}) and the oxidized form (Fe^{3+}) are in the solution, so all we have to do is provide a source and sink for electrons. All we need is a piece of platinum, as shown in Figure 11.4.

Suppose the solution is quite reduced, with $a_{Fe^{2+}}/a_{Fe^{3+}} = 10$. The half-cell reactions are (both written as reductions)

$$2H^+ + 2e = H_2(g) \qquad\qquad \mathcal{E}^\circ = 0.0\,V$$

and

$$Fe^{3+} + e = Fe^{2+} \qquad\qquad \mathcal{E}^\circ = 0.769\,V$$

The complete cell is

$$Fe^{3+} + \tfrac{1}{2}H_2(g) = Fe^{2+} + H^+$$

The Nernst equation may be written for either the complete cell or the Fe half-cell, giving the same answers. Thus

$$\mathcal{E} = Eh = \mathcal{E}^\circ - \frac{0.05916}{1}\log\left(\frac{a_{Fe^{2+}}}{a_{Fe^{3+}}}\right)$$
$$= 0.769 - 0.05916 \times \log(10)$$
$$= 0.710\,V$$

So the Eh of this relatively reduced solution is 0.710 V. If the solution is quite oxidized, with $a_{Fe^{2+}}/a_{Fe^{3+}} = 0.1$,

$$\mathcal{E} = Eh = \mathcal{E}^\circ - \frac{0.05916}{1} \log\left(\frac{a_{Fe^{2+}}}{a_{Fe^{3+}}}\right)$$

$$= 0.769 - 0.05916 \times \log(0.1)$$

$$= 0.828 \text{ V}$$

and the Eh of this more oxidized solution is 0.828 V.

To summarize, you may think of Eh of a solution as either a cell potential or a half-cell potential. It is the potential of a cell having one electrode that responds reversibly to a redox couple (such as Fe^{2+}/Fe^{3+}) or couples in the solution and the SHE as the other electrode. Or it is the half-cell potential of an electrode responding reversibly to a redox couple or couples in the solution. You may use any kind of electrode as the other side of the cell, as long as you correctly deduce the half-cell potential of the electrode that is responding to conditions in your solution. Half-cell potentials are *defined* in terms of the SHE by our IUPAC conventions.

Therefore, we have an index of redox conditions, only assuming that reduced and oxidized species in solution can readily exchange electrons at a platinum surface. There are some practical difficulties in this respect, as discussed below.

11.8.2 Redox Couples Other Than Iron

But suppose our sample solution in Figure 11.4 contains not only the ferrous/ferric redox couple, but also Mn^{2+}/Mn^{4+}, H_2S/SO_4^{2-}, CH_4/CO_2, and others. Even if all these redox couples are at equilibrium, each will have a different activity ratio. To which does our platinum electrode respond? In theory, it responds to all of them simultaneously, and all result in the same Eh. Each couple has a different activity ratio, but each also has a different value of $\mathcal{E}^\circ$, and the resulting Eh for the solution must be the same for each redox couple if they are at equilibrium, and if each reacts with the platinum electrode.

11.8.3 Some Practical Difficulties

No Unique *Eh* Values

However, in practice the platinum electrode does not respond to all redox couples equally. In fact, the only ones it responds to well are Fe and Mn, and then only if concentrations are high enough. Sulfur, carbon, and many other redox pairs simply do not give up or take up electrons easily at the platinum surface, so measured Eh values primarily reflect the ferrous/ferric ratio in solutions. Of course, if equilibrium prevails, this should suffice – the activity ratios of all other couples could be calculated if the ferrous/ferric (and pH) ratio is known. However, at Earth surface conditions in natural environments, equilibrium often (in fact, usually) does not prevail, and the only way to really know the activities of many redox pairs is to analyze the solution for both parts of the pair.

This is strikingly illustrated by the data collected by Lindberg and Runnells (1984), some of which are summarized in Figure 11.5 These authors examined over 150,000

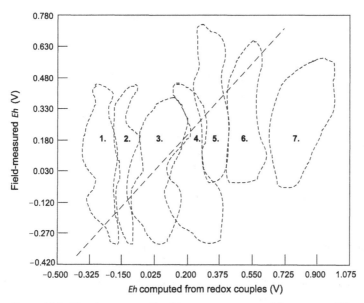

Figure 11.5 *Eh* values, generalized into areas, computed from seven different redox couples compared with the *Eh* value measured in the field. Numbered areas refer to 1. HS^-/SO_4^{2-}; 2. $HS^-/S(s)$; 3. $Fe^{2+}/Fe(OH)_3(s)$; 4. Fe^{3+}/Fe^{2+}; 5. NH_4^+/NO_2^-; 6. NO_2^-/NO_3^-; and 7. $O_2(aq)/H_2O$. The dashed line represents perfect agreement. Modified from Lindberg and Runnells (1984).

groundwater analyses from the USGS database, plus values from the literature, from which they selected 611 analyses of acceptable quality. The ionic species activities were then calculated (this speciation modeling is discussed in Chapters 10 and 13) and theoretical *Eh* values calculated for several redox couples. Comparison with the measured *Eh* illustrates how unreliable these measurements are.

Natural solutions that have not internally equilibrated cannot be said to have a unique *Eh*. Nevertheless, *Eh* measurements are often useful in a qualitative sense and may be fairly accurate in some situations, such as acid Fe-rich mine drainage systems (see the coal mine problem on page 288). But even if *Eh* measurements were not useful at all, the concept of *Eh* is firmly established in the literature of natural environments, where it is used primarily in discussing *models* or hypothetical situations.

Reference Electrodes

Another practical difficulty is that the hydrogen electrode is a rather delicate apparatus, really suitable only for laboratory use. How do we get it into the field, to take *Eh* measurements in natural environments? We don't. Other electrodes can be designed that have fixed potentials, of values independent of what solution they are put into, called reference electrodes. Their potentials with respect to the SHE can be measured, so that field measurements of cells composed of a platinum electrode and a reference electrode can be made, and the readings can then be corrected to what they would have been had a SHE been used, giving the *Eh*.

Redox Reactions

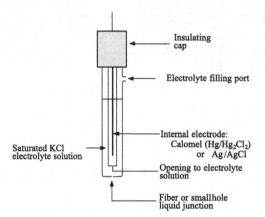

Figure 11.6 A silver/silver chloride reference electrode.

Two common reference electrodes are the calomel (Hg/Hg_2Cl_2) and silver/silver chloride (Ag/AgCl) electrodes. In both, the activities of all active parts of the electrode are fixed, so the electrode potential is fixed at a given temperature. For example, the Ag/AgCl electrode half-cell reaction is

$$AgCl + e = Ag(s) + Cl^-$$

Both Ag and AgCl are present as solid phases so $a_{Ag} = a_{AgCl} = 1$, and these are immersed in a solution saturated with solid KCl, which fixes a_{Cl^-} at a constant value (Figure 11.6). With all reactants and products of the half-cell reaction having fixed activities, $\mathcal{E}^\circ_{Ag/AgCl\ half-cell}$ has a fixed value, which is 0.222 V. That is, a cell composed of the Ag/AgCl electrode on one side and the SHE on the other will record a constant potential of 0.222 V. When used as a reference electrode, immersed in a solution containing ferrous and ferric ions and connected to a platinum electrode, for example, the cell reaction is

$$Fe^{3+} + Ag(s) + Cl^- = Fe^{2+} + AgCl$$

and the standard cell potential of this cell is

$$\mathcal{E}^\circ = \mathcal{E}^\circ_{Fe^{2+}/Fe^{3+}} - \mathcal{E}^\circ_{Ag/AgCl}$$

$$= 0.769 - 0.222$$

$$= 0.547\,V$$

whereas if measured against SHE, $\mathcal{E}^\circ_{Fe^{2+}/Fe^{3+}}$ is 0.769 V. *Eh* measurements are defined as observed cell voltages using the SHE as reference, so if Ag/AgCl is used instead, 0.222 V must be *added* (0.547 + 0.222 = 0.769) to the observed readings. For calomel reference electrodes, 0.268 V must be added.

11.9   *Eh*–pH Diagrams

The only other intensive variable of comparable significance in aqueous systems is pH. It too is a function of the bulk composition at a given *T* and *P*, but both *Eh* and pH are closely related to a large number of important reactions. Therefore, it proves natural to use both as variables in diagrams of systems at fixed T and P, and *Eh*–pH diagrams have become a standard method of displaying and interpreting geochemical data. In the following sections we outline the theoretical basis for calculating these diagrams.

In this section, we will calculate portions of a simple *Eh*–pH diagram for the system $Mn–H_2O$. This illustrates most of the problems encountered in calculating such diagrams. If you wish to add components such as CO_2 or H_2S, the methods are similar, and details are provided by Garrels and Christ (1965).

11.9.1   General Topology of *Eh*–pH Diagrams

First, let us examine the completed *Eh*–pH diagram for $Mn–H_2O–O_2$ in Figure 11.7. There are typically four different types of boundaries shown on these diagrams. The top line, labeled O_2/H_2O, represents conditions for water in equilibrium with O_2 gas at 1 atm. Above this line, a P_{O_2} greater than 1 atm is required for water to exist, so that, because the diagram is drawn for a pressure of 1 atm, water is not stable above this line. Similarly, the bottom line, H_2O/H_2, represents conditions for water in equilibrium with H_2 gas at 1 atm. Below this line, P_{H_2} values greater than 1 atm are required for water to exist; that is, at 1 atm water is not stable. Therefore, the water stability field is between these two lines.

The second type of boundary separates the stability fields of minerals or solid phases such as hausmannite (Mn_3O_4) and pyrochroite [$Mn(OH)_2$]. These are true phase boundaries: hausmannite is thermodynamically unstable below the hausmannite/pyrochroite boundary and pyrochroite is unstable above it. Thus these first two kinds of boundary

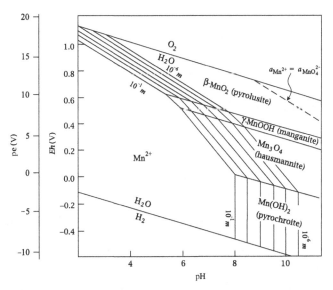

Figure 11.7 *Eh*–pH and pe –pH relations in the system $Mn–H_2O–O_2$ at 25 °C, 1 bar. Mn^{2+} activities and stability fields of Mn-oxide minerals are included. The pe and *Eh* axes are related by the formula pe $= 5040\, Eh/T$.

represent thermodynamic stability fields for different substances. Notice that on this diagram they all have the same slope (equal to the Nernst slope).

The remaining two kinds of lines are not stability boundaries at all but refer to particular concentrations of dissolved ions. For example, the vertical lines within the pyrochroite stability field represent contoured solubilities of pyrochroite as Mn^{2+} concentrations varying from 10^{-1} to 10^{-6} m. Finally, the dashed boundaries between aqueous species, such as that between Mn^{2+} and MnO_4^{2-}, indicate where the activities of the two species are exactly equal. To the right of this line, Mn^{2+} remains present in the solution, but at a lower activity than MnO_4^{2-}, and vice versa.

11.9.2   Sample Calculations

It is possible to look up half-cell potentials for many reactions in physical chemistry textbooks and compilations of electrochemical data. However, it is usually a better procedure to choose Gibbs energy data and use those to calculate Eh and $\mathcal{E}°$ for the reactions of interest.

As the first example, we will calculate the two boundaries for the stability field of water.

For the boundary $H_2O(l)/O_2(g)$, the half-cell reduction reaction is

$$4H^+ + O_2(g) + 4e = 2H_2O(l) \tag{11.15}$$

for which $n = 4$. Using the tabulated $\Delta_f G°$ for water (all others being zero), we find

$$\Delta_r G° = 2(-237.129) - 0$$
$$= -474.258 \text{ kJ mol}^{-1} \tag{11.16}$$

Because

$$\Delta_r G° = -n\mathcal{F}\mathcal{E}°$$

then

$$\mathcal{E}° = -(-474,258/(4 \times 96,485))$$
$$= 1.23 \text{ V}$$

From the Nernst equation, Equation (11.14),

$$Eh = \mathcal{E}° - 2.30259\frac{RT}{n\mathcal{F}} \log[1/(f_{O_2} \cdot a_{H^+}^4)]$$

Setting $f_{O_2} = 1$ atm, and recalling that $pH = -\log a_{H^+}$, gives the equation for the boundary in terms of Eh and pH.

$$Eh = 1.23 - 0.0592pH \tag{11.17}$$

For the boundary $H_2(g)/H_2O(l)$,

$$2H^+ + 2e = H_2(g)$$

with

$$\Delta_r G^\circ = 0$$
$$= -n\mathcal{F}\mathcal{E}^\circ$$

and

$$Eh = 0 - 2.30259 \frac{RT}{n\mathcal{F}} \log(f_{H_2}/a_{H^+}^2)$$

or, with $f_{H_2} = 1$ atm, and $n = 2$,

$$Eh = -0.0592 \text{pH} \tag{11.18}$$

For boundary $Mn(OH)_2/Mn_3O_4$,

$$Mn_3O_4(c) + 2H_2O(l) + 2H^+ + 2e = 3Mn(OH)_2(c)$$

with

$$\Delta_r G^\circ = -94,140 \text{ J mol}^{-1}$$
$$= -n\mathcal{F}\mathcal{E}^\circ$$

Then

$$\mathcal{E}^\circ = -(-94,140/2 \times 96,485)$$
$$= 0.488 \text{ V}$$

and, with $n = 2$,

$$Eh = 0.488 - 0.0592 \text{pH} \tag{11.19}$$

For solubility of Mn_3O_4 as Mn^{2+},

$$Mn_3O_4(c) + 8H^+ + 2e = 3Mn^{2+} + 4H_2O$$

and

$$\Delta_r G^\circ = -349,616 \text{ J mol}^{-1}$$

Then

$$\mathcal{E}^\circ = 352,711/(2 \times 96,485)$$
$$= 1.812 \text{ V}$$

and

$$Eh = \mathcal{E}^\circ - \frac{0.0592}{2} \log \left(\frac{a_{Mn^{2+}}^3}{a_{H^+}^8} \right)$$

or

$$Eh = 1.812 - 0.237 \text{pH} - 0.0887 \log a_{Mn^{2+}} \tag{11.20}$$

This is plotted for selected values of Mn^{2+} activity ranging from 10^{-1} to 10^{-6} in Figure 11.7.

For the equal activity contour of Mn^{2+} and MnO_4^{2-},

$$MnO_4^{2-} + 8H^+ + 4e = Mn^{2+} + 4H_2O$$

with

$$\Delta_r G° = -675,916 \, J \, mol^{-1}$$

Then

$$\mathcal{E}° = 672,369/4 \times 96,485$$
$$= 1.751 \, V$$

and

$$Eh = 1.751 - 0.1182pH - 0.0148 \log(a_{Mn^{2+}}/a_{MnO_4^{2-}})$$

and where the activities of these two aqueous species are equal, this reduces to

$$Eh = 1.751 - 0.1182pH \tag{11.21}$$

This boundary lies at high Eh and pH and is illustrated in Figure 11.7.

11.10   Oxygen Fugacity

As mentioned in Section 11.8, there are two methods in common use to represent the same fundamental variable – the oxidation state of a system. It is time to look at the second method, oxygen fugacity. This is a convenient parameter because any reaction that involves a change in oxidation state (any redox reaction) can be written so as to include oxygen as a reactant or product, whether or not oxygen is actually involved in the reaction. We gave an example in Chapter 9 (Section 9.4.2) of the oxidation of magnetite (in which some of the iron occurs as Fe^{3+} and some as Fe^{2+}) to hematite (in which all of the iron occurs as Fe^{3+}). This reaction often occurs in systems which contain no oxygen molecules at all. Nevertheless, the calculated f_{O_2} ($10^{-68.40}$) for magnetite–hematite equilibrium is a perfectly valid thermodynamic parameter, and the redox reaction involving $O_2(g)$, namely

$$6 \, Fe_2O_3(s) = 4 \, Fe_3O_4(s) + O_2(g)$$

is simpler than the equivalent reaction in Eh mode:

$$3 \, Fe_2O_3(s) + 2 \, H^+ + 2e = 2 \, Fe_3O_4(s) + H_2O$$

We will discuss the relative merits of the two methods a little more further on.

11.10.1   Calculation of Oxygen Fugacity–pH Diagrams

Because f_{O_2} and Eh are both indicators of the same thing – oxidation state – it is possible to draw $\log f_{O_2}$–pH diagrams that are analogous to the Eh–pH calculations we have outlined above. To illustrate this we will construct a $\log f_{O_2}$–pH diagram for the same $Mn–H_2O–O_2$ system at $25 °C$ as has already been described. The completed diagram is shown in

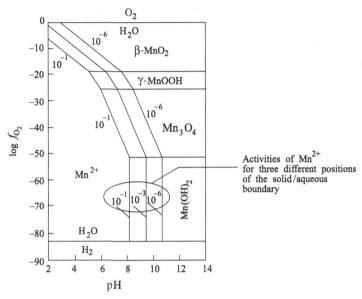

Figure 11.8 The $\log f_{O_2}$–pH diagram for the system Mn–H_2O–O_2 at 25 °C, 1 atm.

Figure 11.8 and should be compared with the analogous Eh–pH diagram of Figure 11.7. The two diagrams are similar except that most of the phase boundaries on the Eh–pH diagram have the Nernst slope, whereas those on the $\log f_{O_2}$–pH diagram have zero slope. Lines on Eh–pH diagrams quite typically have non-zero slopes because hydrogen ions and electrons are so commonly involved in half-cell reactions. Reactions balanced with oxygen instead of electrons require H^+ ions much less frequently, and reactions that contain no hydrogen ions have zero slope.

The method of calculating $\log f_{O_2}$–pH boundaries is illustrated with three examples. All other boundaries are derived in the same way. Our examples include the boundaries for water stability and for coexisting minerals, as well as the aqueous solubility contours of a mineral. Notice that half-cell reactions are not involved in these calculations.

For the water stability boundaries, the dissociation reaction of water is

$$2H_2O = O_2(g) + 2H_2(g) \tag{11.22}$$

To calculate the equilibrium constant for this reaction,

$$\Delta_r G° = -2(-237,129)$$
$$= 474,258 \ \text{J} \, \text{mol}^{-1}$$
$$= -RT \ln K$$

and

$$K = 10^{-83.1}$$
$$= f_{H_2}^2 f_{O_2} \tag{11.23}$$

The upper boundary occurs at 1 atm $O_2(g)$ pressure or $\log f_{O_2} = 0$. The lower boundary is at 1 atm $H_2(g)$ pressure; from the equilibrium constant (11.23), this corresponds to $\log f_{O_2} = -83.1$. As noted before, water can exist under conditions outside of these boundaries, but only if the pressure of oxygen or hydrogen is greater than 1 atm.

For the boundary $Mn(OH)_2/Mn_3O_4$,

$$\tfrac{1}{2}O_2(g) + 3Mn(OH)_2 = Mn_3O_4 + 3H_2O \tag{11.24}$$

and

$$\Delta_r G^\circ = -143{,}093 \text{ J mol}^{-1}$$
$$= -RT \ln(1/f_{O_2}^{1/2})$$

Hence

$$\log f_{O_2} = -50.14 \tag{11.25}$$

For solubility of Mn_3O_4 as Mn^{2+},

$$3H_2O(l) + 3Mn^{2+} + \tfrac{1}{2}O_2(g) = Mn_3O_4 + 6H^+ \tag{11.26}$$

with

$$\Delta_r G^\circ = 114{,}223 \text{ J mol}^{-1}$$
$$= -RT \ln\left(\frac{a_{H^+}^6}{f_{O_2}^{1/2} a_{Mn^{2+}}^3}\right)$$

or

$$\log f_{O_2} = 40 - 12\text{pH} - 6\log a_{Mn^{2+}} \tag{11.27}$$

This is plotted for selected values of Mn^{2+} activity ranging from 10^{-1} to 10^{-6} in Figure 11.8.

11.10.2 Interrelating *Eh*, pH, and Oxygen Fugacity

The obvious similarity between the *Eh*–pH and $\log f_{O_2}$–pH diagrams of Figures 11.7 and 11.8 suggests that it should be possible to convert directly from one set of coordinates to the other. This can be done using the half-cell reaction

$$4H^+ + O_2(g) + 4e = 2H_2O(l) \tag{11.15}$$

and its related Nernst equation,

$$Eh = \mathcal{E}^\circ - \frac{RT}{n\mathcal{F}} \ln[1/(f_{O_2} a_{H^+}^4)] \tag{11.28}$$

This equation can be used to interrelate the three variables *Eh*, pH, and f_{O_2}. The calculations are shown in Section 11.9.2, leading to Equation (11.17) for the *Eh* of solutions having an f_{O_2} of 1 atm. However, instead of 1 atm, we can just as easily insert any other f_{O_2} value, leading to the more general relation

$$Eh = 1.23 + 0.0148 \log f_{O_2} - 0.0592 \text{pH} \tag{11.29}$$

(a)   (b)

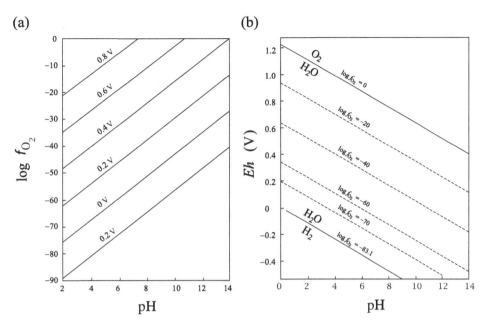

Figure 11.9 (a) *Eh* contours (in volts) on a $\log f_{O_2}$–pH diagram. (b) $\log f_{O_2}$ contours on an *Eh*–pH diagram. Calculated for 25 °C, 1 atm.

Figure 11.9 shows *Eh* contours calculated from (11.29) drawn on a $\log f_{O_2}$–pH diagram, and $\log f_{O_2}$ contours on an *Eh*–pH diagram. The controlling factor is the oxidation state of the system, which in turn is controlled by the bulk composition, and f_{O_2}, f_{H_2}, *Eh*, pe, and all other related variables are simply different ways of quantifying the same thing.

11.11   Summary

On a broad scale the Earth shows a large range of redox conditions, from the highly reduced Ni–Fe core through various silicate layers up into the zone of free water and eventually into the oxygen-rich atmosphere. Therefore, an indicator of the redox state is among the more important of the variables manipulated by geochemists. Like pH, it is an important parameter because it is intimately linked to a large number of reactions of interest to anyone trying to understand the Earth, but again like pH, it is actually no more or less fundamental than any other intensive parameter. For a closed system at a given *T* and *P*, it is completely determined by the bulk composition, as are all intensive parameters, and changes in redox state are accomplished by changing bulk composition.

The measurement of redox conditions by means of a cell voltage, where one electrode has a fixed reference potential and the other is expected to react reversibly with natural systems, is attended by a number of problems. The platinum electrode works well only under certain conditions. It is difficult to get the electrode into reducing environments without allowing some oxidation, and the method is restricted to ambient conditions except in research laboratories. We put up with these problems because there is little choice.

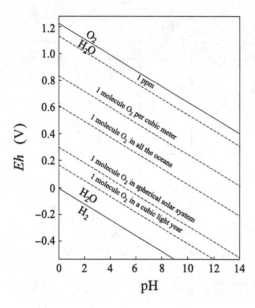

Figure 11.10 An *Eh*–pH diagram with contours of the oxygen concentrations that would result in the redox conditions shown. Obviously, redox conditions over most of the diagram represent solutions that contain no oxygen molecules at all.

1 molecule of O_2 gives $f_{O_2} = 10^{-65}$ bars

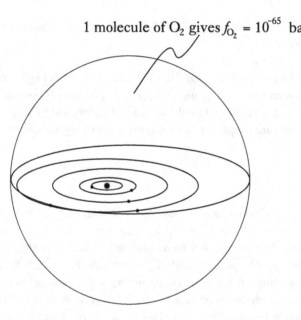

Figure 11.11 An oxygen fugacity $f_{O_2} = 10^{-65}$ bars is equivalent to one molecule of O_2 in a sphere large enough to contain the orbit of Pluto.

Oxygen fugacity, on the other hand, although a much simpler concept, can be directly measured only at high temperatures in research laboratories, a fact that might seem to rule out its use at Earth-surface conditions, but actually does not. Oxygen fugacity can be used (as opposed to *measured*) under any redox conditions, including for systems that contain no oxygen molecules whatsoever, as illustrated in Figures 11.10 and 11.11 (they may contain oxygen in combined form, such as H_2O, of course). Oxygen fugacity is an index of redox conditions, being in some cases but not always an approximation to the partial pressure of $O_2(g)$ in a system. It can be calculated because *any* redox reaction can be written in such a way as to include oxygen, whatever is actually happening in the system. In Chapter 1 we emphasized that we would be describing a *model* of chemical reactions, not real reactions. Parameters such as oxygen fugacity illustrate this point. In a sense, we have oxygen in our models, though perhaps not in our systems.

Eh has been important both in the measuring and in the reporting of redox conditions, but an argument can be made that it should be used only in the measurement and not the reporting of redox conditions; that is, that however measurements are *made*, results should be *reported* as f_{O_2} or $\log f_{O_2}$ values. There would be two advantages to this. First, the use of *Eh* entails the use of a relatively complex set of conventions, which are quite difficult to remember unless used continuously. Second, and more importantly, *Eh* is less useful than f_{O_2} because it is so commonly linked in reactions with pH (giving the "Nernst slope"). This means that *a value for Eh without an accompanying value for* pH *is usually meaningless.* This is illustrated by any *Eh*–pH diagram, in which you can see that an *Eh* of 0.0 V, for example, indicates much more reducing conditions at pH 2 than it does at pH 10. The conversion from an *Eh*–pH point to a $\log f_{O_2}$ value is very simple – Equation (11.29). The very small values of f_{O_2} generally obtained at low temperatures should not be a hindrance to its use, which would greatly simplify the reporting and interpretation of any redox conditions.

Exercises

E11.1 Construct a $\log f_{O_2}$–pH diagram for the sulfide species H_2S, HS^-, and S^{2-} as well as sulfate SO_4^{2-}, having a total sulfur concentration sufficiently low that native sulfur does not appear (less than about 0.01 molal), at 150 °C. The sulfide–sulfate reactions with $\log K$ values are shown in Table 11.1.

Table 11.1. Reactions and *log K* values for some sulfide–sulfate reactions at 150 °C. Data from The Geochemist's Workbench (2015).

Reaction	$\log K_{150}$
$H_2S(aq) + 2 O_2(g) = 2 H^+ + SO_4^{2-}$	79.0464
$HS^- + 2 O_2(g) = H^+ + SO_4^{2-}$	85.8364
$S^{2-} + 3.5 O_2(g) + H_2O = 2 H^+ + 2 SO_4^{2-}$	151.2715
$H_2S + 2 O_2(g) = H^+ + HSO_4^-$	83.3181
$HSO_4^- = SO_4^{2-} + H^+$	-4.2717
$H_2S = HS^- + H^+$	-6.79

Plot the galena/anglesite boundary on the diagram. The relevant reaction is

$$PbSO_4 = PbS + 2 O_2(g) \qquad \log K = -82.982$$

E11.2 Uranium can occur in U^{4+}, U^{5+}, and U^{6+} oxidation states. Minerals that contain uranium in its 4+ state, such as uraninite (UO_2) and coffinite ($USiO_4$), are very insoluble, so U is normally transported in solutions in which most of it is in the 6+ or possibly 5+ state, and precipitation occurs when U is reduced to the 4+ state. Thus redox reactions are important in the formation of uranium ore deposits.

Table 11.2. Extra data for Exercise E11.2.

Species	$\Delta_f G°$ $(kJ\,mol^{-1})$
$UO_2CO_3^\circ(aq)$	-1567.2
$UO_2(CO_3)_2^{2-}(aq)$	-2164.1

(a) Calculate the standard potential for the following half-cells. These are required for the Nernst equations in part (b). (See Table 11.2.)

$$UO_2^{2+}(aq) + 2e = UO_2(s)$$

$$UO_2CO_3^\circ(aq) + 2H^+ + 2e = UO_2(s) + CO_2(g) + H_2O(l)$$

$$UO_2(CO_3)_2^{2-}(aq) + 4H^+ + 2e = UO_2(s) + 2CO_2(g) + 2H_2O(l)$$

(b) Calculate the pH for equal activities of $UO_2^{2+}(aq)$ and $UO_2CO_3^\circ(aq)$, and for $UO_2CO_3^\circ(aq)$ and $UO_2(CO_3)_2^{2-}(aq)$, which are boundaries 4 and 5.

The relevant reactions are

$$UO_2CO_3^\circ(aq) + 2H^+ = UO_2^{2+}(aq) + CO_2(g) + H_2O(l)$$

$$UO_2(CO_3)_2^{2-}(aq) + 2H^+ = UO_2CO_3^\circ(aq) + CO_2(g) + H_2O(l)$$

(c) Write the Nernst equations for boundaries 1, 2, and 3, and derive an equation in Eh and pH for each. All aqueous species have an activity of 10^{-6} (concentration $10^{-6}\,m$), the fugacity of $CO_2(g)$ is constant at 0.1 bar, and water and uraninite have an activity of 1.0.

(d) Construct an Eh–pH diagram for uranium–water–CO_2 using these equations. Figure 11.12 shows the arrangement of the fields, but not the correct position of the boundaries, as it was drawn using other data. Plot f_{O_2} contours of 10^{-20}, 10^{-40}, and 10^{-60} bar across the diagram.

E11.3 An aqueous solution has a pH of 6.0 and is in equilibrium with hematite, $Fe_2O_3(s)$, and magnetite, $Fe_3O_4(s)$.

(a) What are the Eh and oxygen fugacity of the solution?

(b) What is the oxygen fugacity of the same solution at $300\,°C$?

E11.4 Iodine in aqueous solutions exists mostly as iodate ion (IO_3^-) in relatively oxidized solutions and as iodide ion (I^-) in more reduced solutions.

(a) What is the valence of iodine in each of these ions?

(b) Calculate the standard electrode potential $(\mathcal{E}°)$ for the iodate–iodide redox couple, plus the Eh of a solution having a pH of 6.0 and equal activities of the two iodine species.

(c) What is the oxygen fugacity of the solution?

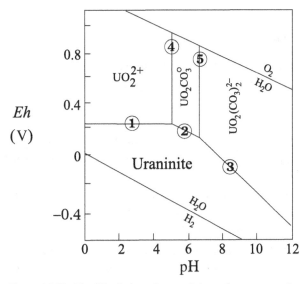

Figure 11.12 *Eh*–pH relations for uraninite and aqueous carbonate species. Note that the diagram shows that uraninite is stable only under quite reducing conditions.

    (d) What ionic species of iodine would you expect in a solution in equilibrium with magnetite?

    (e) Sketch the iodate–iodide boundary on an *Eh*–pH diagram. Sketch a log activity vs. *Eh* diagram through this boundary to show the nature of the boundary.

E11.5 An electrochemical cell like that in Figure 11.2, but using zinc instead of iron, is called a Daniell cell. This cell was actually widely used as a source of electrical power in the nineteenth century. Do a calculation to show why zinc was favored over iron in making this cell.

E11.6 For some reason students often conclude that an $Eh > 0$ means oxidizing conditions, and $Eh < 0$ means reducing conditions. This is not true. What does $Eh = 0$ really mean?

E11.7 Calculate the activity of Fe in coexisting magnetite and hematite at 700 °C, 1 atm. Calculate a_{Fe} for the other iron-mineral pairs $FeO-Fe_3O_4$ and $FeO-Fe$ at the same T, P. This should be sufficient to demonstrate the use of a_{Fe} as a redox indicator.

E11.8 Of course, our new a_{Fe} indicator is limited to systems containing iron. On the other hand, a_{Fe} can also be used as a sulfidation indicator. Calculate a_{Fe} in coexisting pyrite and pyrrhotite in which a_{FeS} is 0.9, at 500 K, 1 bar.

E11.9 Calculate and draw the *Eh*–pH diagram for the iron–water system, using aqueous ion activities of 10^{-6} (concentrations of $10^{-6} m$). See whether you can write the relevant half-cell reactions (shown below) simply by looking at what lies on opposite sides of the boundaries (e.g., Fe^{2+}/hematite) and adding the water, hydrogen ions, and electrons needed to balance. It helps to know that in hematite (Fe_2O_3) both iron atoms are ferric (Fe^{3+}), while in magnetite (Fe_3O_4) one Fe is ferrous (Fe^{2+}) and two are ferric. These are the reactions you should write:

$$Fe^{3+} + e = Fe^{2+}$$

$$2\,Fe^{3+} + 3\,H_2O = Fe_2O_3 + 6\,H^+$$

$$Fe_2O_3 + 6\,H^+ + 2\,e = 2\,Fe^{2+} + 3\,H_2O$$

$$3\,Fe_2O_3 + 2\,H^+ + 2\,e = 2\,Fe_3O_4 + H_2O$$

$$Fe_3O_4 + 8\,H^+ + 2\,e = 3\,Fe^{2+} + 4\,H_2O$$

(a) Calculate and draw Fe^{3+}/hematite, Fe^{2+}/hematite, and Fe^{2+}/magnetite boundaries for aqueous ion activities of 10^{-4} and 10^{-8}. Make them lighter than the 10^{-6} boundaries.

(b) The diagram shows the expected Eh–pH relations between ferrous iron and hematite. What usually precipitates when Fe^{2+} oxidizes, however, is not hematite but ferric hydroxide, $Fe(OH)_3(s)$.

   (i) Write a balanced reaction between hematite and ferric hydroxide, and show that ferric hydroxide is metastable with respect to hematite.

   (ii) Calculate and draw the Fe^{2+}/$Fe(OH)_3$ boundary for $a_{Fe^{2+}} = 10^{-6}$. The half-cell reaction is

$$Fe(OH)_3(s) + 3\,H^+ + e = Fe^{2+} + 3\,H_2O$$

(c) On the same diagram, draw the H_2S/SO_4^{2-} boundary.

$$SO_4^{2-} + 10\,H^+ + 8\,e = H_2S(aq) + 4\,H_2O$$

(d) A groundwater sample smelling of rotten eggs (H_2S) contains 6000 ppm SO_4^{2-} at an Eh of -0.1 V and a pH of 4.0. Could the SO_4^{2-} and H_2S be in equilibrium, or is sulfate being reduced, or is H_2S being oxidized? Bear in mind that acid water at atmospheric pressure can contain no more than about 0.1 molal H_2S. A simple statement of what is happening is not good enough; you have to show why your answer is right.

E11.10 Calculate the standard potential for the half-cell reaction

$$Fe(OH)_3(s) + 3\,H^+ + e = Fe^{2+} + 3\,H_2O(l)$$

What would be the complete cell reaction and the complete cell voltage, using the standard hydrogen electrode?

E11.11 A sample of swamp water has an Eh of 0.533 V, a pH of 3.85, and an Fe^{2+} concentration of 475 ppm. Should ferric hydroxide, $Fe(OH)_3(s)$, precipitate? What would be the equilibrium Eh at this pH ? The atomic weight of Fe is 55.847 $g\,mol^{-1}$.

E11.12 What is the f_{O_2} of the swamp water in Exercise E11.11?

E11.13 Calculate the standard half-cell potential

$$MnO_2(s) + 4\,H^+ + 2\,e = Mn^{2+} + 2\,H_2O(l)$$

Calculate the activity of Mn^{2+} in a solution having pH $= 7.5$ and $Eh = 0.6$ V.

E11.14 What is the f_{O_2} of this solution, i.e, one having pH $= 7.5$ and $Eh = 0.6$ V? Use

$$4\,H^+ + O_2(g) + 4\,e = 2\,H_2O(l)$$

If metallic Mn were present, would it oxidize to $MnO_2(s)$?

E11.15 An aqueous solution has a pH of 6.0 and is in equilibrium with hematite [$Fe_2O_3(s)$] and magnetite [$Fe_3O_4(s)$]. What is the Eh and oxygen fugacity of the solution?

E11.16 What is the oxygen fugacity of the same solution at $300°C$?

E11.17 Iodine in aqueous solutions exists mostly as iodate ion (IO_3^-) in relatively oxidized solutions, and as iodide ion (I^-) in more reduced solutions. What is the valence of iodine in each of these ions? Calculate the standard electrode potential ($\mathcal{E}°$) for the iodate–iodide redox couple, and the Eh of a solution having a pH of 6.0, and equal activities of the two iodine species.

E11.18 Calculate the oxygen fugacity of the solution in Question E11.17. What ionic species of iodine would you expect in a solution in equilibrium with magnetite?

Additional Problems

A11.1 Rounded, fist-sized nodules are found over wide areas of the Pacific Ocean floor. They are composed mostly of Mn oxides and hydroxides, and in some areas they contain valuable quantities of Cu and Zn. How they form has been a long-standing problem. One idea is that they simply precipitate very slowly from the ocean water.

Test this idea by constructing an Eh–pH diagram (suggested ranges pH 4 to 14, Eh -0.6 V to $+1.2$ V) showing the stability fields of the stable Mn phases pyrolusite, manganite, hausmannite, and pyrochroite, assuming an average Mn^{2+} concentration of 2 ppb (parts per billion).

The pH of the sea is 8.1. Compare the results you get by doing the following.

(a) Assuming the sea is in equilibrium with the atmosphere, which has an oxygen fugacity of about 0.2 atm.

(b) An empirical upper limit to observed Eh values in natural waters is

$$Eh = 1.04 - 0.059\text{pH}$$

That is, observed values never plot on the 0.2 bar contour of the diagram, as they should, because the electrodes do not respond well to oxygen. Assume that the Eh is controlled by this empirical upper limit.

(c) Using Eh values actually measured in marine waters, that is, 0.25 to 0.4 V, averaging about 0.35 V.

This diagram can, strictly speaking, be used only to see whether the sea is over- or under-saturated with the mineral phases used in the calculations. However, Mn nodules consist not of these phases, but of minerals such as birnessite and todorokite. These phases are extremely fine-grained and poorly crystallized, and are probably metastable with respect to the phases used in the calculations. Therefore they probably have more positive free energies of formation. What effect will this have on your conclusions? In other words, could a conclusion that Mn is or is not precipitating be reversed by using phases having more positive free energies?

Table 11.3. Extra data, not in Appendix B. (MnO_2 in Appendix B is the same as β-MnO_2.)

Mineral	Formula	$\Delta_f G°$ (kJ mol^{-1})
Manganite	γ-MnOOH	-557.309
Pyrochroite	$Mn(OH)_2$	-616.303

The relevant reactions are as follows (* indicates done in the text.)

1. β-$MnO_2 + 4H^+ + 2e = Mn^{2+} + 2H_2O$
2. γ-$MnOOH + 3H^+ + e = Mn^{2+} + 2H_2O$
3.* $Mn_3O_4 + 8H^+ + 2e = 3Mn^{2+} + 4H_2O$
4. $Mn^{2+} + 2OH^- = Mn(OH)_2$
5. β-$MnO_2 + H^+ + e = \gamma MnOOH$
6. 3γ-$MnOOH + H^+ + e = Mn_3O_4 + 2H_2O$
7.* $Mn_3O_4 + 2H_2O + 2H^+ + 2e = 3Mn(OH)_2$
8.* $MnO_4^{2-} + 8H^+ + 4e = Mn^{2+} + 4H_2O$

A11.2 Once you get accustomed to doing the sort of exercises found in this book you may feel that you know the subject. Actually this is the easy part. Out there in the field, nobody tells you what reactions to consider. You have to figure out what nature is doing before you can do your calculations and plot your diagrams. A good example of how difficult this can be is the work of Barnes *et al.* (1964) on the water draining from abandoned coal mines in Pennsylvania. This acid mine drainage (AMD) causes considerable environmental damage because of iron precipitation, acidity, and dissolved toxic metals. The photo on the front cover shows the typical yellow–orange to red–orange precipitation. Remediation of AMD sites is a large and complex subject.

The overall process is complex, with many different reactions, catalyzed by acidophilic bacteria, but is commonly simplified to three reactions. First, sulfide minerals, usually pyrite, are dissolved:

$$2FeS_2 + 7O_2 + 2H_2O = 2Fe^{2+} + 4SO_4^{2-} + 4H^+$$

See Exercise E9.20 for more detail on this reaction. Then ferrous iron is oxidized to ferric iron:

$$4Fe^{2+} + O_2 + 4H^+ = 4Fe^{3+} + 2H_2O$$

which results in precipitation of ferric hydroxide, giving the orangey color:

$$4Fe^{3+} + 12H_2O + 4Fe(OH)_3(ppd) + 12H^+$$

Or you might consider these two reactions to be combined:

$$4\,Fe^{2+} + 10\,H_2O + O_2 = 4\,Fe(OH)_3(ppd) + 8\,H^+$$

This process requires oxygen and creates acidity (H^+), which is not a problem for reactions occurring at the Earth's surface where oxygen is abundant. Deep within a flooded coal mine, however, pyrite still dissolves, but there is little or no dissolved oxygen. What reaction would occur there? Perhaps, with little or no dissolved oxygen, oxygen might come from the water, resulting in

$$FeS_2 + 8\,H_2O = Fe^{2+} + 2\,SO_4^{2-} + 2\,H^+ + 7\,H_2$$

and, if so, what happens to the hydrogen? Or maybe it is some other reaction. Various possibilities arise.

Careful sampling of the waters at depth in the flooded mine workings (a difficult procedure in itself) and field measurement of Eh gave the measurements in Table 11.4 in the Loree #2 shaft near Wilkes-Barre, Pennsylvania.

Table 11.4. Field measurements at the Loree #2 shaft.

Sample no.	Depth (ft.)	Eh (V)	pH	Fe^{2+} (ppm)	SO_4^{2-} (ppm)
1	224	+0.322	3.41	34	1260
2	292	+0.533	3.55	478	3320
3	467	+0.568	3.36	488	3650
4	751	−0.025	4.0	1073	5800
5	808	−0.103	3.72	1463	6720

Samples 2 and 3 showed a yellow turbidity, the rest were clear. Samples 4 and 5 smelled of rotten eggs, the rest were odorless. Construct an Eh–pH diagram and use it to answer the following questions.

(a) Is the yellow turbidity due to fine particles of $Fe(OH)_3$? If so, is the mine water more or less in equilibrium with it or is it actively precipitating?
(b) Why is sample no. 1 clear?
(c) Is pyrite stable in any of these samples?
(d) In samples nos. 4 and 5 which contain H_2S, are H_2S and SO_4^{2-} apparently at equilibrium, or is sulfur being oxidized or reduced?
(e) Above what pH might pyrite and ferric hydroxide coexist?

The following half-cell reactions will be useful:

$$Fe^{2+} + 3H_2O = Fe(OH)_3 + 3H^+ + e$$

$$FeS_2 + 8H_2O = Fe^{2+} + 2SO_4^{2-} + 16H^+ + 14e$$

$$H_2S + 4H_2O = 10H^+ + SO_4^{2-} + 8e$$

$$FeS_2 + 11H_2O = Fe(OH)_3 + 2SO_4^{2-} + 19H^+ + 15e$$

Strictly speaking, you should draw several diagrams for the different water compositions, or several sets of boundaries on one diagram. You must experiment to determine how many boundaries are necessary to answer the questions. For example, you could safely draw one $Fe^{2+}/Fe(OH)_3$ boundary to consider both sample no. 2 and sample no. 3.

A11.3   In contrast to the case of lead in hydrothermal solutions, which is transported as chloride complexes and deposited as sulfide, gold is transported at somewhat higher temperatures as sulfide complexes and is deposited as the native metal by processes which reduce the sulfide activity in the solution. In order to know how cooling, mixing, boiling, reaction with wall rocks, or other processes might be effective in precipitating metallic minerals, you must know the nature of the complexes involved. Metals in solution mostly as chloride complexes will react to these processes quite differently from metals dissolved as sulfide or bisulfide complexes.

The Rotokawa geothermal system in New Zealand is an active gold depositing environment (Krupp and Seward, 1987). To better understand how changes in oxidation state and pH affect the solubility of gold, we calculate a $\log f_{O_2}$–pH diagram and use it to plot a representative fluid from the geothermal system. To do this we need equilibrium constants for several reactions. These are given in Table 11.5, for a temperature of 320 °C.

Table 11.5. Some equilibrium constants at 320 °C.

	Reaction	$\log K_{320}$
1	$H_2S(aq) + 2O_2 = HSO_4^- + H^+$	50.986
2	$HSO_4^- = H^+ + SO_4^{2-}$	−6.608
3	$H_2S(aq) + 2O_2 = SO_4^{2-} + 2H^+$	44.378
4	$H_2S(aq) = HS^- + H^+$	−7.850
5	$HS^- + 2O_2 = SO_4^{2-} + H^+$	52.228

The sulfide complexation of gold is expressed by the reaction given by Krupp and Seward,

$$Au(s) + H_2S(aq) + HS^- = Au(HS)_2^- + \tfrac{1}{2}H_2(g); \quad \log K = -1.03$$

Note that this is a redox reaction between Au and Au^+, using $H_2(g)$ as the reductant. It is more common to use oxygen. We can either convert all the aqueous sulfide–sulfate reactions above to the corresponding H_2 versions, and use $\log f_{H_2}$–pH space, or convert the gold reaction to its O_2 form and use $\log f_{O_2}$–pH space. We do the latter. The necessary relationship is

$$H_2O(l) = H_2(g) + \tfrac{1}{2}O_2(g); \quad \log K_{320} = -16.952$$

or $\log f_{H_2} = -\tfrac{1}{2}\log f_{O_2} - 16.952$.

We will model the behavior of Sample RK4a from Table 2 of Krupp and Seward (1987), which they say is representative of the deep fluid, i.e., the fluid which deposits gold as it rises and boils.

(a) Rework the equilibrium constant expressions for the aqueous sulfide–sulfate reactions into the form $\log f_{O_2} = a \cdot \text{pH} + b$, where a and b are constants, and plot the boundaries of the H_2S, HS^-, HSO_4^- and SO_4^{2-} fields in $\log f_{O_2}$–pH space.

(b) Plot the position of the 100 ppb contour of $Au(HS)_2^-$. For 100 ppb Au, the equations work out to be

$$\log f_{O_2} = 2\,\text{pH} - 51.446 \text{ in the } HS^- \text{ field}$$

$$\log f_{O_2} = -2\,\text{pH} - 20.046 \text{ in the } H_2S \text{ field}$$

where ppb is parts per billion, or micrograms per kg (μg/kg).

(c) Calculate the f_{O_2} of sample RK4a from the values in Table 11.6. Plot its position on your $\log f_{O_2}$–pH diagram.

(d) What is the concentration of gold (as $Au(HS)_2^-$) in μg/kg in sample RK4a?

(e) Why does the contour in the H_2S field have a negative slope? Do contours for greater Au values lie above or below this contour?

(f) How can you justify using equilibrium thermodynamics on a geothermal system which is so far from being in a state of equilibrium?

More extensive modeling can proceed from here to investigate the effects of cooling, mixing, or pH change. Krupp and Seward conclude that most of the gold will precipitate when the solution cools to below 200 °C.

A11.4 Calculate the half-cell potentials of the silver/silver chloride standard reference electrode and the sulfate–H_2S reaction. Combine these to calculate the standard potential of a cell having these two electrodes (a) if the Ag/AgCl electrode is written as a reduction; (b) if it is written as an oxidation.

Table 11.6. Some compositions from sample RK4a in Table 2 of Krupp and Seward (1987). The pH is the value at aquifer temperature, not after quenching and cooling the sample. Concentrations are in mmol/kg, $T = 318\,°C$ and pH = 5.75.

Na^+	19.40
K^+	3.47
Ca^{2+}	0.0407
SO_4^{2-}	0.0565
SiO_2	11.164
H_2CO_3	278.1
H_2S	6.74
H_2	1.85
CH_4	8.65

A11.5 What would be the observed cell potential in cell (b) in the previous question (with the sulfur electrode written as a reduction) if $a_{SO_4^{2-}} = 0.001$ and $a_{H_2S(aq)} = 0.01$ at a pH of 5.0, assuming that the sulfate–sulfide reaction responded reversibly? (Don't forget that the potential of the Ag/AgCl electrode is constant – all activity terms in its half-cell reaction are fixed, including that of Cl^-).

What would be the Eh of this solution?

A11.6 What is the f_{O_2} of this solution?

A11.7 What is the vapor pressure of liquid mercury at 25 °C? At 100 °C?

A11.8 Calculate the half-cell potential of the Hg^{2+}/Hg_2^{2+} and $Hg_2^{2+}/Hg(l)$ redox couples, and sketch an Eh–pH diagram showing these boundaries. (Note: the first boundary is for $a_{Hg^{2+}}^2 = a_{Hg_2^{2+}}$; the second is for $a_{Hg_2^{2+}} = a_{Hg(l)}$).

Which species should predominate in the ocean, which has a pH of 8.1 and an Eh of 0.35 V?

A11.9 Convert one of these half-cell reactions to its equivalent oxygen fugacity form, and sketch it on a $\log f_{O_2}$–pH diagram.

A11.10 Calculate the standard half-cell potential for

$$Zn^{2+} + SO_4^{2-} + 8\,H^+ + 8e = ZnS(s) + 4\,H_2O(l)$$

where the ZnS is sphalerite, and write the equation of this boundary in terms of Eh and pH for a solution having $a_{Zn^{2+}} = 10^{-6}$ and $a_{SO_4^{2-}} = 10^{-5}$. If this solution has an Eh of 0 V and a pH of 5, will ZnS precipitate?

A11.11 Calculate the standard half-cell potential of the reaction

$$Au^{3+} + 2\,e = Au^+$$

and sketch the Au^+/Au^{3+} boundary on an Eh–pH diagram. What do you conclude about which ion is likely to be more abundant in natural solutions? Recall that the equations for the upper and lower boundaries of water stability are

$$Eh = 1.23 - 0.0592\,pH \quad (f_{O_2} = 1.0\,bar)$$

$$Eh = 0.00 - 0.0592\,pH \quad (f_{H_2} = 1.0\,bar)$$

Box 11.1   You should by now be convinced that we use equations which are derived for idealized equilibrium conditions to better understand real, highly irreversible processes like ore formation and acid mine drainage. Thermodynamics would be of very limited use in understanding natural processes without good field observations and good thermochemical data, but together they are an indispensable tool in helping to develop such understanding.

Thermodynamic theory is not difficult. Finding out what nature is doing or has done in the distant past is difficult. Geochemists call that doing fieldwork.

12 Phase Diagrams

12.1 What Is a Phase Diagram?

A phase diagram in the general sense is any diagram that shows what phase or phases are stable as a function of some chosen system variable or variables. Therefore, the Eh–pH, $\log f_{O_2}$–pH, and activity–activity diagrams we have been looking at are a kind of phase diagram. However, if you mention the subject of phase diagrams to a petrologist, a metallurgist, or a ceramic scientist, they will immediately think of a particular type of diagram that is of great usefulness in these subjects. In these sciences, the compositions of phases and their relationships during phase changes, particularly solid $\rightarrow$ liquid and liquid $\rightarrow$ solid changes, are of particular importance, so diagrams that depict this information as a function of temperature and pressure have come to be the subject of "phase diagrams."

12.1.1 Thermodynamics and Phase Diagrams

Though it is true that phase relations can always be described in terms of the thermodynamic principles and equations we have been discussing, and that any phase diagram can in principle be calculated given the appropriate data, the emphasis in this chapter changes from one of calculating what we want to know from numbers in tables of data, to one of simply representing experimentally derived facts in diagrammatic form. The reason for this is that once we get into systems more complex than a single component, and especially when high-temperature melt phases are involved, the calculations are often not possible because the data are not available, or, even if they are available, the results are not very accurate, because they are very sensitive to small inaccuracies in the data. Therefore, in this book, although we will show the relationship between functions such as G and our diagrams, this will be in an illustrative rather than a quantitative way.

12.1.2 Phase Diagrams as Models

Metallurgists and ceramicists quite often deal with simple two- and three-component systems and use phase diagrams to represent their experimental results on the phase relations in these systems. The diagrams therefore truly represent their systems. Petrologists, on the other hand, are interested in the origins of natural rocks, which commonly have 10 or more important components. Systems this complex cannot be represented in simple diagrams and, in fact, can hardly even be thought about in a quantitative way. Experiments can and have been done using natural rocks, but the results are complex and may not be generally applicable. Therefore, petrologists use simpler systems such as those having two and three components to better understand the principles involved and to investigate simple models of the complex systems in nature.

Phase diagrams represent *equilibrium* relationships. Once these have been depicted, simple *processes* such as melting and crystallization can be considered, but, because as represented on diagrams these involve continuous successions of equilibrium states, they are *reversible processes* in the sense of Section 2.6.1.

12.2   The Phase Rule

The phase rule is a simple but profound relationship, derived by Gibbs (1961a), relating the composition of a system, the number of phases it has, and something called the variance of the system. It requires no thermodynamic data, and although simple in principle, and easily applied to the simple systems usually used to explain it, it can be surprisingly difficult to use when considering geological systems. It is absolutely essential in discussing phase diagrams, and understanding why they look the way they do.

Phase Rules

The phase rule used or implied in modeling systems having aqueous solutions is somewhat different from the one derived by Gibbs. For example the system $NaCl–H_2O$ has two components, but this choice of components is not capable of defining the amounts of the aqueous *species* in the system. To do this a different choice of components, a different basis, is required. A typical choice of *basis species* for this system would be Na^+, Cl^-, $NaCl°$, H^+, OH^-, $NaOH°$, and $HCl°$. See Chapter 11, The Phase Rule, in the online resources.

12.2.1   System Variance or Degrees of Freedom

We discussed phases and components in Chapter 2. There remains the concept of *variance* or *degrees of freedom*. A single homogeneous phase such as an aqueous salt solution, say NaCl in water, has a large number of properties, such as temperature, density, NaCl molality, refractive index, heat capacity, absorption spectra, vapor pressure, conductivity, partial molar entropy of water, partial molar enthalpy of NaCl, ionization constant, osmotic coefficient, ionic strength, and so on. We know, however, that these properties are not all independent of one another. Most chemists know instinctively that a solution of NaCl in water will have all its properties fixed if temperature, pressure, and salt concentration are fixed. In other words, there are apparently three independent variables for this two-component system, or three variables that must be fixed before all variables are fixed. Furthermore, there seems to be no fundamental reason for singling out temperature, pressure, and salt concentration from the dozens of properties available – it's just more convenient; any three would do. The number of variables (system properties) that must be fixed in order to fix *all* system properties is known as the system variance or degrees of freedom.

Now consider two phases at equilibrium, say solid NaCl and a saturated salt solution. Again, intuition or experience tells us that we no longer have three independent variables, but two, because, for example, we cannot choose the composition of the salt solution once T and P have been fixed – it is fixed for us by the solubility of NaCl in water. If we then consider the possibility of having a vapor phase in equilibrium with the salt and the solution, we see that we lose another independent variable because we can no longer choose the pressure on the system independently once the temperature has been chosen – it is fixed by the vapor pressure of the system. So it would seem that, in general, we restrict the number of independent variables in a system by increasing the number of phases at equilibrium.

12.2.2   Mathematical Analogy

The variance of a chemical system is exactly analogous to the variance of a system of linear equations. For example, for the function

$$x + y + z = 0$$

if we choose $x = 2$, $y = 2$, then z is fixed at -4. The equation could be said to have a variance of two, because two variables must be fixed before all variables are fixed. Three variables minus one relationship between them (one equation) leaves two degrees of freedom. If in addition to this function we have another one involving the same variables, such as

$$2x - y + 4z = -19$$

we now have three variables and two functional relationships, and we are free to choose only one of the three variables, the other two then being fixed. For example, if we choose $x = 2$, then there is no further choice: $y = 3$ and $z = -5$. If we choose $x = 3$, then $y = 2.6$ and $z = -5.6$. This situation can be said to be *univariant* or to have one degree of freedom.

And, of course, if we have a third functional relationship, for example,

$$-3x + 2y - 7z = 35$$

then we have no choice: x, y, and z are fixed at 2, 3, and -5, respectively, and the situation is *invariant*.

The reason why the linear equations analogy for phase relationships is so exact is that there is in fact a thermodynamic equation for each phase (see below), and each of these equations has a number of independent variables equal to the number of components in the system plus two. And this, in turn, is because each component represents a degree of freedom (we can add or subtract each component), and there are two more because we defined our systems at the beginning as being able to exchange energy in only two ways – heat and one kind of work.[1] If the number of components is c, then the total number of independent system properties is $c + 2$. If there are p phases in the system, and each phase represents one equation, then there are p equations in $c + 2$ variables, or $c + 2 - p$ degrees of freedom. This is the phase rule:

$$f = c - p + 2 \tag{12.1}$$

where f is the number of degrees of freedom.

12.2.3 Derivation

A more concise derivation uses the Gibbs–Duhem equation,

$$0 = \mathbf{S}\,dT - \mathbf{V}\,dP + \sum_{i=1}^{c} n_i\,d\mu_i \tag{4.75}$$

From this equation we can see the number of independent intensive variables in any homogeneous phase. There are c terms containing μ, i.e., c independent compositional intensive variables, plus two other intensive terms, T and P, for a total of $c + 2$ intensive variables. In a single homogeneous phase, these $c + 2$ variables are linked by one equation (4.75), so only $c + 2 - 1$ of them are independent. If there are p phases, there are still only $c + 2$ intensive variables, because they all have the same value in every phase (at equilibrium), but now there is one equation (4.75) for each phase. Each additional equation reduces the number of independent variables by one, so there are now $c + 2 - p$ independent intensive variables. These independent intensive variables are called degrees of freedom, f, so

$$f = c - p + 2 \tag{12.2}$$

which is the phase rule. Because we usually consider systems at some fixed values of P and T, this "uses up" two degrees of freedom, so the phase rule becomes

$$f = c - p \tag{12.3}$$

which is sometimes called the "mineralogical phase rule."

To the extent that natural systems approach equilibrium, they obey the phase rule. You might reflect now and then on why natural systems should care about the results of this piece of mathematical reasoning.

12.3 Unary Systems

. . . .

Figure 12.1 shows a typical although hypothetical unary (one-component) diagram for compound α (α stands for the formula of some compound, such as NaCl or $CaCO_3$). Although the diagram shows three different *phases* (solid, liquid, and gas), all three have the same composition (whatever the chemical composition of α is), so the system is unary. This simple diagram contains a surprising amount of information, but you must know how

---

[1] If we included other kinds of work in our model, there would be an extra degree of freedom for each.

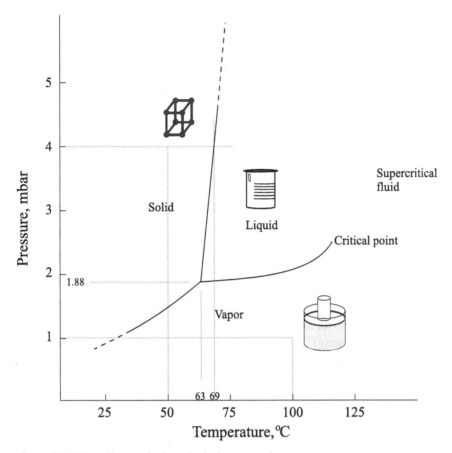

Figure 12.1 Phase diagram for hypothetical compound α.

to "read" the diagram. First, note that the diagram contains labeled *areas*, *lines* that separate the areas, and *points*. Every location on the diagram has a pair of *x–y* coordinates, that is, a pressure and a temperature. For example, a pressure of 4 mbar and a temperature of 50 °C are the coordinates of a point in the area marked "Solid." Under these conditions, the stable form of compound α is observed to be a crystalline solid. If the pressure on α is reduced to 1 mbar, and the temperature increased to 100 °C, the stable form of α is gaseous. Similarly, for any combination of pressures and temperatures within the area marked "Liquid," the stable form of α is liquid. The phase diagram is in fact a record of these experimental observations about the form of α under various conditions of *P* and *T*. As mentioned earlier, the vast majority of phase diagrams record the results of experiments – they are not usually the result of theoretical calculations. They are more often a *source* of thermodynamic data than the result of using such data.

Obviously, within these areas, *P* and *T* could be changed considerably without changing the nature of the phase, although the *properties* of the phase (its density or heat capacity, say) would certainly change with *P* and *T*. It appears, then, that for α, and for any pure compound, we must choose *two* variables in order to define the state of the compound.

Thus to answer the question "what is the density (or heat capacity, refractive index, entropy, ...) of α?" we must first specify two variables – the P and the T we are interested in. One is not enough – at 4 mbar, α can have quite a range of densities, but at 4 mbar, 50 °C, its density is fixed and determinable, as are all its other properties. So we say that in each of its three forms – solid, liquid, and gas – α has two *degrees of freedom* – two variables must be specified before *all* are specified. These two variables are in practice usually T and P, but in principle any two would do. The phase rule summarizes all this discussion by simply saying

$$f = c - p + 2$$
$$= 1 - 1 + 2$$
$$= 2$$

With α at 4 mbar and 50 °C, consider that we raise the temperature gradually. Nothing much happens, except that the properties of α change continuously, until we reach 69 °C, which is the temperature of the boundary between the solid and liquid fields. At this T, solid α is observed to begin to melt, and at this T, any proportions of solid and liquid α are possible (i.e., almost all solid α with a drop of liquid; or almost all liquid α with a tiny amount of solid; or anything in between). However, if the temperature is held very slightly above the melting T, α becomes completely liquid. The solid/liquid boundary line then is a locus of T–P conditions that permit the coexistence of solid and liquid α. It records the melting temperature of α as a function of pressure.

Note too that because it is a line rather than an area, or because there are two coexisting phases rather than one, we now have only one degree of freedom. In other words, at 4 mbar we now have no choice of temperature. If solid and liquid coexist at equilibrium, the temperature must be 69 °C – it is chosen for us. We can still choose whatever P we like (within certain limits), but once we have exercised our one degree of freedom and chosen a pressure, the temperature and all properties of the two phases of α are fixed. Again, we note that the one degree of freedom can be any property, not just T or P. We might choose a certain value for the entropy of solid α, for example; we would then find that there was only one T and P at which solid α with this particular S_α could coexist with liquid α. The phase rule agrees, saying

$$f = c - p + 2$$
$$= 1 - 2 + 2$$
$$= 1$$

Similar comments apply to the boundary between the fields of "Liquid" and "Vapor," which records the *boiling temperatures* of α, and the boundary between the "Solid" and "Vapor" fields, which records the *sublimation temperatures* of α. Where these three boundaries come together at about 63 °C, 1.88 mbar (a *triple point*), the three phase fields come together, and solid, liquid, and gaseous α can coexist in any proportions at this particular T and P. Note that for the coexistence of these three phases, we have lost another degree of freedom. In fact we have no choice at all – if we want three phases to coexist, the

T and P must be 63 °C and 1.88 mbar. As the number of coexisting phases increases, the number of degrees of freedom decreases. Negative numbers of degrees of freedom are not possible, so in a one-component system the phase rule predicts that the *maximum* number of phases at equilibrium is three.

$$f = c - p + 2$$
$$= 1 - 3 + 2$$
$$= 0$$

12.3.1   Gibbs Energy Sections

Despite the fact, mentioned in Section 12.1.1, that phase diagrams are for the most part experimentally derived, they are controlled by and must conform to fundamental thermodynamic relationships. Understanding phase diagrams is enhanced by examining the relationships between the diagrams and the underlying thermodynamics.

From our study of thermodynamics in previous chapters, we know that the stable state of a system under given conditions is that state having the lowest value of the Gibbs energy, $\mathbf{G}$ (or G). If a system does not have the lowest possible value of $\mathbf{G}$, a spontaneous process will take place (according to our model) until this lowest value has been achieved. Also, we know that if two phases are in equilibrium in a unary system, the Gibbs energy of the component is the same in each phase (Section 6.2.1; Figure 6.1). Therefore, the phase boundaries in Figure 12.1 are places where G_α is the same in two phases, as shown in Figure 12.2. Note too that we may calculate and plot the Gibbs energy (and other properties) of a liquid phase in regions where it is not the stable phase. When we say, for example, that at 4 mbar, 50 °C in the solid stability field, $G_\alpha^{\text{solid}} < G_\alpha^{\text{liquid}}$, we imply that, if liquid α could exist at 4 mbar, 50 °C, its G would be greater than that of G_α^{solid}. We could, in fact, plot the values of G for all possible phases over all parts of the diagram. If we did so and looked at a part near the solid/liquid boundary, we would see something like Figure 12.3.

G–T Sections

Figure 12.4 shows a section through Figure 12.1 at a pressure of 2 mbar. At temperatures below 64 °C at 2 mbar pressure, α is solid, and the Gibbs energy of this solid (G_α^{solid}) is shown by the line labeled "Solid". Naturally, as we don't *know* the absolute Gibbs energy of any substance, we cannot place any absolute numbers on the G-axis. However, we *do* know the slope of this line (the slope is $(\partial G_\alpha / \partial T)_P = -S_\alpha$, and we know S for most compounds), so we could establish some arbitrary energy divisions on the G-axis and plot a line with the correct slope. This line would have a gentle downward curvature because S gradually increases with T, but to a first approximation it is a straight line. This line continues to the melting temperature, 64 °C, at which point it intersects another line giving the values of G_α^{liquid}. This line has a steeper slope, because the entropy of a liquid is always greater than the entropy of a solid of the same composition. At the intersection, $G_\alpha^{\text{solid}} = G_\alpha^{\text{liquid}}$, as required by phase equilibrium theory (Sections 6.2.1 and 6.3).

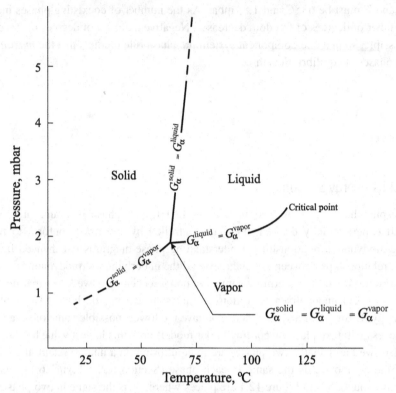

Figure 12.2 Gibbs energy relationships in the phase diagram for compound α.

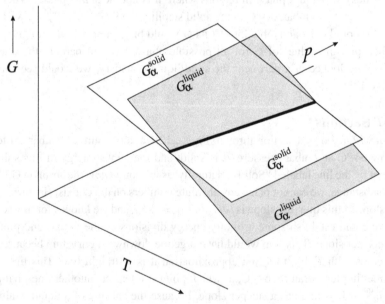

Figure 12.3 $G-T-P$ diagram for part of Figure 12.1. The heavy line at the intersection of the G_α^{solid} and G_α^{liquid} surfaces is the melting curve.

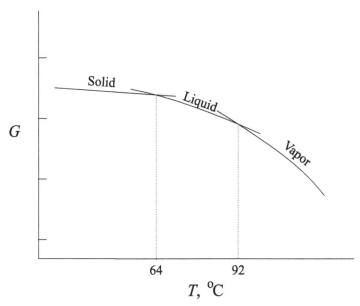

Figure 12.4 *G–T* section through Figure 12.1 at 2 mbar.

The G_α^{liquid} line then continues with a gentle downward curvature through the liquid stability region at 2 mbar until it reaches another phase boundary, the boiling curve, at 92 °C. Here it intersects the G_α^{vapor} curve, which has a still steeper slope, because the entropy of gases is always much greater than that of liquids.

Note the similarity of this diagram to Figure 6.6, where we considered *G–T* sections in a quantitative way, to calculate the positions of phase boundaries.

G–P Sections

Figure 12.5 shows a *G–P* section through Figure 12.1 at a temperature of 69 °C. At pressures below 1.89 mbar, α is gaseous, and the Gibbs energy of this gas is shown by the line labeled "Vapor". The slope of the line is $(\partial G_\alpha / \partial P)_T = V_\alpha$, and as the molar volume of gases is large, the line has a steep slope. This line intersects another line, giving the values of G_α^{liquid}, having a smaller positive slope, because $V_\alpha^{\text{liquid}} < V_\alpha^{\text{gas}}$. This line continues, again with slight downward curvature because the molar volume of the liquid decreases slightly with increasing pressure, until it reaches the freezing curve at 4 mbar, where it intersects the line giving G_α^{solid}. Note the similarity between this diagram and Figure 6.2.

12.3.2   Some Important Unary Systems

Substances whose phase relations are interesting for various reasons include carbon (C), iron (Fe), water (H_2O), silica (SiO_2), aluminum silicate (Al_2SiO_5), and calcium carbonate ($CaCO_3$).

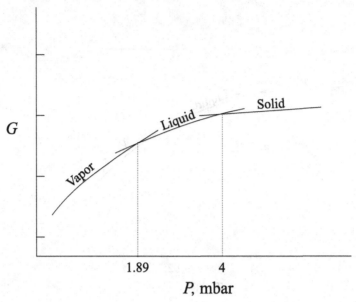

Figure 12.5 *G–P* section through Figure 12.1 at 69 °C.

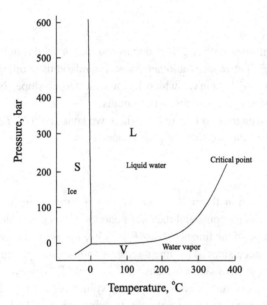

Figure 12.6 Phase diagram for H_2O at relatively low pressures. The solid/liquid boundary is very steep, but in fact has a negative slope.

H_2O

The phase diagram for water at relatively low pressures is shown in Figure 12.6. Water is a most unusual substance. It is one of the very few compounds that expands when it freezes, meaning that ice floats. Most substances have solid forms that are denser than their corresponding liquids, and hence will sink during freezing. The fact that ice floats in water is shown in Figure 12.6 by the fact that the liquid/solid boundary (the freezing/melting curve) has a negative slope. In our "typical" unary system (Figure 12.1), this curve has

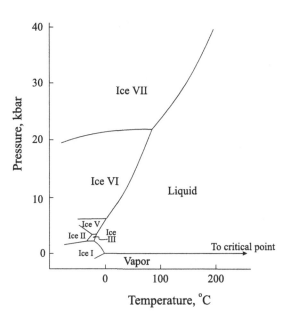

Figure 12.7 Phase diagram for H_2O at high pressures. Ice IV, not shown, is a metastable form of ice in the region of ice V.

a positive slope. In both cases (Figures 12.1 and 12.6), the denser phase lies at higher pressures, as required by Le Chatelier's principle. The unusual thing is that in the H_2O system, the denser phase is the liquid.

The other term in the slope expression ($dP/dT = \Delta S/\Delta V$) is ΔS, which is invariably greater in the liquid than in the solid; therefore, the volume change, ΔV, determines whether dP/dT will be positive or negative.

Figure 12.7 shows the same system over a much greater range of pressures. The striking thing about this diagram is the large number of polymorphs of ice, each with its own stability field. These polymorphs give rise to several *triple points*, showing that the solid–liquid–vapor triple point shown in Figure 12.1, which every unary system has, is often not the only one. We came across this phenomenon (a triple point generated by solid polymorphs) previously (Section 6.4, Figure 6.5). Note the fact that liquid water will freeze (to ice VII) at about 24 kbar at the boiling temperature (100 °C). Note too that the negative slope of the freezing curve (between "Ice I" and "Liquid") extends to only about 2 kbar.

SiO_2

Silica, one of the most common compounds on Earth, has a number of interesting and complex phase relations, shown in Figure 12.8.

12.4 Binary Systems

12.4.1 Types of Diagrams

When we consider the phase relations in systems having two components instead of one, we add one dimension to our diagrams. That is, in unary diagrams all phases have the same composition, so we don't need an axis showing compositions – we can use both

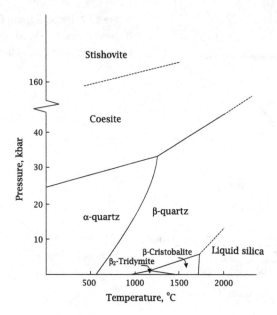

Figure 12.8 Phase diagram for SiO_2.

dimensions available on a sheet of paper for physical parameters, and we choose T and P. With two components, we find that phases commonly contain different proportions of these components – they have different compositions. Since this is of great interest, we use one dimension for composition, leaving only one other for either T or P. Most commonly temperature variations are of more interest, so diagrams showing phase relations on a $T–X$ diagram are very common.[2] This relationship is illustrated in Figure 12.9. In this book, we concentrate on $T–X$ sections, but it is well to realize that there are other varieties of diagrams. Not only do we have $P–X$ sections, but $P–T$ or isoplethal (constant-composition) sections, as well as various projections.

12.4.2 The Melting Relations of Two Components

Suppose that you now understand unary phase relations very well, but have never encountered binary systems, and you are given the following problem. There are two minerals, A and B. We know the melting point of each mineral, T_{m_A} and T_{m_B}, at atmospheric pressure. We grind samples of A and B together in various proportions, say 25% A, 75% B; 50% A, 50% B; and 75% A, 25% B, and we perform experiments to determine the melting temperature of these mixtures. Your job is to draw a diagram *predicting* the most likely results. The diagram should show temperature as the vertical axis and composition as the horizontal axis, and of course the known melting temperatures of the pure minerals A and B should be plotted on the vertical axes at each end of the composition axis.

---

[2] We use the "X" in the expressions $T–X$ or $P–T–X$ to mean "composition" generally, whether measured as mole fractions or weight percent, or in some other way.

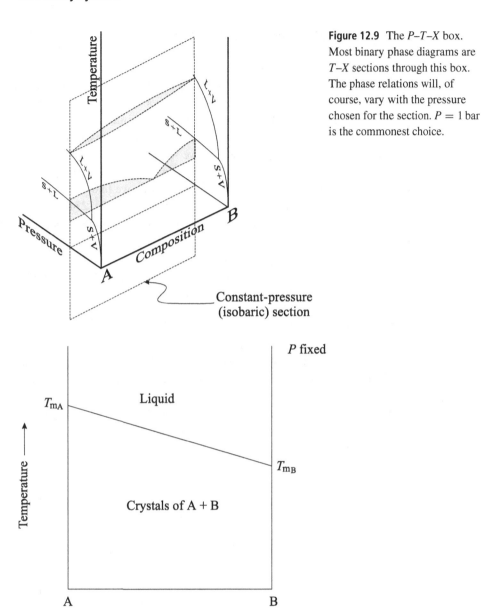

Figure 12.9 The *P–T–X* box. Most binary phase diagrams are *T–X* sections through this box. The phase relations will, of course, vary with the pressure chosen for the section. $P = 1$ bar is the commonest choice.

Figure 12.10 An uneducated guess as to the melting temperatures of mixtures of minerals A and B.

It seems very likely that your guess would look like Figure 12.10. In other words, you would probably suppose that the melting temperature of mixtures of A and B would be some kind of average of the melting temperatures of the pure compounds, much in the way that volumes are averages as shown in Figure 7.5(a). But binary systems are not quite that simple. Figure 12.10 is thermodynamically impossible, even if A and B were not separate phases but formed a solid solution, but we will not bother to prove this. Suffice it to say that experiments on hundreds of binary systems have never given results consistent with Figure 12.10.

What *does* happen depends on what compounds A and B actually are. Let's suppose that A is the component $CaMgSi_2O_6$, and B is the component $CaAl_2Si_2O_8$. The stable forms of these components at ordinary temperatures are the minerals diopside ($CaMgSi_2O_6$) and anorthite ($CaAl_2Si_2O_8$), so we will represent the component $CaMgSi_2O_6$ by the symbol Di and the component $CaAl_2Si_2O_8$ by the symbol An. Diopside melts at 1392 °C, and anorthite melts at 1553 °C. We perform the experiments mentioned above; that is, we grind up both samples into fine powders, then mix the powders in various proportions and heat them up in separate experiments and observe what happens at various temperatures. What a surprise – we find that *all mixtures begin to melt at the same temperature!* And when we analyze the composition of the first liquid to form, we find that *the first liquid to form in all mixtures has the same composition!* The temperature is 1274 °C (called the *eutectic temperature*), and the composition is 42% An, 58% Di (called the *eutectic composition*).

On heating to still higher temperatures, another surprise awaits us. For those mixtures having *more* than 42% An, temperatures above 1274 °C result in disappearance of all diopside in the mixtures—we are left with only liquid plus anorthite crystals. For those mixtures having *less* than 42% An, temperatures above 1274 °C result in the disappearance of all anorthite in the mixtures – only liquid plus diopside crystals are left. So *below* 1274 °C, only two phases coexist – crystals of anorthite and diopside. *Above* 1274 °C, again only two phases coexist – either liquid and anorthite, or liquid and diopside. Only at exactly 1274 °C are three phases observed to coexist at equilibrium – anorthite, diopside, and liquid. And note that in the binary system, we have melting far below the melting temperature of either of the pure components. These relationships are summarized in Figure 12.11. They may seem strange at first, but, as we will see, they are one of a rather small set of relationships that satisfy the phase rule.

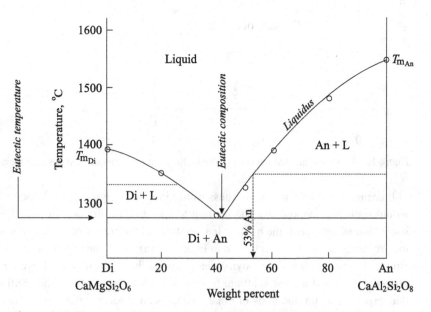

Figure 12.11 The system Di–An at 1 bar pressure. Two representative tie-lines are shown.

The Isobaric Phase Rule

But, first, we must mention a slight modification of the regular phase rule, Equation (12.1). As shown in Figure 12.9, the experiments we are discussing at a fixed pressure of 1 bar can be represented on a plane or section through P–T–X space. The general phase rule (11.1) applies to this P–T–X space. The fact that we confine ourselves to a fixed P plane within this space means that we have "used" one of our degrees of freedom – we have chosen $P = 1$ bar, and the same would be true for any other constant-P section (or constant-T section, for that matter). Therefore *on our T–X plane* the phase rule is

$$f = c - p + 1 \tag{12.4}$$

This shows that the maximum number of phases that can coexist at equilibrium in a binary system at an arbitrarily chosen pressure (or temperature) is three ($p = 3$ for $c = 2, f = 0$), which is consistent with our observations.

12.4.3   Reading the Binary Diagram

The main features of the phase relations in Figure 12.11 follow directly from this fact. During the heating of our mixture of Di and An crystals, we have two phases, and

$$
\begin{aligned}
f &= c - p + 1 \\
&= 2 - 2 + 1 \\
&= 1
\end{aligned}
$$

This means that, to fix all the properties of both kinds of crystals, we need only choose the temperature (pressure being already fixed at 1 bar). However, when the first drop of liquid forms, $p = 3$ (diopside crystals, anorthite crystals, and liquid), and $f = 0$. Another word for $f = 0$ is *invariant*. When $p = 3$ on an isobaric plane, we have *no* choice as to T, P, or the compositions of the phases – they are all fixed. This explains why all mixtures begin to melt at the same temperature, and why the liquid formed is always of the same composition no matter what the proportions of the two kinds of crystal. No other arrangement would satisfy the phase rule.

A line on a phase diagram joining points representing phases that are at equilibrium with each other is called a *tie-line*. Each of the two-phase regions in Figure 12.11 (Di+L; An+L; Di+An) is filled with imaginary tie-lines joining liquid and solid compositions, or two solid compositions, that are at equilibrium. Only two of these tie-lines are shown. Consider the tie-line at 1350 °C in the region labeled An+L. One end of the line is on the curved line representing liquid compositions (called the *liquidus*), and the other end is on the vertical line representing 100% An composition. The composition scale across the bottom of the diagram applies at any temperature, so we can get the liquid composition by dropping a perpendicular from the liquidus to the composition scale, showing that the liquid composition at 1350 °C in equilibrium with pure anorthite crystals is 53% component An, 47% component Di. The composition of the solid phase is given by the other end of the line, which is at 100% An. In each of these two-phase regions, such as An+L, $f = 1$, which means that once we have chosen the temperature, say 1350 °C, all properties of all phases are fixed. Therefore, all proportions of Di and An in this region will have the same liquid and solid compositions. In other words, *any* starting mixture of diopside and anorthite crystals having more than 53% anorthite would, when heated to 1350 °C, consist of a liquid of composition 53% An, 47% Di, plus crystals of pure anorthite. Mixtures having between about 20% An and 53% An would be completely liquid at this temperature, and mixtures having 0 to 20% An would consist of pure diopside crystals plus a liquid of composition 20% An, 80% Di.

By imagining tie-lines across the An + L region at successively higher temperatures, we see that the composition of the liquid in equilibrium with anorthite crystals gets progressively richer in component An. Similarly, the tie-lines in the Di + L region show that the liquid gets richer in Di as the temperature increases.

Because the temperature of the three-phase tie-line is fixed, it follows that both above and below this temperature there must be regions having only two phases. We already know that below the three-phase line the two phases are Di and An. Above the three-phase line one of the phases must be liquid, because melting has started. Therefore, the liquid can coexist with only one other phase, obviously in this case either Di or An, but not both. As *T* increases, the proportion of liquid must increase, eventually becoming 100% liquid. This simple analysis is sufficient to explain the main features of the diagram. "Reading" binary diagrams consists largely of distinguishing between one phase regions, which have no tie-lines (e.g., the "Liquid" region), two-phase regions, which have tie-lines joining two phases at equilibrium, and three-phase tie-lines, which separate two-phase regions, and join three phases at equilibrium.

12.4.4   A More General Example

The system Di–An is misleadingly simple in two respects. For one thing, the diagram shows that both diopside and anorthite remain pure while heated in contact with the other component until the melting temperature is reached (1392 °C for Di, 1553 °C for An). Actually, phases (in theory) never remain perfectly pure when in contact with other phases – some mutual solution always takes place, although as in the case of Di and An it is sometimes small and does not show on the diagram. A more realistic case is shown in Figure 12.12. The diagram is essentially the same as the Di–An diagram, except that there is a field of A_{ss} and of B_{ss}, where the subscript ss stands for solid solution.

The other respect in which the Di–An diagram is misleading is that, in fact, it is not strictly speaking a true binary system. This somewhat surprising statement cannot be fully explained without discussing ternary systems. Suffice it to say that just because you choose two components does not necessarily mean that you have a binary system. *To be truly binary, all compositions of all phases must lie on the plane of the diagram.* This must be the case in simple systems such as Cu–Au and with single solution phases such as liquids; but with complex components such as Di and An, although the *bulk composition* must lie on the plane of the diagram, the compositions of coexisting *phases* may each lie "off the plane" of the diagram. Careful work has shown that, in the system Di–An, diopside crystals are not pure but contain some Al. This means that, because bulk compositions lie on the Di–An plane, phases coexisting with diopside must be somewhat deficient in Al. To portray

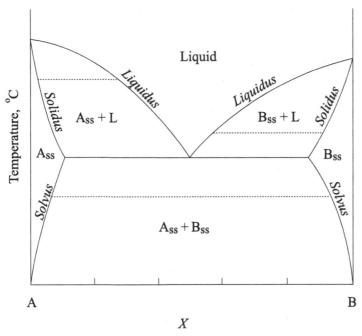

Figure 12.12 A more representative binary system. The difference is that both components show solid solution fields. Three representative two-phase tie-lines are shown.

this in a diagram, one needs a three-component triangle. Just remember that not all choices of two components are binary systems – some are planes within ternary systems.

Solid Solutions

There is no difference in principle between a solid solution and a liquid or gaseous solution. Substances dissolve into one another, like sugar into tea, or like oxygen into nitrogen, because the Gibbs energy change of such a process is negative – they are spontaneous processes. Consider the system Di–An at a temperature of 1600 °C (Figure 12.11). At this temperature, both pure Di and pure An are liquid phases. If one gram of Di liquid and one gram of An liquid were mixed together, they would dissolve into one another to form a homogeneous liquid solution, represented by a point in the middle of the diagram on the 1600 °C isotherm. If pure diopside crystals are mixed with pure anorthite crystals at 1200 °C, on the other hand, nothing happens – they do not dissolve into each other.

Components A and B in Figure 12.12, on the other hand, behave differently. Liquid A and liquid B still mix to form a homogeneous solution, but when solid A and solid B are mixed together, they dissolve into one another to a limited extent. Salt will dissolve into water, but not without limit – it will dissolve only until the water becomes saturated. Similarly, solid B will dissolve into solid A, but not without limit. It dissolves into A until A is saturated with B, and at the same time A dissolves into B until B is saturated with A. The saturation limit of each component in the other is shown by a line called the *solvus*. The existence of a solvus shows that A and B exhibit *limited miscibility* in the solid state. They exhibit *complete miscibility* in the liquid state. "Miscibility" does not really mean "mixability," although they sound similar. "Mixability," if it is a word, just means that things can be mixed together – mutual dissolution is not implied. "Miscibility" means the ability to dissolve into something else.

Figure 12.12 shows a eutectic, but the two solid phases in equilibrium with the liquid are not pure A and pure B; A contains some B in solid solution (A_{ss}) and B contains some A in solid solution (B_{ss}). Similarly, at temperatures above the eutectic, the liquid is not in equilibrium with pure A or pure B, but with A_{ss} and B_{ss}. The compositions of the solid solutions in equilibrium with liquid are given by lines called the *solidus*.

12.4.5   Freezing Point Depression

Figure 12.11 shows that mixtures of diopside and anorthite become completely liquid at temperatures lower than the melting temperature of either pure diopside or pure anorthite. This is also shown by the more general system in Figure 12.12 and is, in fact, an extremely common feature of binary systems. It is called *freezing* (or *melting*) *point depression* and is, in fact, why we put salt on icy roads in winter. The melting temperature of ice is lowered in the presence of the second component (salt, NaCl), so the ice melts and the resulting salty water corrodes our cars. But why is the freezing point depressed?

The answer is found in the basic thermodynamic relationships between the phases. Figure 12.4 shows the absolute Gibbs energies of the solid, liquid, and vapor phases of our compound α as a function of temperature at a pressure of 2 mbar. Let's call compound α component A (just as we called component $CaMgSi_2O_6$ Di). Figure 12.4 therefore shows G_A^{solid}, G_A^{liquid}, and G_A^{vapor} as functions of T. If we now add a second component B to A, what happens to these Gibbs energies? To start with the simplest case, we will suppose that B does not dissolve into solid A or into vapor A, but does dissolve into liquid A (if you add NaCl to H_2O, the salt will not dissolve into ice or into steam, but it will dissolve into liquid water). Therefore, the curves for G_A^{solid} and G_A^{vapor} will not change, because A^{solid} and A^{vapor} are unchanged in the presence of B. But what is the Gibbs energy of component A in a liquid containing both A and B?

The answer is shown in Figure 7.4, which shows that, when B dissolves into A, the molar Gibbs energy of component A in the solution at a given concentration (which we call μ_A), is *lower* than the molar Gibbs energy of pure A (G_A°). This relationship is quite general and without exception, because otherwise A and B would not form a solution. We will be mentioning this relationship at various points throughout this chapter. The consequence of the fact that G_A^{liquid} is lowered but G_A^{solid} is not is shown in Figure 12.13. The shaded surface in this figure represents the free energy curve from Figure 7.4, extended into a range of temperatures. It shows the lowering of the total Gibbs energy of the liquid phase as component B is added. At the arbitrary amount of 10% B, a tangent surface to the free energy surface extends back to the 0% B plane, which is analogous to the tangent at $X_B = 0.4$ in Figure 7.4. The trace of this tangent surface on the G–T section for component A gives μ_A, the molar Gibbs energy of A in the solution containing 10% B, 90% A. It of course lies below the curve of G_A^{liquid} for pure A. But because the curve for G_A^{solid} has not moved, the *intersection* of the G_A^{solid} and μ_A^{liquid} curves is moved to a lower temperature. The intersection of the μ_A^{liquid} and G_A^{solid} curves is the point where these two quantities are equal, and it defines the temperature at which A in the solid state and A in the liquid state

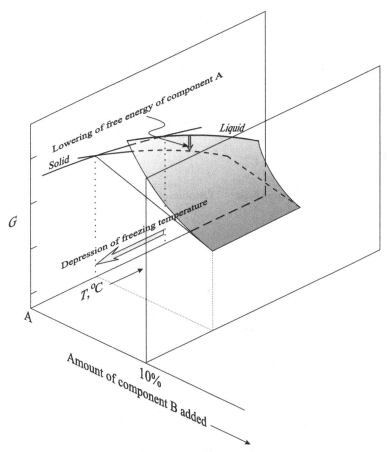

Figure 12.13 The G–T plane is taken from Figure 12.14. Component B enters the liquid phase and causes a lowering of μ_A^{liquid}, which in turn causes a depression of the freezing temperature.

are in equilibrium. For pure A, this is the melting or freezing temperature; for the system A–B, it defines a point on the liquidus of A and is the result of freezing point depression.

This relationship is shown again in Figure 12.14, this time including the vapor curve. If the vapor curve does not move (no B dissolves into vapor A), depression of G_A^{liquid} results in a raising of the boiling temperature as well as a lowering of the freezing temperature. This is also an extremely common effect.

12.4.6   Freezing Point Elevation

But suppose our simplifying assumption that no B enters the solid phase is not true? There is no difference in principle between the thermodynamics of solid and liquid solutions, so if B dissolves into solid A the curve for G_A^{solid} will be lowered for the reasons just discussed. Normally, B is less soluble in solid A than in liquid A, so the amount of lowering is less for the solid phase, and the freezing point is still lowered. This is shown by systems like that in Figure 12.12, where the liquidus of A slopes downward, even though B is shown as entering both the liquid and the solid phases of A.

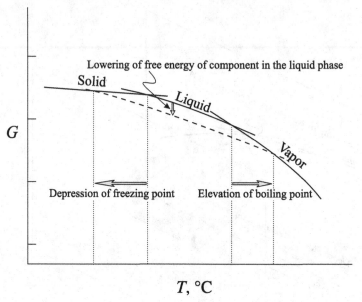

Figure 12.14 *G–T* section, showing lowering of the Gibbs energy of A in the liquid phase, causing depression of the freezing point and elevation of the boiling point.

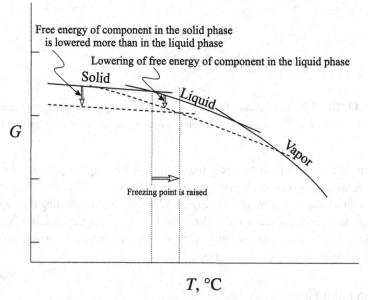

Figure 12.15 *G–T* section, showing a greater lowering of the solid Gibbs energy than liquid Gibbs energy, causing elevation of the freezing point.

However, what of the possibility that the G_A^{solid} curve might be lowered *more* than the G_A^{liquid} curve? This would happen if more B dissolved into solid A than into liquid A and would result in a *freezing point elevation* as shown in Figure 12.15. This explains an important feature of many binary systems.

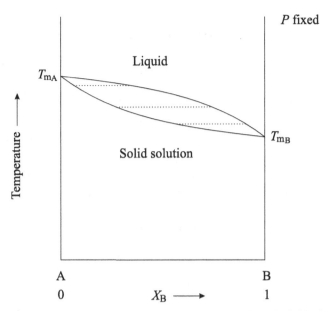

Figure 12.16 A binary system A–B showing complete miscibility both in the solid state and in the liquid state. Three representative tie-lines are shown.

12.4.7   Systems Having Complete Solid Miscibility

A and B in Figure 12.12 show limited solid miscibility, but some important systems show *complete* miscibility in the solid state, giving rise to a diagram that looks quite different, as shown in Figure 12.16. In a sense, it is simpler than the ones we have looked at so far – in fact it looks rather like Figure 12.10, except that the "melting line" in Figure 12.10 is a *melting loop* in Figure 12.16. But the most important difference is that in Figure 12.16 A and B *dissolve completely into one another in the solid state*. This takes some getting used to. We are quite familiar with sugar dissolving into tea, but the idea of placing two solid objects together and observing one disappear into the other is not something in our experience. But this is just another example of something that thermodynamics says *should* happen but in fact does not, because of energy barriers. The thermodynamic model does not consider these barriers, and hence does not always work. These solid solutions do exist, however, because they do not form from solids dissolving into one another at low temperatures. They form at high temperatures, sometimes over long periods of time, and then cool down in their mutually dissolved state. Many important alloys and mineral groups are such complete solid solutions, including the feldspars, olivines, and some pyroxenes and amphiboles.

Note that, in Figure 12.16, T_{mA} is lowered by adding B, but T_{mB} is raised by adding A. This is because more B enters liquid A than solid A, but more A enters solid B than liquid B, and the free energy consequences of this are shown in Figures 12.14 and 12.15, respectively.

The most important mineralogical example of this type of system is the plagioclase feldspar system, which is shown in Figure 12.17. Plagioclase is a mineral whose

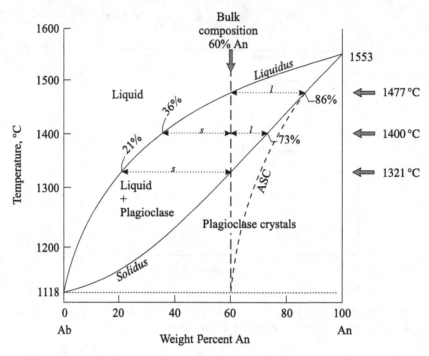

Figure 12.17 The plagioclase–feldspar system at 1 bar pressure. The curve labeled ASC is the average solid composition during fractional crystallization of the 60% An bulk composition.

composition may vary from virtually pure albite ($NaAlSi_3O_8$), or component Ab, to almost pure anorthite ($CaAl_2Si_2O_8$), or component An, depending on the composition of the liquid from which it crystallizes. The melting behavior of a complete solid solution such as this is a simple melting loop – a combined liquidus and solidus that goes from one pure component over to the other. The melting loop is filled with imaginary horizontal tie-lines, three of which are shown in Figure 12.17. They indicate the compositions of liquids, on the liquidus, and the compositions of plagioclase crystals, on the solidus, which are in equilibrium with each other.

12.4.8 Equilibrium vs. Fractional Cooling and Heating

Binary phase diagrams show phase compositions that are at equilibrium – they show what you would obtain if you heated a bulk composition to a certain temperature and waited long enough for equilibrium to be attained. The time required to reach equilibrium after a change in temperature or pressure varies greatly with the system, but equilibrium is *never* achieved instantaneously. Therefore, if we use the diagram to consider what would happen during continuous cooling or heating of a given bulk composition, we cannot be considering what would *really* happen in our system during cooling or heating but are considering *model* processes, as usual.

There are any number of models of processes we could devise involving phase changes in binary systems, but two are especially common – complete equilibrium (reversible) processes, and "surface equilibrium" (perfect fractional) processes. We will discuss only cooling processes. Heating processes are the exact reverse of cooling processes in the equilibrium case, but they are not always the exact reverse in the case of fractional processes.

Perfect Equilibrium Crystallization

Suppose we had a liquid having a composition of 60% An, 40% Ab at a temperature of about 1600 °C (Figure 12.17). On cooling this liquid, nothing much happens (except that the properties of the liquid, such as its density, refractive index, entropy, Gibbs energy, etc., etc., change) until it reaches a temperature of 1477 °C, the liquidus temperature for this composition. At this point, the bulk composition is still 100% liquid, but the first tiny crystal of plagioclase appears. Its composition, given by the solidus, is 86% An, 14% Ab. As cooling continues, plagioclase crystals continue to form, and previously formed crystals change their composition so that all crystals always have the equilibrium composition, with no compositional gradients. When the temperature reaches 1400 °C, the liquid has composition 36% An, and the crystals 73% An. (These compositions are obtained by dropping a perpendicular line from the point of interest to the compositional axis at the bottom of the diagram.) When the composition of the solids reaches the bulk composition of 60% An, the liquid must disappear, and this happens at a temperature of 1321 °C. Further cooling results in no further changes in composition of the crystals.

Perfect Fractional Crystallization

Maintaining perfect equilibrium while cooling is one end of a complete spectrum of possibilities. The other end of the spectrum is that crystals form, but always completely out of equilibrium. This end of the spectrum involves an infinite number of cases and so is rather difficult to discuss in a finite number of words. A subset of these possibilities is the case where crystallization produces crystals in equilibrium with the liquid, as required by the diagram, but, after forming, they do not react with the liquid in any way. This is called surface equilibrium (because the liquid is at all times in equilibrium with the surface of the crystals) or fractional crystallization, and is a model process just as much as is equilibrium crystallization. It is also used in connection with liquid–vapor processes (fractional distillation; fractional condensation), as well as isotope fractionation processes.

There are two ways of imagining a process of perfect fractional crystallization.

- As soon as a tiny crystal forms, it is removed from the liquid. This might be by reaching into the liquid with a pair of tweezers and physically removing the crystal, or the crystal might immediately sink to the bottom or float to the top of the liquid, where it becomes covered by other crystals and is removed from contact with the liquid.
- As soon as a tiny crystal with a composition given by the solidus forms, it is covered by a layer of another composition, given by the solidus at a slightly lower temperature. Successive layers are formed, each controlled by the position of the solidus, but, after forming, the various layers do not homogenize in the slightest. The result is a compositionally zoned crystal.

Note that crystals are removed from contact with the liquid in both cases. This is the essential element of fractional crystallization.

Considering the same bulk composition, 60% An, the cooling history is the same as before until the first tiny crystal forms at 1477 °C, having a composition of 86% An. On further cooling, the liquid composition follows the liquidus, as before, and any *new* crystals that form have compositions given by the solidus at that temperature; but previously formed crystals, being removed from contact with the liquid, do not change their original compositions. The net result is that at any temperature below 1477 °C, the *average* composition of all solids formed is more An-rich than would be the case in equilibrium crystallization, that is, more An-rich than the solidus at that temperature. Because of this, at each temperature below 1477 °C, there must be a larger proportion of liquid of Ab-rich composition to balance the solid composition, that is, to give the known bulk composition. Therefore, whereas in equilibrium crystallization the last drop of liquid must disappear at 1321 °C, in fractional crystallization it does not, and in fact liquids continue to exist right down to pure Ab composition, where the last liquid crystallizes as pure albite. This is the important aspect of fractional crystallization from a petrological point of view – that a given bulk composition can generate a much wider range of liquid compositions, and hence a wider range of igneous rocks, than can equilibrium crystallization.

It is possible to calculate the average composition of the solids during fractional crystallization, but we will not do this. Just note that a curve indicating the average composition of all solids generated must begin at 1477 °C on the solidus, and it must end at 1118 °C at a bulk composition of 60% An, when the last liquid disappears. This curve is shown in Figure 12.17, labeled "ASC." For equilibrium cooling, the "ASC" curve is, of course, the same as the solidus.

12.4.9   The Lever Rule and Mass Balances

Phase diagrams contain information not only about phase compositions and their temperatures and pressures, but about the *proportions* of phases for a given bulk composition. This is done using what is called the lever rule. Look at the three tie-lines we have just been discussing in Figure 12.17. Consider first the line extending from the liquidus (36% An) to the solidus (73% An), at 1400 °C. This line is composed of two parts. One part, labeled l, represents the proportion of liquid, and the other part, labeled s, represents the proportion of solids. The fraction (by weight) of liquid in the bulk composition is thus $l/(l + s)$, and the fraction of solids is $s/(l + s)$. The easiest way to measure the lengths of l and s is probably by comparing the compositions of the end points of the tie-line at the liquidus and solidus with the bulk composition. Thus the s portion of the tie-line has a length of $60 - 36 = 24\%$, and the l portion of the tie-line has a length of $73 - 60 = 13\%$. The total length of the tie-line is $73 - 36 = 37\%$. Therefore the proportion or fraction of solid in the total bulk composition is $24/37 = 0.65$, and the fraction of liquid is $13/37 = 0.35$, and of course $0.65 + 0.35 = 1.0$. If we had a bulk composition weighing 10 g, then at 1400 °C, 1 bar, it would be made up of $0.65 \times 10 = 6.5$ g of crystals (73% An composition), and $0.35 \times 10 = 3.5$ g of liquid (composition 36% An). This lever rule can be used in any two-phase region, given the bulk composition, and the lengths of the lines can be measured in percentage composition as we have done, or in millimeters, or inches, or any other units.

Note that, at the intersection of the bulk composition line and the liquidus, the *s* portion of the line reduces to zero, because there is 100% liquid, and at the intersection of the bulk composition line and the solidus, the *l* portion of the line reduces to zero. This provides a way of remembering which side of the tie-line represents which phase.

Mass Balances

Because we know the proportions and compositions of the phases, it is a simple matter to combine these to calculate the bulk composition. But, as we *know* the bulk composition (we needed it to get the phase proportions), the calculation is circular. Nevertheless, it is a useful check on our reading and construction of diagrams. For example, at 1400 °C in Figure 12.17, the mass balance is

(solid fraction × solid composition)

$+$ (liquid fraction × liquid composition) = bulk composition

$$(0.65 \times 73) + (0.35 \times 36) = 60$$

where 60 is the bulk composition in percentage of An.

12.4.10   Binary *G–X* Sections

The fact that the Gibbs energy of solutions is represented by a convex-downward curve (sometimes called a festoon) was introduced in Chapter 7 and Figure 7.4. Both solid solutions and liquid solutions are represented by such curves, and understanding of binary diagrams is increased by constructing such curves on *G–X* sections. Each *G*-curve moves upward with decreasing temperature ($\partial G/\partial T = -S$), but the liquid curve moves upward faster than the solid curve, because the entropy of liquids is greater than that of solids. The stable phase for any bulk composition is always indicated by the lowest *G*, either on a solid or liquid curve or on a tangent joining two such curves.

A Binary Eutectic System

Consider first the Di–An system in Figure 12.18, an example of a simple binary eutectic system. The *T–X* section is shown at the bottom of the diagram, and *G–X* sections through the system at various temperatures are shown above it. In understanding this diagram, it is important to remember that the *G* of a mixture of crystals that are completely immiscible (show no mutual solid solution) is simply the weighted average of the *G* values of the two pure end-members, which in this case appears as a straight line joining G_{Di} and G_{An}. This straight line appears on all sections, whether the solids are stable phases or not. The line labeled "mixture of Di and An crystals" in Figure 12.18 represents this situation.

When components *do* mutually dissolve to form a solution, Equation (7.38) still works for volumes if the solution is ideal, but even for ideal solutions, it does not work for *G*. The Gibbs energy of the solution must be less than the weighted average of the *G* values of the two pure end-members for mutual solution to take place, and it is represented by the

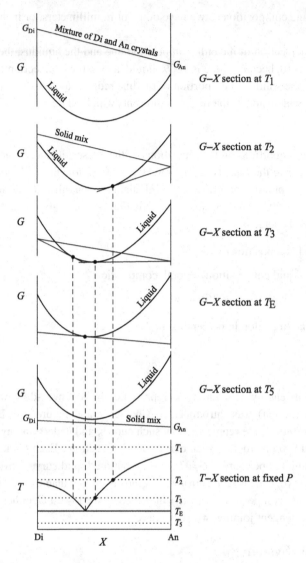

Figure 12.18 *G–X* sections through a binary eutectic diagram.

"festoon," or convex-downward loop. Therefore, the liquid solution formed when liquid Di and liquid An are mixed together is represented by such a loop. It is shown in all sections, even when the liquid is not the stable phase.

Understanding these *G–X* sections is helped by realizing the following.

- The line representing the *G* of the mixture of solid Di and solid An is shown in every section regardless of whether diopside or anorthite is or is not stable. If they are not stable, the line represents *metastable* Gibbs energies.
- The line (festoon) representing the *G* of the complete liquid solution between Di and An is shown in all sections regardless of whether the liquid is stable or not. If the liquid is not stable, its curve represents *metastable* Gibbs energies.
- The stable phase or phases in each section must have the lowest free energy available.

Starting with the G–X section at T_1 at the top of the diagram, we see that for every bulk composition between Di and An, the liquid Gibbs energy is everywhere *lower* than the Gibbs energy of a mixture of solid Di and solid An. In other words, the stable phase across the diagram is liquid. On the T–X section, note that, at T_1, we are above the melting temperatures of both components, and in the field of liquid at all compositions.

As we cool from T_1 to T_2, the Gibbs energy of liquids and solids (whether stable or metastable) increases, but that of liquids increases more, so that, at T_2, the G of liquid An has become greater than that of solid An, but the G of liquid Di remains less than that of solid Di. At some temperature between T_1 and T_2 we must have passed a point where $G_{An}^{liquid} = G_{An}^{solid}$, that is, the melting temperature of An. At T_2, the stable form of pure Di is liquid, but the stable form of pure An is solid. The lowest Gibbs energy available to the system as we go from Di toward An is that of liquid, but just after passing the minimum on the curve, the lowest Gibbs energy available is that of neither liquid nor a mixture of crystals, but a mixture of liquid and An crystals. In this mixture, the Gibbs energy of component An in the crystals must be the same as the Gibbs energy of component An in the liquid.

Recall from Figure 7.4 that the tangent to a Gibbs energy curve of a solution has intercepts giving the chemical potential of each component in the solution. Therefore, a tangent to the liquid curve that has an intercept on the An axis at the Gibbs energy of solid An will indicate that liquid composition in which $\mu_{An}^{liquid} = \mu_{An}^{solid}$. That tangent point is, of course, at the composition of the liquidus at that temperature, as shown by the dotted line joining the G–X section at T_2 with the T–X section. As the temperature falls below T_2, that tangent, rooted on the An axis at the free energy of solid An (G_{An}^{solid}), moves to greater Di compositions, because the liquid loop is moving up with respect to the G_{An}^{solid} point.

At T_3, solid diopside is now the stable phase on the Di side of the diagram, and the tangent situation described above holds for both components. With falling temperature, the two tangent points move toward each other, becoming one tangent at T_E, the eutectic temperature. Note that at T_E there must be only one tangent because μ_{Di} and μ_{An} must be the same in all three phases.[3]

On further cooling, the tangent breaks away from the liquid curve and becomes a straight line below the liquid curve, giving the Gibbs energy of a mixture of diopside and anorthite crystals just as in the section at T_1. The difference is that now it is completely *below* the liquid curve, and therefore a mixture of crystals is the stable configuration of the system.

A Melting Loop System

The story is rather similar for G–X sections through a melting loop diagram at various temperatures (Figure 12.19). However, instead of dealing with the intersection of a solution curve, or festoon, and a straight line, we have the intersection of two solution curves – one for the liquid solution and one for the solid solution. If this seems confusing, go back to Chapter 7 and recall why a Gibbs energy curve for a solution must be convex downward. Then remember that this applies whether the solution is solid, liquid, or gaseous. Finally, remember that in these sections (Figure 12.19) we plot the positions of both solution curves

---

[3] If you think about this statement, and look at Equations (7.26), you will see why we said (in Section 12.4.4) that phases never remain absolutely pure when heated together, at least according to our model.

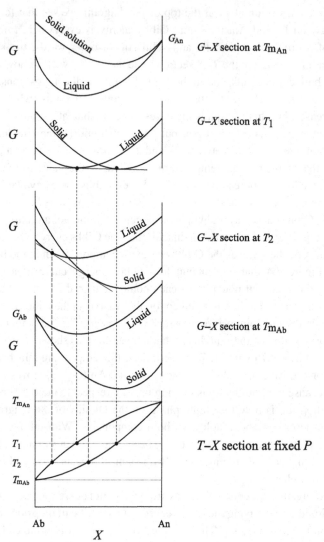

Figure 12.19 *G–X* sections through a melting loop diagram.

in every section, regardless of whether the solution is the stable phase or not. The point is to determine which parts of which curves give, or combine to give, the lowest Gibbs energy available to the system at each composition across the system.

At the top of the diagram, the section is drawn at the melting temperature of anorthite crystals. Therefore, the liquid and solid curves join at the An axis, because $G_{An}^{solid} = G_{An}^{liquid}$. Going toward component Ab, the liquid curve is everywhere below the solid curve, showing that liquid is everywhere the stable phase at this temperature. Similarly, the section at T_{mAb} shows that $G_{Ab}^{solid} = G_{Ab}^{liquid}$, and that at compositions toward component An, the solid curve is below the liquid curve, showing that, at this temperature, a solid solution of Ab and An is the stable phase. At intermediate temperatures, the two curves intersect. The liquid curve is lower on one side, and the solid curve is lower on the other side. Intermediate compositions have the lowest possible Gibbs energy only by being a mixture of solid and liquid, and, because the chemical potentials of both Ab and An must be the same in both phases, the compositions of the two solutions at equilibrium must be given by the only common tangent to the two curves at each temperature.

The sections in Figure 12.19 have been reassembled into a three-dimensional view in Figure 12.20. The only major advantage of this is that you can now see the relative rises of the liquid and solid loops with decreasing temperature, as indicated by the dotted lines on

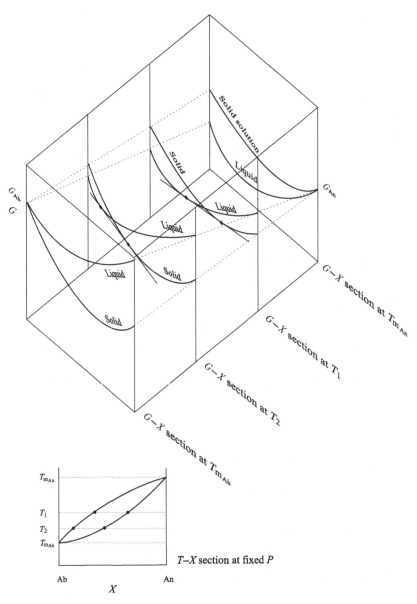

Figure 12.20 A perspective view of G–X sections through a melting loop diagram.

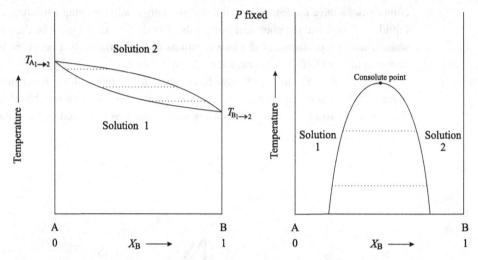

Figure 12.21 The two basic binary diagram elements. In the phase transition loop (left diagram) solution 1 and solution 2 can be solid and solid, solid and liquid, or liquid and vapor, respectively. $T_{A_{1 \to 2}}$ and $T_{B_{1 \to 2}}$ are melting temperatures, boiling temperatures, or polymorphic phase transition temperatures for pure A and B, respectively. Three representative tie-lines are shown. In the solvus (right diagram), solution 1 and solution 2 can be two solids or two liquids. Two representative tie-lines are shown.

the sides of the box. The line representing liquids has a steeper slope than that representing solids for the same reasons as in Figures 12.4 and 6.6.

12.4.11 Binary Diagram Elements

Binary phase diagrams can become quite complex, but the complexities are nothing but the elements of simpler diagrams, combined in such a way as to satisfy the phase rule. There are essentially only two elements (Figure 12.21), both of which contain two-phase tie-lines, and hence are elements controlling the compositions of coexisting phases.

• The phase transition loop, which separates two different kinds of solutions. This can be a melting loop as shown previously in Figure 12.16, separating a liquid solution from a solid solution, the end points being melting temperatures. But it can also be a boiling loop, separating a gas or vapor from a liquid, the end points being boiling temperatures, and it can also be a polymorphic or solid–solid phase transition loop separating two solid solutions having different structures, the end points being polymorphic transition temperatures. Phase transition loops occur simply because solutions cannot change to other solutions with no change in composition.[4]

---

[4] This can be proven using thermodynamics, but we will not bother. The exception to this rule is a solution having a maximum or minimum in temperature, where it can melt or boil to another solution having the same composition (see Figure 12.22, lower left, upper right.)

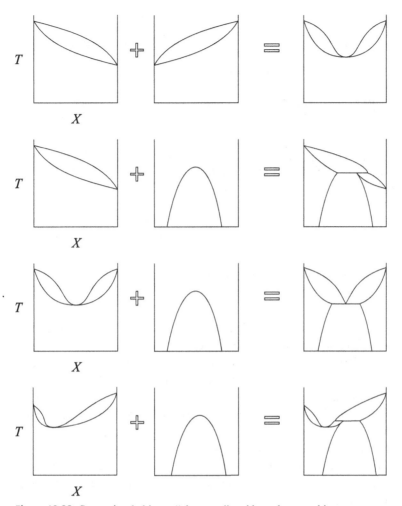

Figure 12.22 Some simple binary "elements," and how they combine.

- The solvus, which separates two solutions of the same kind, such as two solid solutions, or two liquid solutions (Figure 12.21). Increasing the temperature normally increases the solubility of one in the other, and the two phases can become identical (the solvus closes) at an *upper consolute point*. As this name implies, solvi can sometimes close downward (with decreasing temperature), but this is rare. Normally the solvus keeps widening downward.

Figure 12.22 shows some examples of how these two elements combine. For example, a binary minimum melting loop (top of Figure 12.22) can be considered to be produced by combining two simple melting loops. A simple *peritectic* can be considered to be what happens when a solvus intersects a simple melting loop, and a eutectic what happens when a solvus intersects a binary minimum melting loop. You can try to make these intersections with other topologies, but they will generally not obey the phase rule. (The difference

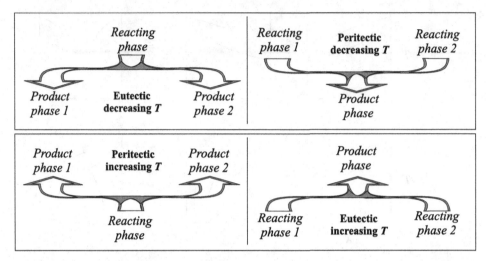

Figure 12.23 The difference between a peritectic and a eutectic.

between a peritectic and a eutectic is illustrated in Figure 12.23. In both, three phases exist together at equilibrium, and so both are represented by a three-phase tie-line. However, the reaction relationships are exactly reversed. What happens at a eutectic during cooling is the same as what happens at a peritectic on heating. Which phases are solids and which are liquids is immaterial.)

Similarly, as conditions change (say, increasing pressure), these configurations can become more complex, but without introducing any new features. For example, at the top of Figure 12.24 is an attempt to portray the effect produced by a solvus moving upward, due to changing conditions. At low pressures, the solvus intersects the melting loop, but does not go through it. If the solvus moves upward in temperature faster than does the melting loop, it must eventually poke its way through the top of the melting loop, as shown. Every such intersection must be accomplished with no more than three coexisting phases, so three-phase tie-lines are produced. Try to imagine the top right diagram on Figure 12.24 as a cross-cutting melting loop and solvus, perhaps with dotted lines completing the individual elements. Then try to satisfy the Phase Rule in some other way; you will find it difficult.

Finally, note that as far as the phase rule is concerned, one phase transition is much like another. Thus an α–β polymorphic transition in a single component behaves just like a melting point or a boiling point when a second component is added. Because one polymorph will in general have a greater capacity for the second component than the other polymorph, a polymorphic phase transition loop is created in the binary system, exactly analogous to boiling and melting loops. These phase transition loops may extend from side to side of the diagram in a completely miscible solid solution, but more likely they will intersect a solvus, as shown at the bottom of Figure 12.24. As before, a three-phase tie-line is created by every such intersection.

It is possible to have a four-phase tie-line in a binary system, but this could only be at a unique temperature and pressure, just like a triple point in a unary system. Binary sections

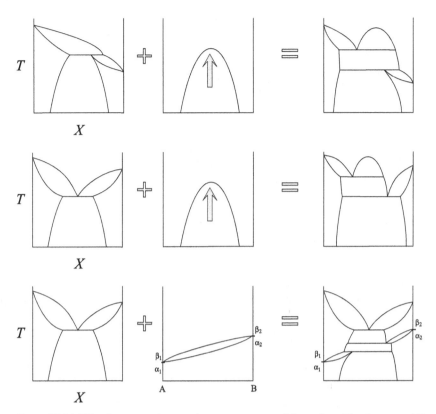

Figure 12.24 What happens when a solvus moves upward through melting loops, and the effect of adding a polymorphic transition.

are not usually drawn for such unique conditions. That is, when we chose our pressure for our T–X section in Section 12.4.2 (the isobaric phase rule), it would be extremely unlikely for this choice to be just the pressure needed for four-phase equilibrium, so three-phase tie-lines are the norm in binary sections.

12.4.12 Cooling Curves

Temperature–time curves are often used as a means of experimentally determining the temperatures of phase changes, and thinking about them can add to your understanding of phase relations. In looking at the four different cooling curves in Figure 12.25, you should imagine that you are in a laboratory, conducting an investigation into the system A–B. One way to proceed would be to prepare a number of different bulk compositions (thoroughly mix various proportions of A and B together), then heat each bulk composition to a number of different temperatures, wait long enough to achieve equilibrium, then observe what phases are present, and measure their compositions. Figure 12.25 (left side) would then represent the results you obtained from a large number of such experiments.

Another way to proceed would be to heat each bulk composition to a temperature sufficiently high to produce a homogeneous liquid, and then to cool slowly while observing

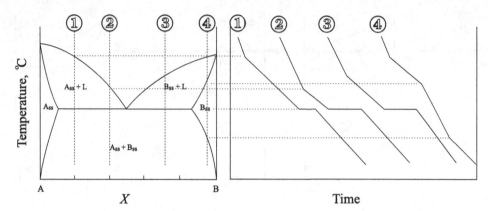

Figure 12.25 Temperature–time curves through a binary eutectic system.

the temperature. When the liquidus temperature is reached, the latent heat of crystallization is released, resulting in a slower rate of cooling. When the eutectic temperature is reached, cooling will cease completely, while three phases coexist. When the liquid disappears, cooling will resume. By observing the inflection points and plateaus in the temperature–time curves, you may deduce the positions of the liquidus and eutectic for the various compositions. You would still need to do more experiments to determine phase compositions, but the cooling-curve method can often give the general shape of a diagram in a relatively short time. Note that composition number 4 does not pass through the eutectic, and hence does not show a temperature plateau.

12.4.13   Intermediate Compounds

In all the binary systems we have considered so far, no compounds are formed *between* the compounds A and B. That is, there are no compounds AB, or A_2B, or A_2B_3, and so on. What happens if these do exist? Consider the binary system A–B that contains the binary compound AB. The simplest possibility is that both A–AB and AB–B are binary systems of the same type, such as simple eutectic systems. Then the two systems are "glued together," as in Figure 12.26.

Another common possibility is that the liquidus for one of the end-member compounds extends completely over the intermediate compound, as in Figure 12.27. When this happens, the compound AB does not melt to a liquid of its own composition – it breaks down at the peritectic temperature to a different compound plus a liquid. This is known as *incongruent melting*. A good example of this is the system $KAlSi_2O_6$–SiO_2 (leucite–silica), which contains the intermediate compound $KAlSi_2O_6 \cdot SiO_2$, or $KAlSi_3O_8$, K-feldspar, shown in Figure 12.28. The large liquidus surface extending over the intermediate compound in these diagrams will often "shrink" with increasing pressure, leading ultimately to the "glued together" type of system (Figure 12.26). In other words, AB may melt incongruently at low pressures and congruently at high pressures.

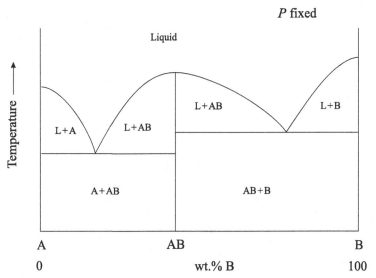

Figure 12.26 Intermediate compound AB divides the binary system A–B into two similar parts. Note that if the composition axis were in mole fraction or mole %, AB would appear midway between A and B, but this is not so when the axis is in weight %.

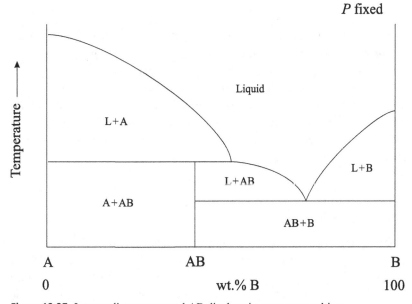

Figure 12.27 Intermediate compound AB displays incongruent melting.

12.5 Ternary Systems

12.5.1 Ternary Compositions

With the addition of a third component, we now need two dimensions to display all possible compositions in a system, and so we lose the ability to display composition and temperature

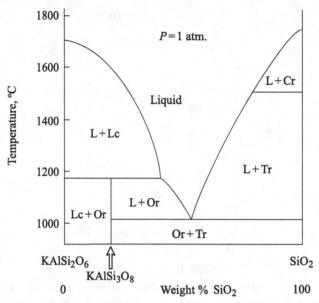

Figure 12.28 The system $KAlSi_2O_6$–SiO_2, with the intermediate compound $KAlSi_3O_8$. Lc, leucite; Or, orthoclase (K-feldspar); Tr, tridymite; Cr, cristobalite; L, liquid.

simultaneously. We can only display compositions at a chosen T and P on a section, or we can *project* compositions from various conditions onto a single plane, as discussed below. The method of depicting compositions within a triangle is shown in Figure 12.29.

Each apex of the triangle represents 100% of one of the components. The proportion of each component in a ternary composition is measured by the distance of a point from the side of the triangle opposite the component in question, as shown. The triangles are usually isometric, but not necessarily. Right-angled triangles are also used in some circumstances, and other shapes could be used if desired.

12.5.2 Sections and Projections

In discussing binary systems we used only binary *sections*, although we mentioned that various kinds of *projections* could be used. We must now expand on this statement. As shown in Figure 12.9, a section shows what you would see after slicing through a P–T–X box, as though with a knife. A projection, on the other hand, is what you see by peering through the box at some angle, usually parallel to one of the axes, and seeing the curves on all sides of the box at the same time. For example, Figure 12.30 shows a P–T *projection* of Figure 12.9. You see the unary phase diagrams for both components superimposed on one another. In a more complete projection, you would also see various curves projected from *within* the box, as well as the curves on the faces of the box. For example, there is a curve joining the critical points of each pure component, which crosses through the box, showing the critical points of binary compositions, which is not shown.

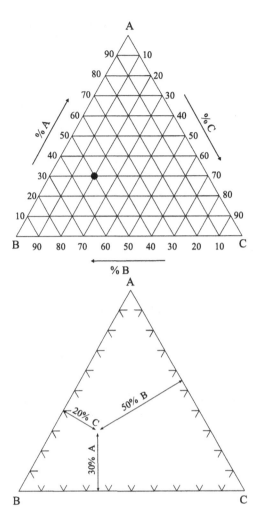

Figure 12.29 Representation of a ternary composition. The dot in the upper triangle represents a composition of 30% A, 50% B, and 20% C, as shown in the lower triangle.

Ternary Projections

In looking at ternary systems, we will start with the projection. To best see the meaning of ternary projections, we start with the oblique view of a simple ternary eutectic system, Figure 12.31. In this figure we see that each side of the compositional triangle has a binary isobaric T–X section constructed on it, perpendicular to the compositional triangle. We see, too, that the liquidus lines of the binaries are joined into surfaces that extend across the ternary space. Each binary eutectic point becomes a ternary *cotectic line* extending into the ternary, down to a *ternary eutectic* point. Points and lines on these surfaces are projected onto a plane surface, as depicted in Figure 12.31. The projection is what you would see if you looked straight down on the three-dimensional object in Figure 12.31, parallel to the temperature axis. The results of such projection are shown in Figures 12.32 and 12.33.

In Figure 12.32 the cotectic lines and *isothermal contours* on the liquidus surfaces have been projected onto a plane surface. The triangle is divided by the cotectic lines into three areas, labeled A + L, B + L, and C + L (where A stands for solid A, etc.), because bulk compositions in these areas will consist of these phases at temperatures below those of

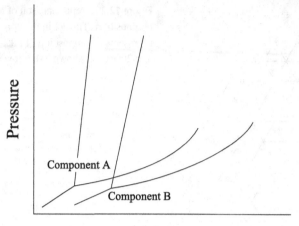

Figure 12.30 A *P–T* projection of Figure 12.9.

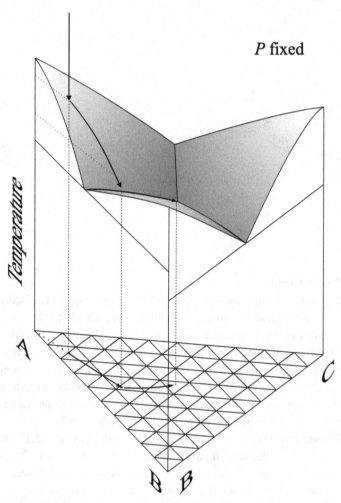

Figure 12.31 A ternary eutectic system A–B–C in an oblique view. The cooling path of a liquid of composition 80% A, 15% B, 5% C is shown.

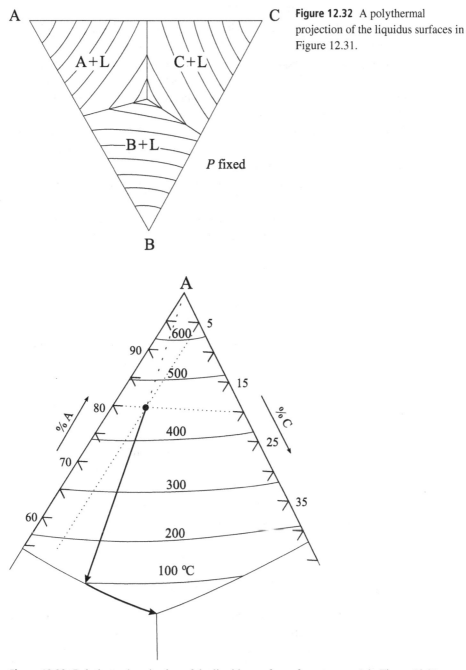

Figure 12.32 A polythermal projection of the liquidus surfaces in Figure 12.31.

Figure 12.33 Polythermal projection of the liquidus surface of component A in Figure 12.31.

the liquidus surfaces. This will become clearer by examining what happens to a liquid of composition 80% A, 15% B, 5% C, as shown in Figures 12.31 and 12.33.

Starting at a temperature well above the liquidus surface, as shown in Figure 12.31, the liquid cools vertically downward until it hits the liquidus surface at a temperature of 450 °C (Figure 12.33). This liquidus surface is the locus of points indicating the first appearance of crystals of composition A, so crystals of A start to separate from the liquid. A tie-line joins the liquid composition to the A-axis. Because composition A is being subtracted from the liquid, the liquid composition *must move directly away from* A, as shown in Figure 12.33. The composition of the liquid stays on the liquidus surface, always on the continuation of a straight line through composition A and the point (80% A, 15% B, 5% C). A continuous series of tie-lines joins the liquid composition to the A-axis (two of which are shown in Figure 12.31). This continues until the liquid composition hits the cotectic line (temperature 100 °C, Figure 12.33), which joins the liquidus surfaces of A and B. On this line, the liquid is simultaneously in equilibrium with crystals of A and crystals of B, so B starts to precipitate. On further cooling, both A and B precipitate, and the liquid composition moves down the cotectic line until it hits the ternary eutectic. At this point, C starts to precipitate, and all three solids precipitate until the liquid is used up. At the ternary eutectic, the number of phases is four (solid A, solid B, solid C, and L), and

$$f = c - p + 1$$
$$= 3 - 4 + 1$$
$$= 0$$

The ternary eutectic is thus an isobaric invariant point, and no temperature change can take place until the liquid is all used up, at which time the crystals will resume cooling.

Ternary Sections

Consider a temperature midway between the melting points of the pure components and the three eutectic temperatures in the system A–B–C, as shown in Figure 12.34. An isothermal plane at this temperature will cut through all three liquidus surfaces. Near each apex, the plane lies below each liquidus surface, so it shows an area of solid plus liquid filled with tie-lines. In the center of the diagram, the plane is everywhere above the liquidus surfaces, so it shows a blank "field of liquid." Sections at successively lower temperatures would show the two-phase fields expanding and coalescing, leaving smaller and smaller liquid fields.

12.5.3 The Granite System

The simple ternary eutectic system discussed above represents just a beginning to the general subject of ternary phase diagrams. Features such as solid solutions, peritectics, intermediate compounds, and so on all introduce complications that we will not discuss. However, as an example of the simple ternary eutectic diagram, let's have a look at the system SiO_2–$NaAlSi_3O_8$–$KAlSi_3O_8$–H_2O, which is used as a model for natural granites. Component SiO_2 is called q, component $NaAlSi_3O_8$ is called ab, and component $KAlSi_3O_8$ is called or. This is because the most common form of SiO_2 is quartz (q), pure $NaAlSi_3O_8$ is the mineral albite (ab), and one of the varieties of $KAlSi_3O_8$ is the mineral orthoclase (or). Quartz is abundant in granites, as is plagioclase, made up mostly of ab, and K-feldspar, made up mostly of or. Granites usually contain several minerals in

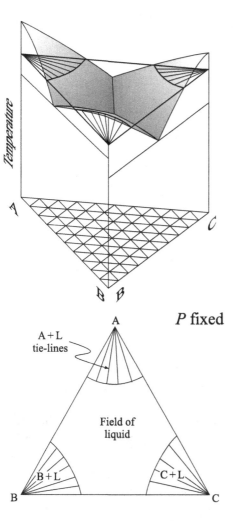

Figure 12.34 An isothermal section through system A–B–C.

addition to quartz, plagioclase, and K-feldspar, but these minerals (that is, quartz, albite-rich plagioclase, and K-feldspar) often account for 80% to 90% of a granite, so the system q–ab–or is quite a useful model in trying to understand the crystallizing or melting histories of granites in general.

The Granite System at 10 kbar

The granite system diagram is shown as a polythermal projection in Figure 12.35.

Compare this diagram with Figures 12.8 and 12.17. In Figure 12.8 you see that the melting point of solid SiO_2 (as β-cristobalite) is about 1700 °C at 1 bar, whereas in Figure 12.35 it is somewhat less than 1100 °C. The melting point of albite at 1 bar in Figure 12.17 is 1118 °C, whereas in Figure 12.35 it is just over 700 °C. Raising the pressure from 1 bar to 10,000 bar increases the melting points, so that cannot be the explanation. What is going on?

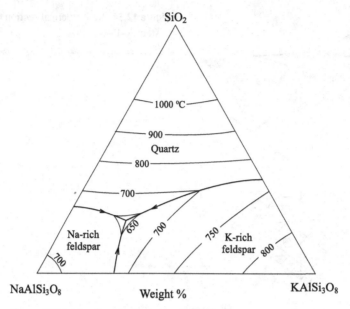

Figure 12.35 The Granite System at 10 kilobars.

The difference lies in the fact that the granite system is not the "dry" system q–ab–or, but the "wet" system q–ab–or–H_2O. The presence of water at high pressures has the effect of substantially lowering the melting temperatures of the pure minerals. Thus Figure 12.35 is not really a ternary projection, but a quaternary projection. All liquids in the diagram have not ternary compositions, but quaternary compositions; that is, they are all saturated with water, and supercritical water is an extra phase that is not shown in the diagram. The presence of the water brings the liquidus temperatures down well into the range of temperatures found in the Earth's crust and means that water is an important component in the history of real granites. However, as water is an extra phase as well as an extra component, we can treat Figure 12.35 exactly like a ternary eutectic and essentially forget about the water.

According to Figure 12.35, crystallization of any bulk composition within the system will generate a final liquid composition at the ternary eutectic at just under 650 °C. Note that the composition of this final liquid (or initial liquid on heating) lies quite close to the albite corner – ternary eutectics do not always occur near the center of the diagram. At other pressures, the position of this eutectic changes. At lower pressures, it moves "northeast," roughly directly away from the albite composition, as shown in Figure 12.36.

A complicating factor here is that the system shows a eutectic only at high pressures. Below about 5 kbar, the eutectic changes to a ternary minimum, as indicated by the change from the circle to plus signs in Figure 12.36. The reason for this is shown in the lower two sequences of diagrams in Figure 12.22. In one of these, a binary eutectic is generated when a more-or-less symmetrical melting loop intersects a solvus. This corresponds to the ab–or–H_2O system at high pressures. In the other, the melting loop

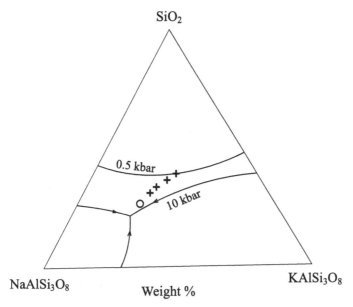

Figure 12.36 The granite system at pressures from 0.5 to 10 kbar. The plus signs show the position of the ternary minimum at 0.5, 2, 3, and 4 kbar. The circle shows the position of the ternary eutectic at 5 kbar.

has a minimum temperature offset to one side, so that even after the solvus intersection has taken place, the minimum is preserved. This corresponds to the system ab–or–H_2O at low pressures. This difference between a eutectic and a minimum is preserved in the ternary system. However, from the point of view of liquid compositions generated during cooling in ternary systems, there is little difference between a ternary eutectic and a ternary minimum. Both represent the final liquid composition for many bulk compositions in the system.

12.5.4   Granite Compositions

This brief explanation is not sufficient for you to understand all the details of this system, but it is sufficient to understand one important result. When the compositions of natural granites are normalized to q–ab–or and plotted in the q–ab–or triangle, a remarkable coincidence of compositions and the ternary minima and eutectics results, as shown in Figure 12.37. As natural granites have undoubtedly crystallized at a variety of pressures, the compositions would be expected to be strung out along the track defined by the ternary minima and eutectics, as they are, *provided that natural granites actually form by crystallizing from silicate liquids*. The demonstration by Tuttle and Bowen (1958) that this was the case provided strong evidence for the magmatic origin of granites. The slight offset of the highest frequency of natural compositions toward the $KAlSi_3O_8$ apex, as well as other aspects of the diagram, have been the subject of much discussion.

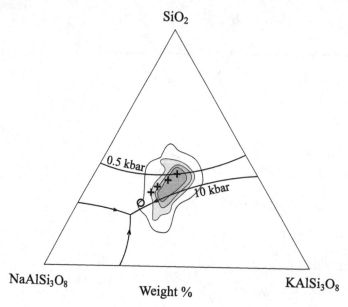

Figure 12.37 The compositions of natural granites superimposed on Figure 12.36. Several hundred individual points are contoured, with the darker colors indicating a higher frequency of points.

12.6   Summary

Phase diagrams are a kind of concise representation of the equilibrium relationships between phases as a function of chosen intensive variables, such as temperature, pressure, composition, pH, oxidation potential, activity ratio, and so on. They are extremely useful, not only in representing what is known about a system, but also in thinking about processes involving phase changes. Most of the diagrams in Chapter 11 are phase diagrams, although the term is most often used in relation to the P–T–X type of diagram discussed in this chapter.

Although in principle phase diagrams can be calculated from thermodynamic data, in complex systems the relationships are generally determined experimentally. The diagrams must nevertheless obey rules established by equilibrium thermodynamics. Therefore an understanding of the material in all the previous chapters is a prerequisite to a real understanding of the simple points, lines, and surfaces found in phase diagrams.

Exercises

E12.1 If the slope of a phase transition of a mineral from phase α to phase β is -21.0 bar deg^{-1} at a temperature of 600 K, the $\Delta_{\alpha \to \beta} V$ of the transition is $+0.150$ cal bar^{-1} mol^{-1}, and $\Delta_f H^\circ_{600}$ of phase α is $-17,000$ cal mol^{-1}, what is $\Delta_f H^\circ_{600}$ of phase β? Sketch and label the phase diagram.

E12.2 The slope of the ice–water phase boundary is $-131.7\,\text{bar deg}^{-1}$. Knowing the heat of fusion of ice ($\Delta_r H^\circ_{\text{ice}\rightarrow\text{water}} = 6010\,\text{J mol}^{-1}$) and the molar volume of water at $0\,^\circ\text{C}$ ($V^\circ_{\text{H}_2\text{O(l)}} = 18.01826\,\text{cm}^3\,\text{mol}^{-1}$), calculate the molar volume of ice at $0\,^\circ\text{C}$.

E12.3 A hypothetical compound β has been found to have at least seven phases, named A, B, ..., G, arranged as shown in Figure 12.38. Point 6 is a critical point. The locations of points 1, 2, ..., 7 and some of the thermodynamic properties have been found experimentally to have the values in Tables 12.1 and 12.2.

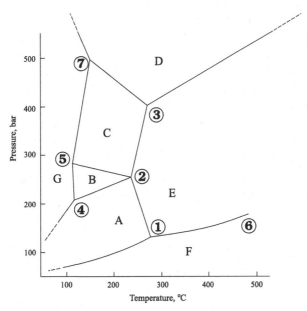

Figure 12.38 Phase diagram for compound β.

Table 12.1. *T* and *P* data.

Point	*T* (°C)	*P* (bar)
1	277	132
2	236	256
3	270	404
6	483	179
7	157	

(a) There are two parts of this diagram that are thermodynamically impossible. Where are they, and why are they incorrect? Sketch possible correct relationships at each location.

(b) Sketch G–P and V–P sections at $250\,^\circ\text{C}$.

(c) Sketch G–T and H–T sections at $150\,\text{bar}$.

(d) Calculate V°_E.

Table 12.2. Data for the four phases.

Phase	$V°$ ($cm^3\,mol^{-1}$)	$S°$ ($J\,mol^{-1}\,K^{-1}$)
A	15.0	22.0
C	11.62	
D	10.0	21.4
E		22.3

(e) Using this result, calculate $S_C°$.

(f) Calculate the pressure at point 7. If it doesn't look about right compared with the diagram, you have made a mistake.

(g) What are the upper and lower limits for possible values for $V_B°$?

(h) Identify each phase as solid, liquid, or gas. Which solids float and which sink in liquid β?

(i) Why are boundaries A/F and E/F curved, while all the others are straight?

E12.4 What are the melting points of pure A and pure B in Figure 12.39? How many polymorphs of component B are there? What is the temperature of transition between them? If the entropy of α is $10\,J\,mol^{-1}\,K^{-1}$, what is the entropy of β? $\Delta_f H_\alpha° = -100\,kJ\,mol^{-1}$, $\Delta_f H_\beta° = -102\,kJ\,mol^{-1}$.

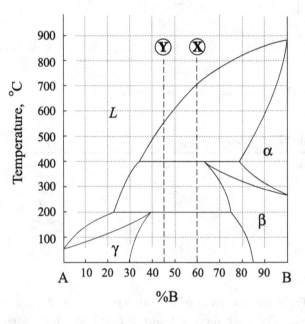

Figure 12.39 System A–B at 1 bar pressure.

E12.5 Figure 12.39 could be described as being made up of a simple melting loop intersected by a miscibility gap, complicated slightly by the polymorphs of component B. If component B had only one crystal form (α), what would the phase diagram look like?

E12.6 If liquid immiscibility developed at higher pressures, what might the diagram in Figure 12.39 look like in the following cases: (a) B having only one crystal form; (b) B having two polymorphs?

E12.7 In Section 12.4.11 it says that there are two basic binary diagram elements. However, note that if the phase transition loop (Figure 12.21, left) becomes detached from the vertical axes on both sides, and if the solvus (Figure 12.21, right) closes downward as well as upwards (i.e., has both upper and lower consolute points – this does happen in some systems), the two elements become indistinguishable. So perhaps there is only one "basic element." Then again, these elements are created by tie-lines, so perhaps the "basic elements" are the two-phase and three-phase tie-lines. Which point of view is right? Under what circumstances does the phase transition loop become detached from the temperature axes?

E12.8 A central part of the novel *Cat's Cradle* by Kurt Vonnegut, Jr., involves the existence of a new polymorphic form of ice (ice IX), which is more stable than ordinary ice (ice I) at atmospheric pressure and has a melting point of $114.4\,°F$ ($45.8\,°C$). When ice IX gets loose it acts as a seed, and all the water on Earth, which is metastable with respect to ice IX, freezes immediately. This, of course, means the end of life on Earth. Is there any thermodynamic reason why ice IX could not exist? If it did exist, is $114.4\,°F$ a reasonable melting point for it? Answer this with the help of a G–T section through system H_2O at 1 bar.

E12.9 Construct the phase diagram for H_2O in the range 0 to 7000 bar (Y-axis), $-40\,°C$ to $20\,°C$ (X-axis), using the data in Table 12.3. Label each phase boundary and triple point with the coexisting phases and degrees of freedom. Show the metastable extensions of the phase boundaries.

Table 12.3. Data for Exercise E12.9 (point f is off the diagram).

Point	Phases	$T\,(°C)$	$P\,(bar)$
a	Ice I, ice III, L	−22	2215
b	Ice III, ice V, L	−17	3530
c	Ice I, ice II, ice III	−35	2170
d	Ice II, ice III, ice V	−24	3510
e	Ice V, ice VI, L	0.2	6380
f	Ice VI, ice VII, L	81.6	22,400

(a) The slope of the ice II/ice V boundary is negative, and the slope of the ice I/ice II boundary is positive (you don't need to know the exact slopes). Sketch a diagram of density vs. pressure (0 to 5000 bar) for H_2O at $-40\,°C$ and at $-10\,°C$, with no numbers on the density axis.

(b) Does ice V sink or float in water?

(c) Sketch a G–P section through the diagram at $-10\,°C$ from 0 to 7000 bar showing ice I, water, ice V, and ice VI. Show the location on this section of the freezing of water to ice VI and ice V as discussed in the quotation from

Bridgman below. Why does water freeze sometimes to ice V and sometimes to ice VI, according to Bridgman?

The following extract is from *The Physics of High Pressure*, by P. W. Bridgman (1958, page 423). It may help to dispel any notion that determining phase relations is a simple and straightforward exercise. (Bridgman was a professor at Harvard University and in 1946 received the Nobel Prize in physics for his research on materials at high pressures.)

The conditions under which the different modifications of ice appear are somewhat capricious, and often inconvenient manipulation is necessary to arrive in the part of the phase diagram desired. The behavior is particularly striking in the neighbourhood of the V–VI–liquid triple point, say, between 0 °C and −10 °C and between 5000 and 6000 bar. With the ordinary form of apparatus, water in a steel piezometer with pressure transmitted to it by mercury, it is very difficult to produce the modification V. For example, if liquid water at −10 °C is compressed across the melting curve, the liquid will persist in the sub-cooled condition for an indefinite time, without freezing to V, the stable form. But if the compression is carried further, to the unstable prolongation of the liquid–VI line, freezing to the form VI will take place almost at once on crossing the line, in spite of the fact that VI is unstable with respect to V. Now in order to make V appear, VI may be cooled 30° or 40°, when it will spontaneously change to V, the stable form. If pressure is now released back to 5000, say, and temperature is raised to the melting line of V, melting takes place at once and so the coordinates of the melting line may be found. Suppose now after the melting is completed the liquid water be kept in the neighbourhood of the melting line for several days, and then the pressure increased again across the melting line at −10 °C; it will be found that the instant the melting line of V is crossed the liquid freezes to V. This suggests that there has persisted in the liquid phase some sort of structure, not detected by ordinary large-scale experiments, favorable to the formation of V. It is now known, of course, from X-ray analysis, that structures are possible in the liquid; this experiment suggests a specificity in these structures that might well be the subject of further study.

The formation of the nucleus of V may be favored by the proper surface conditions. Thus if there is any glass in contact with the liquid, either a fragment of glass wool purposely introduced into the liquid, or by enclosing the water in a glass instead of a steel bulb, freezing to V takes place at once without the slightest hesitation immediately on carrying the virgin liquid across the freezing line.

Additional Problems

A12.1 (a) Describe in detail the equilibrium cooling history of a liquid of composition 6 in Figure 12.40.

   (b) Describe in detail the perfect fractional cooling history of a liquid of composition 2 in Figure 12.40. This diagram looks frighteningly complex at first, but it contains nothing more than the diagram elements already discussed, and working through such a diagram is not more difficult than using a simpler diagram; it just takes longer.

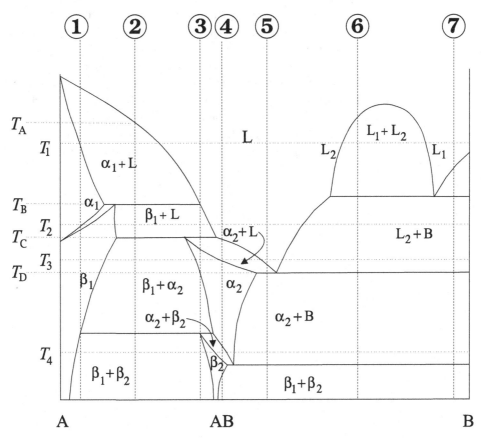

Figure 12.40 The T–X phase diagram for the system A–B.

A12.2 You have performed a series of experiments in the lab involving the cooling of various mixtures of compound A and compound B at 1 atm. The results of the experiments are given in Table 12.4, in which T represents the temperature at which you first observed crystals in the cooling mixtures.

Table 12.4. The results of some cooling experiments.

Run no.	Composition (wt.%)	T (°C)
1	100% compound A	63
2	100% compound B	88.9
3	90% compound B	85.4
4	75% compound B	75
5	60% compound B	63.9
6	45% compound B	50.5
7	30% compound B	42.9
8	15% compound B	54.4

(a) Construct a phase diagram using as axes T (°C) and wt.% B, and label all the phase fields. Assume that there is no appreciable solid solution in the solid phases.

(b) What are the eutectic temperature and composition?

(c) Construct a schematic cooling curve (temperature vs. time) for run no. 4.

(d) Consider 1.5 g of a system containing 35 wt.% compound A and 65 wt.% compound B at a temperature of 60 °C. What phases are present at equilibrium, and how much of each phase is there? Note that this information is contained in the diagram, even though no such experiment was conducted.

13 Affinity and Extent of Reaction

13.1 Introduction

In this chapter we will see how equilibrium thermodynamics is used in dealing with reacting systems, rather than just systems at equilibrium. Realistically, this should mean including the science of chemical kinetics with our thermodynamics, and we should also include other factors, such as fluid flow, temperature and pressure gradients, and adsorption on mineral surfaces, to build increasingly realistic models of complex natural phenomena involving the movement and chemical reactions of fluids in soils and rocks in the Earth's crust.

That is all a bit too ambitious for this book.[1] Computational models including all these subjects are at the forefront of research in several areas. What we will attempt in this chapter is to show how equilibrium thermodynamics simulates chemically reacting systems *without* using kinetics. Process modeling in thermodynamics has some points in common with kinetics, because both sciences consider the problem of chemical reactions proceeding from start to finish, and both use the extent of reaction variable ξ. But although thermodynamics uses the extent of reaction variable it does not use a real time variable, with which it is closely connected in kinetics. The use of a real time variable is what distinguishes kinetics from thermodynamics.

13.2 Quasistatic Processes

The term thermodynamic constraint is particularly appropriate for the extent of reaction variable ξ, because for any equilibrium state of a system which does not appear on the stable equilibrium G–T–P surface, the question arises as to what prevents that system from spontaneously changing toward the stable equilibrium state. In real systems the reason is usually that there is a kinetic barrier, an activation energy which must be overcome for the change to take place (see Lambert (1998) for an entertaining discussion of this point). However, there is no term or variable in thermodynamics for activation energy. Activation energy as well as the enthalpy and entropy of activation are variables in transition state theory and reaction kinetics, but these quantities are not relevant to the present discussion.

---

[1] An introduction to these complications is Zhu and Anderson (2002).

We have equations like (4.48) which enable us to calculate energy changes in processes involving changes in temperature and pressure,

$$d\mathbf{G} = -\mathbf{S}\,dT + \mathbf{V}\,dP \tag{4.48}$$

or changes in entropy and volume,

$$\Delta U = T\,\Delta S - P\,\Delta V \tag{4.7}$$

But now we have to deal with irreversible processes, which we know are characterized by complete disequilibrium. How can we do this, while maintaining metastable equilibrium? This is the role of the extent of reaction variable ξ and quasistatic processes.

13.2.1 Reversible vs. Quasistatic Processes

The term quasistatic has various meanings as used by other authors, but in this chapter it refers to an irreversible process carried out in a number of steps, usually a very large number of very small steps ($d\xi$). Both reversible and quasistatic processes are idealized and are represented by continuous functions, but there is an important difference. As described in Section 2.6.1, a reversible process refers to a continuous series of *stable* equilibrium states, which may be plotted on a stable equilibrium surface having only the normal two independent constraint variables such as T and P or $\mathbf{S}$ and $\mathbf{V}$ as in Figure 4.7. Equilibrium states having three constraint variables can be plotted on a *metastable* equilibrium surface such as path A → B in Figure 4.11, but the third constraint is held constant (a point amplified in Section 13.5.2). A quasistatic process is a series of metastable equilibrium states along an irreversible path such as represented by path B → C in Figure 4.11. After every increment of a quasistatic process the system is in a metastable equilibrium state having three constraints, but the third constraint is incrementally changed. Each point in a reversible process represents a system in a balanced state; it has no tendency to change in one direction or another. On a metastable equilibrium surface or in a quasistatic process the system has a definite tendency to change to a lower energy state but is prevented from doing so by a (third) constraint. Increments in this constraint variable, called the extent of reaction variable ξ, result in a quasistatic process.

13.3 The Extent of Reaction Variable

Consider a generalized chemical reaction

$$a\mathrm{A} + b\mathrm{B} = c\mathrm{C} + d\mathrm{D} \tag{13.1}$$

where A, B, C, and D are chemical formulae, and a, b, c, and d are the stoichiometric coefficients. We pointed out in Section 9.3 that, when this reaction reaches equilibrium,

$$c\mu_\mathrm{C} + d\mu_\mathrm{D} = a\mu_\mathrm{A} + b\mu_\mathrm{B}$$

and

$$\Delta_\mathrm{r}\mu = c\mu_\mathrm{C} + d\mu_\mathrm{D} - a\mu_\mathrm{A} - b\mu_\mathrm{B} \tag{13.2}$$
$$= 0$$

Equation (13.1) can be generalized to

$$\sum_i v_i M_i = 0 \tag{13.3}$$

where M_i are the chemical formulae and v_i are the stoichiometric coefficients, with the stipulation that v_i is positive for products and negative for reactants. Equation (13.2) can then be generalized to

$$\sum_i v_i \mu_i = 0 \tag{13.4}$$

Up to now, we have been most interested in reactions that reach equilibrium. Now let's look at what happens before that point is reached, that is, while the reaction is taking place. Let's say that the reaction proceeds from left to right as written. It doesn't matter for the moment whether all the reactants and products are in the same phase (a *homogeneous* reaction) or in different phases (a *heterogeneous* reaction). During the reaction, A and B disappear and C and D appear, but the *proportions* of A:B:C:D that appear and disappear are fixed by the stoichiometric coefficients. If the reaction is

$$A + 2B \rightarrow 3C + 4D \tag{13.5}$$

then for every mole of A that reacts (disappears), 2 moles of B must also disappear, while 3 moles of C and 4 moles of D must appear. This is simply a mass balance, independent of thermodynamics or kinetics, and can be expressed as

$$\frac{dn_A}{v_A} = \frac{dn_B}{v_B} = \frac{dn_C}{v_C} = \frac{dn_D}{v_D} = \frac{dn_i}{v_i} \tag{13.6}$$

where, if all reactants and products are pure phases, the differentials dn_A, dn_B, and so on can refer to a change in the amount of A, B, and so on, of any convenient magnitude, not necessarily an infinitesimal change. We can then represent every term in (13.6) by increments in a single variable, $d\xi$, so

$$\frac{dn_A}{v_A} = \frac{dn_B}{v_B} = \frac{dn_C}{v_C} = \frac{dn_D}{v_D} = d\xi \tag{13.7}$$

from which it appears that

$$\frac{dn_A}{d\xi} = v_A; \quad \frac{dn_B}{d\xi} = v_B; \quad \ldots; \quad \frac{dn_i}{d\xi} = v_i \tag{13.8}$$

where our new variable ξ is called the extent of reaction variable, and in this case represents an arbitrary number of moles. Equation (13.8) says that in reaction (13.5), $dn_A/d\xi = -1$, $dn_B/d\xi = -2$, $dn_C/d\xi = 3$, and so on, which simply means that for every mole of A that disappears, 2 moles of B also disappear, 3 moles of C appear, and so on.

13.3.1 Using the Extent of Reaction Variable

Figure 4.8(b) shows the G–T–P surfaces for the stable and metastable forms of a particular reaction. In Figure 13.1 we show the same thing, but the surfaces represent (metastable) reactants and (stable) products. The reactants and products can represent any chemical

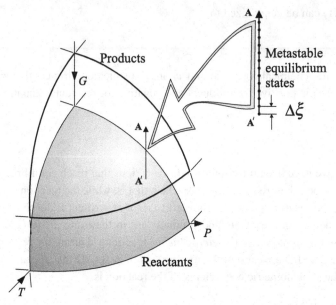

Figure 13.1 A sequence of metastable equilibrium states for the reaction A′ → A at constant T, P. The extent of reaction variable is ξ. Compare this with Figure 4.8.

reaction whatsoever. An irreversible reaction between these two states is represented by A′ → A at a fixed T and P (A′ → A here and B → C in Figure 4.11 represent the same idea). At A′, which is metastable because of some unspecified constraint, we release this constraint momentarily and allow the reaction to proceed irreversibly by an amount $\Delta\xi$ (measured in moles), forming some products. Then we reapply the constraint, and the system settles down into its new metastable state between A′ and A, with both reactants and a little bit of products coexisting. Then we release the constraint momentarily again, another $\Delta\xi$ of reaction occurs, and we reapply the constraint again. Note the difference between this and a reversible reaction. The reversible reaction is also a continuous succession of equilibrium states, but they are stable equilibrium states, having only the normal two constraints such as T and P, with no third constraint and no tendency to change from this state. In Figure 13.1 a reversible reaction would lie entirely on the "Products" surface, where there is no need for a third variable. On the "Reactants" surface or any surface in between, the extent of reaction variable would have a constant value.

Well, that's fine in the abstract, but how do we represent this arrow in thermodynamics? And why would we want to do it? How we do it is simplicity itself and illustrates once again the difference between reality and our model or simulation of reality.

Aragonite–Calcite Example

Let's consider one of the simplest kinds of reaction, a polymorphic change such as aragonite → calcite. Aragonite on museum shelves actually does not change to calcite at all, but we can do it mathematically with ease. From (13.7) we have

$$\frac{dn_A}{\nu_A} = \frac{dn_B}{\nu_B} = d\xi$$

where, if A is aragonite and B is calcite, then $\nu_A = -1$ and $\nu_B = 1$. Thus

$$dn_{\text{aragonite}} = -d\xi \qquad (13.9)$$

$$dn_{\text{calcite}} = d\xi \qquad (13.10)$$

where $n_{\text{aragonite}}$ is some number of moles of aragonite, and similarly for calcite. If $n°$ is the number of moles of each to start with, then integrating these equations from $n°$ to some new value of n gives

$$\int_{n°}^{n} dn_{\text{aragonite}} = -\int d\xi$$

$$n_{\text{aragonite}} - n°_{\text{aragonite}} = -\Delta\xi \qquad (13.11)$$

and similarly

$$n_{\text{calcite}} - n°_{\text{calcite}} = \Delta\xi \qquad (13.12)$$

Equations (13.11) and (13.12) can also be rewritten

$$n_{\text{aragonite}} = n°_{\text{aragonite}} - \Delta\xi \qquad (13.13)$$

$$n_{\text{calcite}} = n°_{\text{calcite}} + \Delta\xi \qquad (13.14)$$

which shows that whatever amounts of each mineral we have to start with, this amount is decreased by $\Delta\xi$ moles for aragonite and increased by $\Delta\xi$ moles for calcite, every time we allow the reaction (the integration) to proceed by $\Delta\xi$. What could be simpler? If we let $n°_{\text{calcite}} = 0$ and $n°_{\text{aragonite}} = 1$, and we proceed from pure aragonite to pure calcite in four steps of $\Delta\xi = 0.25$ moles, then after one step $n_{\text{aragonite}} = 0.75$ moles, $n_{\text{calcite}} = 0.25$ moles, and so on, and we obtain the result shown in Figure 13.2. We could, of course, use

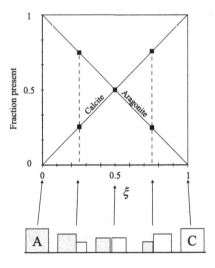

Figure 13.2 The irreversible reaction aragonite (A) → calcite (C) considered as a function of the extent of reaction variable ξ.

as many small steps as we like, changing aragonite into calcite quasistatically, although this never happens in nature.

A striking feature of the reaction path calculations we have been considering is that in a sense they are independent of the Gibbs energy which we know controls which way these reactions proceed. In other words, by changing the sign in equations like (13.9) and (13.10) we can use exactly the same methods to change calcite into aragonite or graphite into diamond, or to form a mineral from its dissolved species in solution, like running a movie backwards. To investigate this further, we must link the extent of reaction variable with the Gibbs energy. The calcite–aragonite case is too simple, so we choose a system that has an equilibrium constant for one reaction, and a manageable number of species, the system N_2–H_2.

13.4 Components and Species Again

In the two-component system N_2–H_2, components N_2 and H_2 refer to the result you would get if you analyzed the system for total nitrogen and total hydrogen. The actual molecular form taken by each element in the system is irrelevant. However, N_2 and H_2 might also refer to the *species* in the system, i.e., diatomic nitrogen molecules and diatomic hydrogen molecules. In real systems, species N_2 and H_2 combine to form ammonia,

$$N_2(g) + 3H_2(g) = 2NH_3(g) \tag{13.15}$$

so that there are at least three major species. In many systems, there may be only a few components, but dozens or hundreds of species.

Equation (4.72)

$$dG = -S\,dT + V\,dP + \sum_i^c \mu_i\,dn_i \tag{4.72}$$

was derived for systems of c independent components. For example, in the two-component system N_2–H_2, $c = 2$, and the final term on the right side would be

$$\mu_{N_2}\,dn_{N_2} + \mu_{H_2}\,dn_{H_2}$$

However, the equation can also be used with i representing some or all of the *species* in the system, rather than independent components, as long as the species are related to one another in balanced chemical reactions, such that the number of independent compositional parameters remains equal to c. Thus the final term on the right side of (4.72) could also read

$$\mu_{N_2}\,dn_{N_2} + \mu_{H_2}\,dn_{H_2} + \mu_{NH_3}\,dn_{NH_3}$$

where N_2 and H_2 are now species and not components. Although we now have three compositional terms, we also have an equation, (13.15), relating them (assuming equilibrium), so there are still only two independent compositional terms.

Changing from components to species in this way provides some flexibility we shall take advantage of shortly. In a closed system, we cannot change the amounts (or mole numbers) of components N_2 and H_2, and the last term in (4.72) is zero. However, the amounts of the species in (13.15) *can* change in a closed system, if the reaction progresses to the left or the right from some metastable state towards the stable equilibrium state. The last term in (4.72) (which would look a bit different in having $\sum_{i=1}^{s}$ rather than $\sum_{i=1}^{c}$, that is, s species rather than c components) would not be zero, even in a closed system, if we were considering chemical reactions progressing towards equilibrium. Furthermore, in such cases this last term must always be negative, no matter in which direction the reaction proceeds, to be consistent with $d\mathbf{G}_{T,P} \leq 0$. One other thing to note is that, although in this case there is one reaction and three species, in other cases you may be considering one out of many simultaneous reactions, and many species. In these cases, all possible reactions must be proceeding towards equilibrium at each increment of ξ, not just the one you happen to be considering.

13.5　The Affinity

Considering a system at constant T and P, Equation (4.72) now shows that

$$d\mathbf{G}_{T,P} = \sum_{i=1}^{s} \mu_i \, dn_i \tag{13.16}$$

where the i are not the c independent components, but the s species we are considering which form from those components in a single reaction. In our ammonia example, Equation (13.15), $s = 3$. Into this we substitute the relation $dn_i = \nu_i \, d\xi$ from (13.8), to get

$$d\mathbf{G}_{T,P} = \sum_{i=1}^{s} (\nu_i \mu_i) d\xi \tag{13.17}$$

On comparing this with (4.53),

$$d\mathbf{G} = -\mathbf{S} \, dT + \mathbf{V} \, dP - \mathcal{A} \, d\xi \tag{4.53}$$

we see that another definition of the affinity is

$$\mathcal{A} = -\sum_{i}^{s} \nu_i \mu_i \tag{13.18}$$

Recalling that our ν_i are positive for products and negative for reactants, the quantity $\sum_i \nu_i \mu_i$ is simply the difference in partial molar Gibbs energy between products and reactants. For (13.5) this is

$$\underbrace{(3\,\mu_C + 4\,\mu_D)}_{\text{products}} - \underbrace{(\mu_A + 2\,\mu_B)}_{\text{reactants}} \tag{13.19}$$

If this sum is zero, the reaction is at stable equilibrium, there is no third constraint, the affinity is zero, the third term on the right side of (4.53) disappears, and both

$$d\mathbf{G} = -\mathbf{S} \, dT + \mathbf{V} \, dP$$

and (4.48)

$$dG = -S\,dT + V\,dP \tag{4.48}$$

apply to the (closed) system. If some constraint prevents reaction (13.5) from proceeding to equilibrium, then the sum in (13.19) is not zero, but may be positive or negative, depending on whether (13.5) wants to go to the left or to the right. If the sum $\sum_i \nu_i \mu_i$ is positive ($\mathcal{A}$ negative), the reaction wants to go to the left as written, $d\xi$ is negative, and $\mathcal{A}\,d\xi$ is positive. If the sum is negative ($\mathcal{A}$ positive), the reaction proceeds to the right, $d\xi$ is positive, and $\mathcal{A}\,d\xi$ is positive. So $\mathcal{A}\,d\xi$ is inherently positive (or zero), and $d\mathbf{G}_{T,P}$ is inherently negative for a spontaneous reaction, which is consistent with (from (4.53))

$$d\mathbf{G}_{T,P} = -\mathcal{A}\,d\xi \tag{13.20}$$

Equation (13.18) shows that the affinity is a $\Delta\mu_{T,P}$ term, giving the "distance" in joules per mole between stable and metastable equilibrium surfaces or states, and the amount of useful work that a chemical reaction could theoretically do as it proceeds toward equilibrium. It is in fact represented by the path $A \rightarrow A'$ in Figures 4.8(b) and 13.1 and path B $\rightarrow$ C in Figure 4.11. It is worth noting, too, that the units of ξ are moles and those of $\mathcal{A}$ are $\mathrm{J\,mol^{-1}}$, so the Gibbs energy in equations such as (13.20) is the total, not the molar, Gibbs energy. A common modeling practice is to define the system as containing one kilogram of water, so that all mole numbers (n_i) become molalities (m_i).

An Example
Equation (4.46)

$$d\mathbf{G}_{T,P} \leq 0 \tag{4.46}$$

clearly implies that $\mathbf{G}_{T,P}$ as a function of some third variable is a U-shaped function, having $d\mathbf{G}_{T,P} < 0$ except at the minimum, where $d\mathbf{G}_{T,P} = 0$, just as Equation (4.14)

$$d\mathbf{S}_{U,V} \geq 0 \tag{4.14}$$

implies that entropy is an *inverted*-U shaped function of some third constraint variable (Figure 4.3), having $d\mathbf{S}_{U,V} > 0$ except at the maximum, where $d\mathbf{S}_{U,V} = 0$. Furthermore, being differentiable, these functions are continuous, so there must be a function of the kind we need, a continuous function describing metastable equilibrium states between a starting composition off the appropriate stable $\mathbf{G}$–T–P surface (e.g., point B in Figure 4.11) and giving the change in Gibbs energy between this state and every state in between as the reaction changes the system composition toward the stable equilibrium state (point C in Figure 4.11).

This function is, in total differential form analogous to Equation (4.23),

$$dG = \left(\frac{\partial \mathbf{G}}{\partial T}\right)_{P,\xi} dT + \left(\frac{\partial \mathbf{G}}{\partial P}\right)_{T,\xi} dP + \left(\frac{\partial \mathbf{G}}{\partial \xi}\right)_{T,P} d\xi \tag{13.21}$$

$$= -S\,dT + V\,dP - \mathcal{A}\,d\xi \tag{4.53}$$

where

$$\mathcal{A} = -(\partial \mathbf{G}/\partial \xi)_{T,P} = -\sum_{i=1}^{s} \nu_i \mu_i \qquad (13.22)$$

where s is the number of species i in a chemical reaction, ν_i are stoichiometric coefficients, negative for reactants and positive for products, $\mathcal{A}$ is the affinity, and ξ is the third constraint variable, introduced by De Donder (1920, 1922, 1927) and De Donder and van Rysselberghe (1936). It appears in all the fundamental equations of thermodynamics if irreversible processes are involved. If ξ is non-zero, G is then a function of three instead of two variables, and represents the Gibbs energy of a metastable equilibrium state.

Equation (13.22) shows that the rate of decrease in Gibbs energy of the system with each reaction increment (that is, $\partial \mathbf{G}/\partial \xi$) will be greater the larger the difference in chemical potential between products and reactants. Because this difference steadily decreases to zero at equilibrium during a reaction, this results in the function implied by Equation (4.46), that is, a continuous U-shaped curve of $\mathbf{G}$ vs. ξ with a minimum at the equilibrium composition where $d\mathbf{G}_{T,P} = 0$. Such curves are shown by many authors, e.g., Denbigh (1981, Figure 18), de Heer (1986, Figure 28), and McQuarrie and Simon (1997, Figure 26.2). Diagrams showing the inverted-U shaped entropy curve (Figure 4.2 in this text) are less common. de Heer (1986, Figure 17b) and Bent (1965, Figure 14.1) show schematic examples.

Now consider the reaction

$$\mathrm{N_2O_4(g)} = 2\,\mathrm{NO_2(g)} \qquad (13.23)$$

If we start with 1 mole of $\mathrm{N_2O_4}$ and zero moles of $\mathrm{NO_2}$, the number of moles of $\mathrm{N_2O_4}$ as the reaction proceeds will be $1 - \xi$ and the number of moles of $\mathrm{NO_2}$ will be 2ξ. The total number of moles in the system is $1 - \xi + 2\xi = 1 + \xi$, so the mole fractions are

$$\left. \begin{array}{l} x_{\mathrm{N_2O_4}} = \dfrac{1-\xi}{1+\xi} \\[3mm] x_{\mathrm{NO_2}} = \dfrac{2\xi}{1+\xi} \end{array} \right\} \qquad (13.24)$$

The goal is to minimize an expression for the Gibbs energy of this solution, so, starting with expressions from Chapter 7,

$$\Delta_{\mathrm{mix}}G_{\mathrm{ideal\ solution}} = G_{\mathrm{ideal\ solution}} - \sum_i x_i G_i^{\circ} \qquad (7.18)$$

$$= RT \sum_i x_i \ln x_i \qquad (7.19)$$

we write

$$G_{\mathrm{ideal\ solution}} = \sum_i x_i G_i^{\circ} + RT \sum_i x_i \ln x_i \qquad (13.25)$$

In the present case, this becomes (for pure substances $G^{\circ} = \mu^{\circ}$)

$$G_{\mathrm{ideal\ solution}} = x_{\mathrm{N_2O_4}}\mu_{\mathrm{N_2O_4}}^{\circ} + x_{\mathrm{NO_2}}\mu_{\mathrm{NO_2}}^{\circ} + RT[x_{\mathrm{N_2O_4}} \ln x_{\mathrm{N_2O_4}} + x_{\mathrm{NO_2}} \ln x_{\mathrm{NO_2}}] \qquad (13.26)$$

To convert molar G to the total $\mathbf{G}$ of the solution, we multiply both sides by the denominator of the mole fraction term, $(1 + \xi)$, so

$$\mathbf{G}_{\text{ideal solution}} = n_{\text{N}_2\text{O}_4} \mu^\circ_{\text{N}_2\text{O}_4} + n_{\text{NO}_2} \mu^\circ_{\text{NO}_2} + RT[n_{\text{N}_2\text{O}_4} \ln x_{\text{N}_2\text{O}_4} + n_{\text{NO}_2} \ln x_{\text{NO}_2}] \quad (13.27)$$

Substituting $n_{\text{N}_2\text{O}_4} = (1 - \xi)$ and $n_{\text{NO}_2} = 2\xi$,

$$\mathbf{G}_{\text{ideal solution}} = (1 - \xi)\mu^\circ_{\text{N}_2\text{O}_4} + 2\xi \mu^\circ_{\text{NO}_2} + RT[(1 - \xi)\ln x_{\text{N}_2\text{O}_4} + 2\xi \ln x_{\text{NO}_2}]$$

$$\mathbf{G}_{\text{ideal solution}} - \mu^\circ_{\text{N}_2\text{O}_4} = \xi(2\mu^\circ_{\text{NO}_2} - \mu^\circ_{\text{N}_2\text{O}_4}) + RT[(1 - \xi)\ln x_{\text{N}_2\text{O}_4} + 2\xi \ln x_{\text{NO}_2}] \quad (13.28)$$

so

$$\mathbf{G}_{\text{ideal solution}} - \mu^\circ_{\text{N}_2\text{O}_4} = \xi(\Delta_r\mu^\circ) + RT[(1 - \xi)\ln x_{\text{N}_2\text{O}_4} + 2\xi \ln x_{\text{NO}_2}]$$

$$= \xi(\Delta_r\mu^\circ) + \Delta_{\text{mix}}\mathbf{G} \quad (13.29)$$

where $\mathbf{G}$ is the total Gibbs energy of the system containing $(1 + \xi)$ moles of N_2O_4 and NO_2 gases, assuming ideality, $\Delta_r\mu^\circ$ is the standard Gibbs energy of reaction at T, and ξ is the extent of reaction variable which can have any value between 0 and 1.0. Note that the $RT[\ldots]$ term is simply the total energy form of (7.19), and can be called $\Delta_{\text{mix}}\mathbf{G}$, the total Gibbs energy of mixing. All we have done is to convert Equation (13.25) into Equation (13.29) by switching to the total Gibbs energy and introducing our variable mole fractions (13.24).

Plotting values of the right side of Equation (13.29) for values of ξ between 0 and 1 results in Figure 13.3. It shows how $\mathbf{G} - \mu^\circ_{\text{N}_2\text{O}_4}$ changes as the reaction proceeds from pure

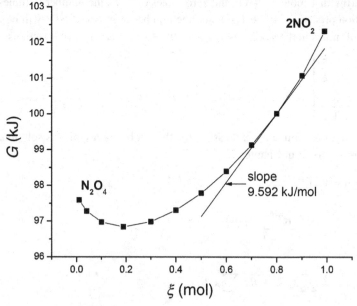

Figure 13.3 The U-shaped curve showing the Gibbs energy for the reaction $\text{N}_2\text{O}_4 = 2\,\text{NO}_2$ as a function of ξ. The minimum occurs at $\xi = 0.1892$ mol. The slope of 9.592 kJ mol^{-1} at $\xi = 0.8$ is $-\mathcal{A}$. $\mathbf{G}$ on the y-axis is actually $\mathbf{G} - \mathbf{G}^\circ_{\text{N}_2\text{O}_4}$. From Anderson (2015).

N_2O_4 to pure NO_2, and as $\mu^\circ_{N_2O_4}$, although an unknown quantity, is certainly a constant, the curve shows the true shape of the Gibbs energy variation. It shows that the Gibbs energy has a minimum value, an extremum, at the equilibrium composition, just as our theory has predicted.

The derivative of Equation (13.28) is

$$\left(\frac{d\mathbf{G}}{d\xi}\right)_{T,P} = 2\,\Delta_f G^\circ{}_{NO_2} - \Delta_f G^\circ{}_{N_2O_4} - RT\ln\left(\frac{1-\xi}{1+\xi}\right) + 2RT\ln\left(\frac{2\xi}{1+\xi}\right) \qquad (13.30)$$

and, using $\Delta_f G^\circ$ data from Appendix B, this becomes

$$= 4.729 - 2.479\ln\left(\frac{1-\xi}{1+\xi}\right) + 4.958\ln\left(\frac{2\xi}{1+\xi}\right)$$

$$= -\mathcal{A} \qquad (13.31)$$

and is shown in Figure 13.4.

Setting this result equal to zero and solving for ξ gives $\xi = 0.1892$ mol at equilibrium. If you do this differentiation in Maxima (2013) or some other program you will find that Equation (13.30) has two more terms, which, however, cancel each other out. Equation (13.31) can be written $d\mathbf{G}_{T,P} = -\mathcal{A}\,d\xi$ (Equation (13.20)), and Maxima will integrate this, but no one ever does this because the result is known to be exactly equal to $\Delta_r G^\circ$ for the reaction, $4.729\,\text{kJ}\,\text{mol}^{-1}$.

All these equations relating ξ to the change in Gibbs energy assume that the reaction proceeds stoichiometrically and ideally. Real chemical reactions never proceed through a series of metastable equilibrium states and cannot be represented by continuous functions like these. The reaction as written is a model reaction, an idealization, like much of the rest of thermodynamics. It does not represent what happens in the real system.

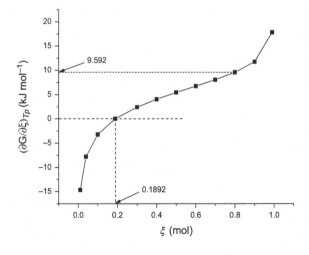

Figure 13.4 The derivative $(\partial\mathbf{G}/\partial\xi)_{T,P}$ of Equation (13.29), which is also $-\mathcal{A}$. The slope 9.592 kJ mol^{-1} is shown graphically in Figure 13.3. From Anderson (2015).

When $\xi = 0.5 \, \text{mol}$ in the reaction thus represented, the mole fractions are

$$x_{NO_2} = \frac{2\xi}{1+\xi} = 0.667$$

$$x_{N_2O_4} = \frac{1-\xi}{1+\xi} = 0.333$$

These species coexist at metastable equilibrium in the model system and are prevented from reacting spontaneously to the stable equilibrium composition of

$$x_{NO_2} = 0.318$$

$$x_{N_2O_4} = 0.682$$

at $\xi = 0.1892 \, \text{mol}$ by the constraint variable. Just as reversible processes are a continuous series of equilibrium states described by continuous functions, so, despite the fact that this U-shaped curve represents an irreversible process, it is continuous and differentiable, and hence must also represent equilibrium states. As mentioned in Chapter 4, only equilibrium states can be plotted on a diagram.

Despite the idealized nature of the procedure, it gives a result consistent with the known value of the equilibrium constant, again as shown by McQuarrie and Simon (1997),

$$K = \frac{x_{NO_2}^2}{x_{N_2O_4}} = \frac{0.318^2}{0.682} = 0.148 \tag{13.32}$$

$$= \exp\left(-\frac{\Delta_r G^\circ}{RT}\right) \tag{13.33}$$

13.5.1 Multicomponent Reactions

In a multicomponent system, there may be many independent reactions proceeding, and each will have its own value of the ξ variable. Only one reaction is considered in Figure 13.3, but other possible reactions in the same system would also proceed with the same increments of ξ and when they all reach equilibrium then so does the system, so that $d\mathbf{G}_{T,P} = 0$. If either NO_2 or N_2O_4 is involved in other reactions the value of ξ at equilibrium for the reaction $N_2O_4 = 2\,NO_2$ will be different from the value it has when it is the only reaction. This is discussed by Raff (2014) for a simplified model system.

Rather than solve the equations required to directly obtain the composition of the equilibrium state for a multicomponent system, a common practice in the geochemical literature is to follow the irreversible reaction of some complex disequilibrium starting composition to its final stable equilibrium composition in a series of small increments of ξ, as discussed first in Section 2.9. After each increment all the species involved in all the reactions are adjusted to their equilibrium activities and any phase changes thus made necessary are performed at that point in the reaction. The bulk composition is unchanged. Each of these intermediate compositions represents a metastable equilibrium state within the overall irreversible reaction, and the series of such states represents an idealized reaction path. As we said in Section 2.9, such computed reaction paths do not necessarily represent what happens in reality, but they usually provide useful information nonetheless, and are routinely performed by petrologists, oceanographers, and others. They show that the highly idealized metastable equilibrium states referred to in discussing Figure 13.3 can actually be realized in these calculated reaction paths. A summary of developments in this field is given by Bethke (2008).

13.5.2   Continuous Functions and Metastable States

As has been pointed out in various places, reversible processes, an essential building block of thermodynamics, are implied by the use of continuous functions in determining differences in state variables between two equilibrium states. Very similar (but not as important) processes involve metastable equilibrium states as shown by path A $\rightarrow$ B in Figure 4.11, where the degree of advancement parameter ξ has a fixed value. This is shown by Equation (13.21), where ξ appears in the subscripts on the first two terms on the right and so is held constant, and if ξ is fixed then $d\xi = 0$ in the final term. Therefore, if ξ fixed, the same equations are valid both for stable and for metastable equilibrium states. On the aragonite surface $\xi = 0$ and on the calcite surface $\xi = 1$ mol. At $\xi = 0.5$ mol equal amounts of the two phases coexist at metastable equilibrium. The real process of aragonite recrystallizing to calcite, if it does, is quite different. Aragonite on a museum shelf will never change to calcite, but in nature the reaction can go either way, and the controlling factors (e.g., temperature, pressure, nucleation, crystal growth) are research topics. To reiterate, thermodynamics does not describe real processes; it simulates them with a series of equilibrium states, either stable or metastable. This is just as true for the irreversible process B $\rightarrow$ C and Equation (4.53) as it is for A $\rightarrow$ B and Equation (4.48).

13.6   Conclusion

The extent of reaction variable ξ calls attention to the importance of metastable equilibrium states in irreversible processes. The importance of both real and thermodynamic metastable equilibrium states tends to be overlooked, but real metastable states are extremely common and thermodynamics must have a way of dealing with them. Because thermodynamic parameters are defined only in equilibrium states, thermodynamic metastable equilibrium states then become an essential part of the thermodynamics of irreversible processes. Many discussions implicitly assume that the term equilibrium is synonymous with stable equilibrium. This would considerably limit the usefulness of thermodynamics, which deals with irreversible as well as reversible processes, so the concept of equilibrium cannot be limited to stable equilibrium states.

13.7   Final Comment

We have emphasized in various places a point of view about thermodynamics which is fairly philosophical, which may also seem to be of not much use to someone interested in complex natural phenomena. That is the idea that, in doing our thermodynamic calculations, we are not really calculating the properties of natural systems, but are instead calculating the properties of simplified models of these systems. The models are mathematical, and the properties and processes in the model include some that have no counterpart in the real world. Nevertheless, model results are useful in understanding the real systems, if the models are properly or appropriately constructed. The calculations must satisfy stringent mathematical relationships, often giving them a gloss of certainty to the untrained eye, but actually they are of use only if the model they constitute is appropriate, i.e., if it has some similarity to the problems of interest.

One reason why it is a good idea to make the distinction between our models and reality is that our "explanations" of thermodynamics (and other exact sciences) are often couched in terms of mathematical planes, surfaces, tangents, and other even more abstract concepts. Students look at these explanations with understanding of a sort, but also with an underlying but usually unexpressed bafflement as to what these shining, unblemished, perfect, mathematical constructs have to do with anything real. It is best to come to terms with this problem by admitting that thermodynamics is in fact all about these mathematical constructs. It is not difficult to understand thermodynamics as the mathematics of certain planes, surfaces, tangents, etc. The more difficult problem is to understand why these mathematical constructs have such direct relevance to our universe, but that is best left to the philosophers.

How do we assure ourselves that our models are appropriate? There is of course no way to be sure. The construction of models useful in understanding nature is the essence of science and relies as much upon creativity and imagination as any painting or musical composition. Unfortunately, the models use the language of mathematics, rather than shapes and colors or musical notes, and are hence not understandable to anyone who does not know the language. This is unfortunate, because mathematics and mathematical models have their own kind of beauty, which easily rivals that of the arts. Although we don't know why the universe should be such that mathematics is so useful in describing it, there is no doubt that it is useful. It seems clear that the more mathematical tools you have in your repertoire, the more adept you will be at fitting mathematics to your observations.

The models used to simulate complex natural phenomena are becoming very complex. Not only is it difficult to create such models, and to compile enough basic data to enable them to work, but also it is becoming increasingly difficult to know how well they work. There is a great deal of uncertainty about what nature is actually doing, so that it's often hard to know just how well model results coincide with natural observations. This means that, in understanding natural systems, insightful field observations are just as important as model construction. And model results may turn out to be reasonably accurate, despite their being based on entirely incorrect ideas. Just because a model "works" does not mean it is right. All this means that there is plenty of scope for thermodynamics and the tools built upon it in the years to come, in our quest to understand our world and how it works.

Exercises

E13.1 (a) Write an equation showing the Helmholtz energy as a total differential, using appropriate constraint variables (also called natural variables).

(b) Do the same for enthalpy.

(c) Do the same for volume.

E13.2 Show that mixing ideal gases does not change the molar internal energy.

E13.3 Consider the reaction

$$N_2(g) + 3H_2(g) = 2NH_3(g)$$

What are the mole fractions in terms of the degree of advancement variable ξ, i.e., similar to Equations (13.24)?

Additional Problems

A13.1 Some data from Appendix B are shown in Table 13.1. For both NO_2 and N_2O_4, do the following.

(a) Calculate $\Delta_f U^\circ$ two different ways:

(i) from $\Delta_f H^\circ - P\Delta_f V^\circ$;

(ii) from $\Delta_f G^\circ + T\Delta_f S^\circ - P\Delta_f V^\circ$.

(b) Calculate $T\Delta_f S^\circ$ two different ways:

(i) from $T\Delta_f S^\circ$;

(ii) from $\Delta_f G^\circ - \Delta_f H^\circ$.

(c) What is the reason for the differences between the two results for the same quantity?

(d) You may never see another reference to $\Delta_f U^\circ$ or $\Delta_f V^\circ$ in the thermodynamic literature. Why is that?

Table 13.1. Some data from Appendix B for 298.15 K and 1 bar.

	$\Delta_f H^\circ$ (J mol^{-1})	$\Delta_f G^\circ$ (J mol^{-1})	S° (J mol^{-1} K^{-1})	Ideal V° (cm^3 mol^{-1})
$NO_2(g)$	33,180	51,310	240.06	24,465.6
$N_2O_4(g)$	9160	97,890	304.29	24,465.6
$N_2(g)$	0	0	191.61	24,465.6
$O_2(g)$	0	0	205.138	24,465.6

Constants and Numerical Values

The SI (Système International) units

Base (fundamental) units

Physical quantity	SI unit	Symbol
Length	meter	m
Mass	kilogram	kg
Time	second	s
Electric current	ampere	A
Temperature	kelvin	K
Amount of substance	mole	mol

Derived SI units

Physical quantity	SI unit	Symbol for SI unit	Unit in terms of base units	Unit in terms of other SI units
Velocity (speed)			m/s	
Acceleration			m/s^2	N/kg
Force	newton	N	$kg\,m/s^2$	J/m
Pressure	pascal	Pa	$kg/(m\,s^2)$	N/m^2
Energy	joule	J	$kg\,m^2/s^2$	$N\,m$
Entropy	joule per kelvin	S	$kg\,m^2/(s^2\,K)$	J/K
Power	watt	W	$kg\,m^2/s^3$	J/s
Momentum			$kg\,m/s$	
Frequency	hertz	Hz	s^{-1}	
Electric charge	coulomb	C	$A\,s$	$V\,F$
Voltage (emf)	volt	V	$kg\,m^2/(A\,s^3)$	$W/A;\ C/F$
Electric resistance	ohm	Ω	$kg\,m^2/(A^2\,s^3)$	V/A
Capacitance	farad	F	$A^2\,s^4/(kg\,m^2)$	C/V

Fundamental physical constants (Cohen and Taylor, 1988)

Quantity	Symbol in this text	Value	Units
Speed of light in vacuum	c	299,792,458	$m\,s^{-1}$
Constant of gravitation	g	6.67259	$10^{-11}\,m^3\,kg^{-1}\,s^{-2}$
Elementary charge	e	1.60217733	$10^{-19}\,C$
Planck constant	h	6.6260755	$10^{-34}\,J\,s$
Avogadro constant	N_A	6.022136	$10^{23}\,mol^{-1}$
Faraday constant	$\mathcal{F}$	96,485.309	$C\,mol^{-1}$
Molar gas constant	R	8.314510	$J\,mol^{-1}K^{-1}$
Boltzmann constant, R/N_A	k	1.380658	$10^{-23}J\,K^{-1}$
Molar volume[a]	V	0.02241410	$m^3\,mol^{-1}$
Molar volume[b]	V	0.0247896	$m^3\,mol^{-1}$

[a]The volume per mole of ideal gas at 101,325 Pa and 273.15 K.
[b]The volume per mole of ideal gas at 10^5 Pa (1 bar) and 298.15 K.

Miscellaneous useful conversions and older units

ln 10	2.302585
$\ln x$	$\ln 10 \times \log_{10} x$
1 cal	4.184 J[a]
R	$1.987216\,cal\,K^{-1}\,mol^{-1}$
$\mathcal{F}$	$96,485.309\,J\,V^{-1}\,mol^{-1}$
	$23,060.542\,cal\,V^{-1}mol^{-1}$
$RT/\mathcal{F}$	0.02569273 V ($T = 298.15$ K)
$2.302585\,RT/\mathcal{F}$	0.0591597 V ($T = 298.15$ K)
1 bar	10^5 pascal
	14.504 psi
	$0.10\,J\,cm^{-3}$
	$0.0239006\,cal\,cm^{-3}$
1 atm	1.01325 bar
	101325 pascal
	14.696 psi
1 cm^3	$0.10\,J\,bar^{-1}$
	$0.0239006\,cal\,bar^{-1}$
1 Å	10^{-8} cm

[a]This is the thermochemical calorie, used in most of physical chemistry. The International Table calorie used in steam tables is 4.1868 J.

APPENDIX B

Standard State Thermodynamic Properties of Selected Minerals and Other Compounds

Table B.1. Inorganic substances. Data from Wagman et al. (1982), with a few additions from other sources – Al species from Drever (1988); silica species and all volume data from Johnson et al. (1992).

Formula	Form	Molecular weight (g mol^{-1})	$\Delta_f H^\circ$ (kJ mol^{-1})	$\Delta_f G^\circ$ (kJ mol^{-1})	S° (J mol^{-1} K^{-1})	C_P° (J mol^{-1} K^{-1})	V° (cm^3 mol^{-1})
Aluminum							
Al	s	26.9815	0	0	28.33	24.35	—
Al^{3+}	aq	26.9815	−531	−485	−321.7	—	−45.3
Al(OH)$^{2+}$	aq	43.9889	−767.0	−693.7	—	—	—
Al(OH)$_2^+$	aq	60.9963	−1010.7	−901.4	—	—	—
Al(OH)$_3^\circ$	aq	78.0037	−1250.4	−1100.7	—	—	—
Al(OH)$_4^-$	aq	95.0111	−1490.0	−1307.0	102.9	—	45.60
Al$_2$O$_3$	α, corundum	101.9612	−1675.7	−1582.3	50.92	79.04	25.575
Al$_2$O$_3 \cdot$ H$_2$O	boehmite	119.9766	−1980.7	−1831.7	96.86	131.25	39.07
Al$_2$O$_3 \cdot$ H$_2$O	diaspore	119.9766	−1998.91	−1841.78	70.67	106.19	35.52
Al$_2$O$_3 \cdot$ 3H$_2$O	gibbsite	156.0074	−2586.67	−2310.21	136.90	183.47	63.912
Al$_2$O$_3 \cdot$ 3H$_2$O	bayerite	156.0074	−2576.5	—	—	—	—
Al(OH)$_3$	amorphous	78.0037	−1276	—	—	—	—
Al$_2$SiO$_5$	andalusite	162.0460	−2590.27	−2442.66	93.22	122.72	51.53
Al$_2$SiO$_5$	kyanite	162.0460	−2594.29	−2443.88	83.81	121.71	44.09
Al$_2$SiO$_5$	sillimanite	162.0460	−2587.76	−2440.99	96.11	124.52	49.90
Al$_2$Si$_2$O$_7 \cdot$ 2H$_2$O	kaolinite	258.1616	−4119.6	−3799.7	205.0	246.14	99.52
Al$_2$Si$_2$O$_7 \cdot$ 2H$_2$O	halloysite	258.1616	−4101.2	−3780.5	203.3	246.27	99.30
Al$_2$Si$_2$O$_7 \cdot$ 2H$_2$O	dickite	258.1616	−4118.3	−3795.9	197.1	239.49	99.30
Al$_6$Si$_2$O$_{13}$	mullite	426.0532	−6816.2	−6432.7	255.	326.10	—
Al$_2$Si$_4$O$_{10}$(OH)$_2$	pyrophyllite	360.3158	−5642.04	−5268.14	239.41	294.34	126.6

Barium							
Ba	s	137.3400	0	0	62.8	28.07	—
Ba^{2+}	aq	137.3400	−537.64	−560.77	9.6	—	−12.9
BaO	s	153.3394	−553.5	−525.1	70.42	47.78	—
BaO$_2$	s	169.3388	−634.3	—	—	66.9	—
BaF$_2$	s	175.3368	−1207.1	−1156.8	96.36	71.21	—
BaS	s	169.4040	−460	−456	78.2	49.37	—
BaSO$_4$	barite	233.4016	−1473.2	−1362.2	132.2	101.75	52.10
BaCO$_3$	witherite	197.3494	−1216.3	−1137.6	112.1	85.35	45.81
BaSiO$_3$	s	213.4242	−1623.60	−1540.21	109.6	90.00	—
Calcium							
Ca	s	40.0800	0	0	41.42	25.31	—
Ca^{2+}	aq	40.0800	−542.83	−553.58	−53.1	—	−18.4
CaO	s	56.0794	−635.09	−604.03	39.75	42.80	—
Ca(OH)$_2$	portlandite	74.0948	−986.09	−898.49	83.39	87.49	—
CaF$_2$	fluorite	78.0768	−1219.6	−1167.3	68.87	67.03	24.542
CaS	s	72.1440	−482.4	−477.4	56.5	47.40	—
CaSO$_4$	anhydrite	136.1416	−1434.11	−1321.79	106.7	99.66	45.94
CaSO$_4$ · 2H$_2$O	gypsum	172.1724	−2022.63	−1797.28	194.1	186.02	—
Ca$_3$(PO$_4$)$_2$	β,whitlockite	310.1828	−4120.8	−3884.7	236.0	227.82	—
Ca$_3$(PO$_4$)$_2$	α	310.1828	−4109.9	−3875.5	240.91	231.58	—
CaCO$_3$	calcite	100.0894	−1206.92	−1128.79	92.9	81.88	36.934
CaCO$_3$	aragonite	100.0894	−1207.13	−1127.75	88.7	81.25	34.150
CaSiO$_3$	wollastonite	116.1642	−1634.94	−1549.66	81.92	85.27	39.93

Table B.1. (continued)

Formula	Form	Molecular weight (g mol⁻¹)	$\Delta_f H°$ (kJ mol⁻¹)	$\Delta_f G°$ (kJ mol⁻¹)	$S°$ (J mol⁻¹ K⁻¹)	$C°_P$ (J mol⁻¹ K⁻¹)	$V°$ (cm³ mol⁻¹)
$CaSiO_3$	pseudo-wollastonite	116.1642	−1628.4	−1544.7	87.36	86.48	—
$CaAl_2SiO_6$	Ca–Al pyroxene	218.1254	−3298.2	−3122.0	141.4	165.7	—
$CaAl_2Si_2O_8$	anorthite	278.2102	−4227.9	−4002.3	199.28	211.42	100.79
$CaTiO_3$	perovskite	135.9782	−1660.6	−1575.2	93.64	97.65	—
$CaTiSiO_5$	sphene	196.0630	−2603.3	−2461.8	129.20	138.95	—
$CaMg(CO_3)_2$	dolomite	184.4108	−2326.3	−2163.4	155.18	157.53	64.365
$CaMgSi_2O_6$	diopside	216.5604	−3206.2	−3032.0	142.93	166.52	66.090
Carbon							
C	graphite	12.0112	0	0	5.740	8.527	5.298
C	diamond	12.0112	1.895	2.900	2.377	6.113	3.417
CO_3^{2-}	aq	60.0094	−677.149	−527.81	−56.9	—	−6.1
HCO_3^-	aq	61.0174	−691.99	−586.77	91.2	—	24.2
CO	g	28.0106	−110.525	−137.168	197.674	29.142	24465.6
CO_2	g	44.0100	−393.509	−394.359	213.74	37.11	24465.6
CO_2	aq	44.0100	−413.80	−385.98	117.6	—	32.8
H_2CO_3	aq	62.0254	−699.65	−623.08	187.4	—	—
CH_4	g	16.0432	−74.81	−50.72	186.264	35.309	24465.6
C_2H_6	g	30.0704	−84.68	−32.82	229.60	52.63	24465.6
CN	g	26.0179	437.6	407.5	202.6	29.16	—
CN^-	aq	26.0179	150.6	172.4	94.1	—	—
HCN	g	27.0259	135.1	124.7	201.78	35.86	—
HCN	aq	27.0259	107.1	119.7	124.7	—	—
Chlorine							
Cl_2	g	70.9060	0	0	233.066	33.907	24465.6
Cl^-	aq	35.4530	−167.159	−131.228	56.5	−136.4	17.3
HCl	aq	36.4610	−167.159	−131.228	56.5	−136.4	17.3
HCl	g	36.4610	−92.307	−95.299	186.908	29.12	24465.6

Copper							
Cu	s	63.5400	0	0	33.15	24.435	—
Cu$^+$	aq	63.5400	71.67	49.98	40.6	—	—
Cu^{2+}	aq	63.5400	64.77	65.49	−99.6	—	—
CuO	tenorite	79.5394	−157.3	−129.7	42.63	42.30	—
Cu$_2$O	cuprite	143.0794	−168.6	−146.0	93.14	63.64	—
CuSO$_4$ · 3H$_2$O	bonattite	213.6478	−1684.31	−1399.36	221.3	205	—
CuSO$_4$ · 5H$_2$O	calcanthite	249.6786	−2279.65	−1879.745	300.4	280	—
CuS	covellite	96.6040	−53.1	−53.6	66.5	47.82	—
Cu$_2$S	chalcocite	159.1440	−79.5	−86.2	120.9	76.32	—
Fluorine							
F$_2$	g	37.9968	0	0	202.78	31.30	—
HF	g	20.0064	−271.1	−273.2	173.779	29.133	—
HF	aq	20.0064	−320.08	−296.82	88.7	—	—
F$^-$	aq	18.9984	−332.63	−278.79	−13.8	−106.7	—
Hydrogen							
H$_2$	g	2.0160	0	0	130.684	28.824	24465.6
H$^+$	aq	1.0080	0	0	0	0	0
OH$^-$	aq	17.0074	−229.994	−157.244	−10.75	−148.5	—
H$_2$O	l	18.0154	−285.830	−237.129	69.91	75.291	18.068
H$_2$O	g	18.0154	−241.818	−228.572	188.825	33.577	24465.6
Iodine							
I$_2$	s	253.8088	0	0	116.135	54.438	—
I$^-$	aq	126.9044	−55.19	−51.57	111.3	−142.3	—

Table B.1. (continued)

Formula	Form	Molecular weight (g mol⁻¹)	$\Delta_f H°$ (kJ mol⁻¹)	$\Delta_f G°$ (kJ mol⁻¹)	$S°$ (J mol⁻¹ K⁻¹)	$C_P°$ (J mol⁻¹ K⁻¹)	$V°$ (cm³ mol⁻¹)
HI	aq	127.9124	−55.19	−51.57	111.3	—	—
IO_3^-	aq	174.9026	−221.3	−128.0	118.4	—	—
IO_4^-	aq	190.9020	−155.5	−58.5	222.0	—	—
Iron							
Fe	s	55.8470	0	0	27.28	25.10	—
Fe^{2+}	aq	55.8470	−89.1	−78.90	−137.7	—	—
Fe^{3+}	aq	55.8470	−48.5	−4.7	−315.9	—	—
$Fe_{0.947}O$	wüstite	68.8865	−266.27	−245.12	57.49	48.12	—
Fe_2O_3	hematite	159.6922	−824.2	−742.2	87.40	103.85	—
Fe_3O_4	magnetite	231.5386	−1118.4	−1015.4	146.4	143.43	—
$Fe(OH)_2$	s	89.8618	−569.0	−486.5	88.	—	—
$Fe(OH)_3$	s	106.8692	−823.0	−696.5	106.7	—	—
FeS	troilite	87.9110	−100.0	−100.4	60.29	50.54	—
FeS_2	pyrite	119.9750	−178.2	−166.9	52.93	62.17	—
$FeCO_3$	siderite	115.8564	−740.57	−666.67	92.9	82.13	—
Fe_2SiO_4	fayalite	203.7776	−1479.9	−1379.0	145.2	132.88	—
Lead							
Pb	s	207.1900	0	0	64.81	26.44	—
Pb^{2+}	aq	207.1900	−1.7	−24.43	10.5	—	—
PbO	yellow	223.1894	−217.32	−187.89	68.70	45.77	—
PbO	red	223.1894	−218.99	−188.93	66.5	45.81	—
PbF_2	s	245.1868	−664.0	−617.1	110.5	—	—
$PbCl_2$	s	278.0960	−359.41	−314.10	136.0	—	—
PbS	galena	239.2540	−100.42	−98.7	91.2	49.50	—
$PbSO_4$	anglesite	303.2516	−919.94	−813.14	148.57	103.207	—
$PbCO_3$	cerussite	267.1994	−699.1	−625.5	131.0	87.40	—
$PbSiO_3$	s	283.2742	−1145.70	−1062.10	109.6	90.04	—

Magnesium

Mg	s	24.3120	0	0	32.68	24.89	—
Mg²⁺	aq	24.3120	−466.85	−454.8	−138.1	—	—
MgO	periclase	40.3114	−601.70	−569.43	26.94	37.15	—
Mg(OH)₂	brucite	58.3268	−924.54	−833.51	63.18	77.03	—
MgF₂	sellaite	62.3088	−1123.4	−1070.2	57.24	61.59	—
MgS	s	56.3760	−346.0	−341.8	50.33	45.56	—
MgCO₃	magnesite	84.3214	−1095.8	−1012.1	65.7	75.52	—
MgCO₃ · 3H₂O	nesquehonite	138.3676	—	−1726.1	—	—	28.018
MgSiO₃	enstatite	100.3962	−1549.00	−1462.09	67.74	81.38	—
Mg₂SiO₄	forsterite	140.7076	−2174.0	−2055.1	95.14	118.49	—

Manganese

Mn	s	54.9380	0	0	32.01	26.32	—
Mn²⁺	aq	54.9380	−220.75	−228.1	−73.6	50	—
MnO₄⁻	aq	118.9356	−541.4	−447.2	191.2	−82.0	—
MnO₄²⁻	aq	118.9356	−653.0	−500.7	59.0	—	—
MnO	manganosite	70.9374	−385.22	−362.90	59.71	45.44	—
Mn₃O₄	hausmannite	228.8116	−1387.8	−1283.2	155.6	139.66	—
Mn₂O₃	s	157.8742	−959.0	−881.1	110.5	107.65	—
MnO₂	pyrolusite	86.9368	−520.03	−465.14	53.05	54.14	—
Mn(OH)₂	amorphous	88.9528	−695.4	−615.0	99.2	—	—
MnS	alabandite	87.0020	−214.2	−218.4	78.2	49.96	—
MnCO₃	rhodochrosite	114.9474	−894.1	−816.7	85.8	81.50	—
MnSiO₃	rhodonite	131.0222	−1320.9	−1240.5	89.1	86.44	—
Mn₂SiO₄	tephroite	201.9596	−1730.5	−1632.1	163.2	129.87	—

Table B.1. (continued)

Formula	Form	Molecular weight (g mol^{-1})	$\Delta_f H°$ (kJ mol^{-1})	$\Delta_f G°$ (kJ mol^{-1})	$S°$ (J mol^{-1} K^{-1})	$C_P°$ (J mol^{-1} K^{-1})	$V°$ (cm^3 mol^{-1})
Mercury							
Hg	l	200.5900	0	0	76.02	27.983	—
Hg	g	200.5900	61.317	31.820	174.96	20.786	—
Hg^{2+}	aq	200.5900	171.1	164.4	−32.2		—
Hg$_2^{2+}$	aq	401.1800	172.4	153.52	84.5	—	—
HgS$_2^{2-}$	aq	264.7180	—	41.9	—	—	—
HgCl$_4^{2-}$	aq	342.4020	−554.0	−446.8	293		—
Hg$_2$Cl$_2$	s	472.0860	−265.22	−210.745	192.5		—
HgO	s, red	216.5894	−90.83	−58.539	70.29	44.06	—
HgO	s, yellow	216.5894	−90.46	−58.409	71.1		—
HgS	cinnabar	232.6540	−58.2	−50.6	82.4	48.41	—
HgS	metacinnabar	232.6540	−53.6	−47.7	88.3		—
Molybdenum							
Mo	s	95.9400	0	0	28.66	24.06	—
MoO$_3$	s	127.9388	−745.09	−667.97	77.74	74.98	—
MoS$_2$	molybdenite	160.0680	−235.1	−225.9	62.59	63.55	—
Nickel							
Ni	s	58.7100	0	0	29.87	26.07	—
Ni^{2+}	aq	58.7100	−54.0	−45.6	−128.9	—	—
NiO	bunsenite	74.7094	−239.7	−211.7	37.99	44.31	—
NiS	s	90.7740	−82.0	−79.5	52.97	47.11	—
Nitrogen							
N$_2$	g	28.0134	0	0	191.61	29.125	—
NO	g	30.0061	90.25	86.55	210.761	29.844	—
NO$_2$	g	46.0055	33.18	51.31	240.06	37.20	—
N$_2$O	g	44.0128	82.05	104.2	219.85	38.45	—

N_2O_4	l	92.0110	−19.50	97.54	209.2	142.7	—
N_2O_4	g	92.0110	9.16	97.89	304.29	77.28	—
N_2O_5	s	108.0104	−43.1	113.9	178.2	143.1	—
N_2O_5	g	108.0104	11.3	115.1	355.7	84.5	—
NH_3	g	17.0307	−46.11	−16.45	192.45	35.06	—
NO_3^-	aq	62.0049	−205.0	−108.74	146.45	−86.6	—
NH_4^+	aq	18.0837	−132.51	−79.31	113.4	79.9	—
NH_4OH	aq	35.0461	−366.12	−263.63	181.21	—	—

Oxygen

O_2	g	31.9988	0	0	205.138	29.355	—
O_2	aq	31.9988	−11.7	16.4	110.9	—	—
OH^-	aq	17.0074	−229.994	−157.244	−10.75	−148.5	—
H_2O	l	18.0154	−285.830	−237.129	69.91	75.291	18.068
H_2O	g	18.0154	−241.818	−228.572	188.825	33.577	24465.6

Potassium

K	s	39.1020	0	0	64.18	29.58	—
K^+	aq	39.1020	−252.38	−283.27	102.5	21.8	9.0
KCl	sylvite	74.5550	−436.747	−409.14	82.59	51.30	—
$KAlSi_3O_8$	sanidine	278.3367	−3959.7	−3739.9	232.88	204.51	—
$KAlSi_3O_8$	microcline	278.3367	−3968.1	−3742.9	214.22	202.38	108.741
$KAlSiO_4$	kaliophilite	158.1671	−2121.3	−2005.3	133.1	119.79	—
$KAlSi_2O_6$	leucite	218.2519	−3034.2	−2871.4	200.08	164.14	—
$KAl_3Si_3O_{10}OH_2$	muscovite	398.3133	−5984.4	−5608.4	306.3	—	14.087

Table B.1. (continued)

Formula	Form	Molecular weight (g mol⁻¹)	$\Delta_f H°$ (kJ mol⁻¹)	$\Delta_f G°$ (kJ mol⁻¹)	$S°$ (J mol⁻¹ K⁻¹)	$C°_P$ (J mol⁻¹ K⁻¹)	$V°$ (cm³ mol⁻¹)
Silicon							
Si	s	28.0860	0	0	18.83	20.00	—
SiO_2	α-quartz	60.0848	−910.94	−856.64	41.84	44.43	22.688
SiO_2	α-cristobalite	60.0848	−909.48	−855.43	42.68	44.18	—
SiO_2	α-tridymite	60.0848	−909.06	−855.26	43.5	44.60	25.740
SiO_2	coesite	60.0848	−906.31	−851.62	40.376	43.51	20.641
SiO_2	amorphous	60.0848	−903.49	−850.70	46.9	44.4	16.1
SiO_2	aq	60.0848	−877.699	−833.411	75.312	318.40	
H_4SiO_4	aq	96.1156	−1449.359	−1307.669	215.132	468.98	
$HSiO_3^-$	aq	77.0922	−1125.583	−1013.783	41.84	−137.24	9.5
Silver							
Ag	s	107.8700	0	0	42.55	25.351	—
Ag^+	aq	107.8700	105.579	77.107	72.68	21.8	—
Ag_2O	s	231.7394	−31.05	−11.20	121.3	65.86	—
AgCl	cerargyrite	143.3230	−127.068	−109.789	96.2	50.79	—
Ag_2S	acanthite	247.8040	−32.59	−40.67	144.01	76.53	—
Ag_2S	argentite	247.8040	−29.41	−39.46	150.6	—	—
Sodium							
Na	s	22.9898	0	0	51.21	28.24	—
Na^+	aq	22.9898	−240.12	−261.905	59.0	46.4	−1.2
NaCl	halite	58.4428	−411.153	−384.138	72.13	50.50	27.015
Na_2SiO_3	s	122.0638	−1554.90	−1462.80	113.85		
$NaAlSiO_4$	nepheline	142.0549	−2092.8	−1978.1	124.3		54.16
$NaAlSi_3O_8$	low albite	262.2245	−3935.1	−3711.5	207.40	205.10	100.07
$NaAlSi_2O_6$	jadeite	202.1397	−3030.9	−2852.1	133.5		60.40
Sulfur							
S	orthorhombic	32.0640	0	0	31.80	22.64	—
S^{2-}	aq	32.0640	33.1	85.8	−14.6		—
HS^-	aq	33.0720	−17.6	12.08	62.8		—
SO_4^{2-}	aq	96.0616	−909.27	−744.53	20.1	−293	—

			$\Delta_f H°$	$\Delta_f G°$	$S°$	C_p	
HSO_4^-	aq	97.0696	−887.34	−755.91	131.8	−84.0	—
S_2	g	64.1280	128.37	79.30	228.18	32.47	—
H_2S	g	34.0800	−20.63	−33.56	205.79	34.23	—
H_2S	aq	34.0800	−39.7	−27.83	121	—	—
SO_2	g	64.0628	−296.830	−300.194	248.22	39.87	—
SO_3	g	80.0622	−395.72	−371.06	256.76	50.67	—
Titanium							
Ti	s	47.9000	0	0	30.63	25.02	
TiO	s	63.8994	−519.7	−495.0	50.0	39.96	
TiO_2	anatase	79.8988	−939.7	−884.5	49.92	55.48	
TiO_2	brookite	79.8988	−941.8	—	—	—	
TiO_2	rutile	79.8988	−944.7	−889.5	50.33	55.02	
Uranium							
U	s	238.0290	0	0	50.21	27.665	
UO_2	uraninite	270.0278	−1084.9	−1031.7	77.03	63.60	
UO_3	orthorhombic	286.0272	−1223.8	−1145.9	96.11	81.67	
U^{3+}	aq	238.0290	−489.1	−475.4	192.	—	
U^{4+}	aq	238.0290	−591.2	−531.0	410.	—	
UO_2^{2+}	aq	270.0278	−1019.6	−953.5	−97.5	—	
Zinc							
Zn	s	65.3700	0	0	41.63	25.40	
Zn^{2+}	aq	65.3700	−155.89	−147.06	−112.1	46	
ZnO	zincite	81.3694	−348.28	−318.30	43.64	40.25	
ZnS	wurtzite	97.4340	−192.63	—	—	—	
ZnS	sphalerite	97.4340	−205.98	−201.29	57.7	46.0	
$ZnCO_3$	smithsonite	125.3794	−812.78	−731.52	82.4	79.71	
Zn_2SiO_4	willemite	222.8236	−1636.74	−1523.16	131.4	123.34	

Table B.2. Data from Shock and Helgeson (1990). (Columns for $\Delta_f G°$ and $\Delta_f H°$ are reversed in comparison with Table B.1, and $\Delta_f G°$ and $\Delta_f H°$ are in J rather than kJ. Note that a large database of data on organic compounds is being assembled by Everett Shock, and is freely available at http://webdocs.asu.edu.)

Formula	Form	Name	$\Delta_f G°$ (J mol^{-1})	$\Delta_f H°$ (J mol^{-1})	$S°$ (J mol^{-1} K^{-1})	$C_P°$ (J mol^{-1} K^{-1})	$V°$ (cm^3 mol^{-1})
n-Alkanes							
CH_4	aq	methane	−34451	−87906	87.82	277.4	37.30
CH_4	g	methane	−50720	−74810	186.26	35.31	24465.6
C_2H_6	aq	ethane	−16259	−103136	112.17	369.4	51.20
C_3H_8	aq	propane	−8213	−127570	141.00	462.8	67.00
C_4H_{10}	aq	*n*-butane	151	−151586	167.44	560.2	82.80
C_5H_{12}	aq	*n*-pentane	8912	−173887	198.74	640.2	98.60
C_6H_{14}	aq	*n*-hexane	18493	−198322	221.33	733.0	114.40
C_7H_{16}	aq	*n*-heptane	27070	−221543	251.04	821.7	130.20
C_8H_{18}	aq	*n*-octane	35899	−248571	266.94	910.4	146.00
1-Alkenes							
C_2H_4	aq	ethylene	81379	35857	120.08	261.5	45.50
C_3H_6	aq	1-propene	74935	−1213	153.55	350.2	61.30
C_4H_8	aq	1-butene	84977	−23577	181.59	438.9	77.10
C_5H_{10}	aq	1-pentene	94014	−46861	209.62	527.6	92.90
C_6H_{12}	aq	1-hexene	101964	−71233	237.65	616.3	108.70
C_7H_{14}	aq	1-heptene	110667	−94851	265.68	705.0	124.50
C_8H_{16}	aq	1-octene	120164	−117654	293.72	793.7	140.30

Species		State					
Alkylbenzenes							
C_6H_6	benzene	aq	133888	51170	148.53	361.1	83.50
$C_6H_5CH_3$	toluene	aq	126608	13724	183.68	430.1	97.71
$C_6H_5C_2H_5$	ethylbenzene	aq	135729	−10460	208.36	504.2	113.80
Alcohols							
CH_3OH	methanol	aq	−175937	−246312	134.72	158.2	38.17
C_2H_5OH	ethanol	aq	−181293	−287232	150.21	260.2	55.08
C_6H_5OH	phenol	aq	−52656	−153302	191.63	315.1	86.17
Ketones							
C_3H_6O	acetone	aq	−161084	−258236	185.77	241.4	66.92
Carboxylic acids							
HCOOH	formic acid	aq	−372301	−425429	162.76	79.5	34.69
CH_3COOH	acetic acid	aq	−396476	−485762	178.66	169.7	52.01
C_2H_5COOH	propanoic acid	aq	−390911	−512414	206.69	234.3	67.90
C_3H_7COOH	butanoic acid	aq	−381539	−535343	234.72	336.8	84.61
C_4H_9COOH	pentanoic acid	aq	−373288	−559359	262.76	432.2	100.50
$C_5H_{11}COOH$	hexanoic acid	aq	−364343	−582789	292.46	523.8	116.55
$C_6H_{13}COOH$	heptanoic acid	aq	−356268	−607015	318.82	612.5	132.30
$C_7H_{15}COOH$	octanoic acid	aq	−348946	−631993	346.85	701.2	148.10
Carboxylate anions							
$HCOO^-$	formate	aq	−350879	−425429	90.79	−92.0	26.16
CH_3COO^-	acetate	aq	−369322	−486097	86.19	25.9	40.50

Table B.2. (continued)

Formula	Form	Name	$\Delta_f G°$ (J mol^{-1})	$\Delta_f H°$ (J mol^{-1})	$S°$ (J mol^{-1} K^{-1})	$C_P°$ (J mol^{-1} K^{-1})	$V°$ (cm^3 mol^{-1})
$C_2H_5COO^-$	aq	propanoate	−363046	−513084	110.88	129.3	54.95
$C_3H_7COO^-$	aq	butanoate	−354008	−535259	133.05	186.2	70.30
$C_4H_9COO^-$	aq	pentanoate	−345598	−562371	160.25	329.7	86.31
$C_5H_{11}COO^-$	aq	hexanoate	−336603	−585300	189.54	418.4	102.21
$C_6H_{13}COO^-$	aq	heptanoate	−327984	−609023	217.57	469.4	118.60
$C_7H_{15}COO^-$	aq	octanoate	−319407	−632746	242.67	558.1	134.40
Amino acids							
$C_2H_5NO_2$	aq	glycine	−370778	−513988	158.32	39.3	43.25
$C_3H_7NO_2$	aq	alanine	−371539	−552832	167.36	141.4	60.45
$C_5H_{11}NO_2$	aq	valine	−356895	−616303	178.24	302.1	90.79
$C_6H_{13}NO_2$	aq	leucine	−343088	−632077	215.48	397.9	107.57
$C_6H_{13}NO_2$	aq	isoleucine	−343925	−631366	220.92	383.3	105.45
$C_3H_7NO_3$	aq	serine	−510866	−714627	194.56	117.6	60.62
$C_4H_9NO_3$	aq	threonine	−502080	−749354	222.59	210.0	76.86
$C_4H_7NO_4$	aq	aspartic acid	−721322	−947132	229.28	127.2	71.79
$C_5H_9NO_4$	aq	glutamic acid	−723832	−970688	294.97	177.0	89.36
$C_4H_8N_2O_3$	aq	asparagine	−538272	−780985	230.96	125.1	77.18
$C_5H_{10}N_2O_3$	aq	glutamine	−529694	−804709	258.99	187.0	94.36
$C_9H_{11}NO_2$	aq	phenylalanine	−207108	−460575	221.33	384.1	121.92
$C_{11}H_{11}N_2O_2$	aq	tryptophan	−112550	−409195	153.13	420.1	144.00
$C_9H_{11}NO_3$	aq	tyrosine	−365263	−658562	190.37	299.2	123.00
$C_5H_{11}NO_2S$	aq	methionine	−502917	−743078	274.89	292.9	105.30
Peptides							
$C_4H_8N_2O_3$	aq	diglycine	−489612	−734878	226.77	158.99	319.11
$C_5H_{10}N_2O_3$	aq	alanylglycine	−488398	−778684	212.13	252.30	398.40
$C_8H_{16}N_2O_3$	aq	leucylglycine	−462834	−847929	303.76	497.06	608.10
$C_4H_6N_2O_2$	aq	diketopiperazine	−240329	−415471	223.84	71.13	321.04

APPENDIX C

Answers to Exercises

Chapter 1

E1.1 The change in potential energy ΔPE is

$$\Delta PE = mg \, \Delta h$$
$$= 0.1 \times 9.81 \times (4 - 2)$$
$$= 1.962 \, \text{kg} \, \text{m}^2/\text{s}^2$$

E1.2 On a frictionless surface, the ball will roll down to the bottom of the valley, then up the other side to a height equal to the height it started from; then back down and back up to where it started; then down again and up and down endlessly. There would be no loss of energy. With a frictional force between the ball and the surface of the valley, the ball will gradually slow down and come to rest at the bottom. This friction generates an amount of heat energy equal to the loss in potential energy which may escape from the system and increase the entropy of the surroundings, which some call the universe.

Chapter 2

E2.1 $\Delta_r V° = -15.867 \, \text{cm}^3 \, \text{mol}^{-1}$.

E2.2 $\Delta_r V° = -7.934 \, \text{cm}^3 \, \text{mol}^{-1}$. Note that you must pay attention to how the equation is written when you calculate reaction properties.

E2.3 $\Delta_r V° = -475.348 \, \text{cm}^3 \, \text{mol}^{-1}$.

E2.4 $\Delta_r V° = -10.915 \, \text{cm}^3$. The properties of aqueous (dissolved) species in the tables are actually *partial molar* properties. This important concept is discussed in Chapter 7.

E2.5 (a) One.
   (b) Two, $MgSiO_3$ and $FeSiO_3$. There could be other choices.
   (c) Two. Mg_2SiO_4 and Fe_2SiO_4.
   (d) Three. MgO, FeO, and SiO_2.
      Illustrate this with the ternary diagram.

Chapter 3

E3.1 See Figure C.1.

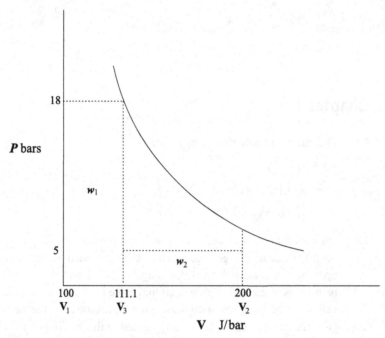

Figure C.1 Problem 3.1 in graphical form.

$$1\,cm^3 = 0.10\,J\,bar^{-1}$$

so

$$\mathbf{V}_1 = 1000 \times 0.1$$
$$= 100\,J\,bar^{-1}$$

$\mathbf{V}_2$ has half the pressure of $\mathbf{V}_1$, so by the ideal gas law it has twice the volume, or $\mathbf{V}_2 = 200\,J\,bar^{-1}$. (Apologies for the fact that in Figure 3.4 the volume at $P = 10$ does not look like twice the volume at $P = 20$.)

(a) The volume ($\mathbf{V}_3$) after the first expansion is

$$P_1\mathbf{V}_1 = P_3\mathbf{V}_3$$
$$20 \cdot 100 = 18 \cdot \mathbf{V}_3$$
$$\mathbf{V}_3 = 111.11\,J\,bar^{-1}$$

So the work done is

$$w_1 = -18(111.11 - 100)$$

$$= -200\,\text{J}$$

$$w_2 = -5(200 - 111.11)$$

$$= -444.5\,\text{J}$$

The total work is

$$w_1 + w_2 = -200 - 444.5$$

$$= -644.5\,\text{J}$$

(b)

$$w = -[18(111.1 - 100) + 16(125 - 111.1) + 14(142.85 - 125)$$

$$+ 10(200 - 142.85)]$$

$$= -1243.6\,\text{J}$$

(c)

$$w_{max} = -\int_{V_1}^{V_2} P\,dV$$

$$= -\int_{V_1}^{V_2} (RT/V)dV$$

$$= -RT\ln(V_2/V_1)$$

$$= -8.31451 \times 298.15 \times \ln(2000/1000)$$

$$= -1718.29\,\text{J}$$

(d) +2200 J.

(e) The essential part of doing these work calculations is to know the gas volume as a function of pressure, that is, the equation of state (EoS) of the gas. We use the equation for an ideal gas ($PV = nRT$) because it is simple. Such equations for real gases are sufficiently hard to use that they become a topic of study in themselves, and attention would be diverted from the simple idea of work being discussed. The *principles* of w and w_{max} are illustrated just as well with an ideal gas as with a real gas.

E3.2 Problem 2.1 found the volume change to be $\Delta_r V° = -15.867\,\text{cm}^3\,\text{mol}^{-1}$. Atmospheric pressure is 1.01325 bar, so

$$w = -P(V_2 - V_1)$$

$$= -1.01325(-15.867)$$

$$= 16.077\,\text{bar}\,\text{cm}^3\,\text{mol}^{-1}$$

$$= 1.608\,\text{J}\,\text{mol}^{-1}$$

E3.3 On page 57 we found $\Delta_r H° = -53{,}480\,\mathrm{J\,mol^{-1}}$, and from the previous question $P\Delta V = 1.608\,\mathrm{J\,mol^{-1}}$. So, by Equation (3.30), we have $\Delta_r U° = -53{,}480 - 1.608 = -53{,}481.6\,\mathrm{J\,mol^{-1}}$. Notice how small the $P\Delta V$ term is at low pressures.

E3.4 $\Delta_r H° = -89{,}475\,\mathrm{J\,mol^{-1}}$.

E3.5 Reaction (2.1): $w_{\mathrm{net}} = \Delta_r G° = -13{,}903\,\mathrm{J\,mol^{-1}}$.
Reaction (2.3): $w_{\mathrm{net}} = \Delta_r G° = -267{,}550\,\mathrm{J\,mol^{-1}}$.

E3.6 $w = -9782.6\,\mathrm{J\,mol^{-1}}$ at 1 bar, $-9912.2\,\mathrm{J\,mol^{-1}}$ at 1 atm. Notice that this is considerably larger than w for reaction (2.5) ($1.608\,\mathrm{J\,mol^{-1}}$) because gases are among the products. It is still much smaller than the maximum useful work, $\Delta_r G° = -267{,}550\,\mathrm{J\,mol^{-1}}$, which is typical of chemical reactions.

Chapter 4

E4.1 The surface shows Gibbs energies (strictly speaking, Gibbs energies of formation) as a function of T and P, so the question refers to $(\partial G/\partial T)_P$, which is $-S$, and $(\partial G/\partial P)_T$, which is V. Both S and V are always positive quantities for pure substances, so the slopes must be as shown.

E4.2 $\Delta_f S° = -757.012\,\mathrm{J\,mol^{-1}\,K^{-1}}$ for $CaAl_2Si_2O_8$. This gives a value of $\Delta_f G° = -4002.2\,\mathrm{kJ\,mol^{-1}}$ when combined with $\Delta_f H°$. Appendix B has $\Delta_f G° = -4002.3\,\mathrm{kJ\,mol^{-1}}$.

E4.3 $\Delta_r H° = -4.9\,\mathrm{kJ\,mol^{-1}}$. Exothermic. $\Delta_r G° = -5.1\,\mathrm{kJ\,mol^{-1}}$. Albite should form spontaneously. Nepheline and quartz together at 25 °C will not react; it is a truly metastable assemblage. They will react at high temperatures.

E4.4 (a) $P = 1.01325\,\mathrm{bar} = 0.101325\,\mathrm{J\,cm^{-3}}$. $V = 0.0224141\,\mathrm{m^3\,mol^{-1}} = 22{,}414.1\,\mathrm{cm^3\,mol^{-1}}$.

$$R = \frac{0.101325 \times 22{,}414.1}{273.15}$$
$$= 8.31451\,\mathrm{J\,mol^{-1}\,K^{-1}}$$

(b) Energy is $\mathrm{kg\,m^3\,s^{-2}}$; pressure is $\mathrm{kg\,m^{-1}\,s^{-2}}$, so energy/pressure is $\mathrm{m^3}$. Therefore $\mathrm{J\,bar^{-1}}$ is a volume.

The SI convention is $1\,\mathrm{J/Pa} = 1\,\mathrm{m^3}$, so

$$1\,\mathrm{J\,bar^{-1}} = 1 \times 10^{-5}\,\mathrm{m^3}$$
$$= 10\,\mathrm{cm^3}$$

so

$$1\,\mathrm{cm^3} = 0.10\,\mathrm{J\,bar^{-1}}$$

E4.5 See Figure C.2 and Table C.1.

The equations give values of $\Delta_f G°$ for each mineral at 2800 bars, but, as the elements cancel out, the difference in these $\Delta_f G°$ values is the same as the difference

Table C.1. Calcite–aragonite equilibrium pressures from data in Table 4.1.

$T(°C)$	25	50	100	150	200	250	300	325
$\Delta_r G°(\text{J mol}^{-1})$	−150	−230	−380	−570	−760	−960	−1160	−1260
P_2 (bars)	3648	3825	4493	5340	6187	7078	7969	8415

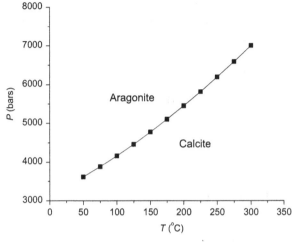

Figure C.2 The aragonite – calcite phase boundary from data in Table 4.1.

in Gibbs energies $G_{\text{calcite}} - G_{\text{aragonite}}$ and is called $\Delta_r G°$. We want the pressure at which this difference is zero. For the reaction in the direction aragonite → calcite and where $P_1 = 2800$ bars, P_2 is the equilibrium pressure, and $\Delta_r G°_{P_2} = 0$,

$$\Delta_r G°_{P_2} - \Delta_r G°_{P_1} = \Delta_r V°(P_2 - P_1)$$

$$P_2 = (-\Delta_r G°_{P_1} + 2800\,\Delta_r V°)/\Delta_r V°$$

E4.6 The reaction is $MgCO_3 + 3\,H_2O = MgCO_3 \cdot 3H_2O$, with

$$\Delta_r G° = -1726.1 - (-1012.1 - 3 \times 237.129)$$

$$= -2.613\,\text{kJ mol}^{-1}$$

which is negative, so nesquehonite is more stable in water.

E4.7 Diaspore > gibbsite > boehmite > corundum is the order of stability, with diaspore the most stable.

E4.8 The reaction is $Fe_2O_3 + H_2O = 2\,FeO(OH)$, with

$$\Delta_r G° = 2 \times -487.02 - (-742.2 - 237.129)$$

$$= 5.289\,\text{kJ mol}^{-1}$$

which is positive, so hematite is more stable in water.

On page 216 we said that $\Delta_r G°$ is not a reliable guide to which way the reaction will go. But the qualifier was "unless the reaction consists only of pure phases." That is the case here, and in the previous questions. A more exact qualifier is "unless all reactants and products are in their standard states." Data are tabulated for substances in their various standard states. For solids and liquids this is the pure phase at 298 K and 1 bar, so, if the reaction involves only pure phases and these conditions, then $\Delta_r G°$ is the same as $\Delta_r G$, and can be used to to tell which way the reaction should go.

Chapter 5

E5.1 From Appendix B,

$$\Delta_r H° = -2590.27 - (-910.94) - (-1675.7)$$
$$= -3.63 \, \text{kJ mol}^{-1}$$
$$= -0.87 \, \text{kcal mol}^{-1}$$

compared with Holm and Kleppa's $-1.37 \, \text{kcal mol}^{-1}$ calculated from their high-temperature measurements.

E5.2

$$\Delta_r H° \text{ at } 25 \,°\text{C} = -870 \, \text{cal mol}^{-1}$$
$$\Delta_r H_T° - \Delta_r H_{T_r}° = -303.1 \, \text{cal mol}^{-1}$$
$$\Delta_{\text{transition}} H = -290 \, \text{cal mol}^{-1}$$

so the calculated $\Delta_r H°$ at 695 °C is $-1463 \, \text{cal mol}^{-1}$, compared with Holm and Kleppa's measured value of $-1990 \, \text{cal mol}^{-1}$. Finding out why there is such a difference might take a great deal of effort, but you would learn a lot about thermochemical data.

Chapter 6

E6.1 If the reaction is written graphite $\rightarrow$ diamond, then

$$\Delta_r G° = 2900 \, \text{J mol}^{-1}$$
$$\Delta_r V° = -0.1881 \, \text{J bar}^{-1} \, \text{mol}^{-1}$$
$$\Delta_r S° = -3.363 \, \text{J mol}^{-1} \, \text{K}^{-1}$$

Letting P_2 be the equilibrium pressure and $P_1 = 1$ bar,

$$\Delta_r G_{P_2}° - \Delta_r G_{P_1}° = \Delta_r V°(P_2 - P_1)$$
$$0 - 2900 = -0.1881(P_2 - 1)$$

$$P_2 = 1 + \frac{2900}{0.1881}$$
$$= 15{,}418 \text{ bars}$$

To determine $\Delta_r G^\circ$ at $T_2 = 100\,^\circ\text{C}$,

$$\Delta_r G^\circ_{T_2} - \Delta_r G^\circ_{T_1} = \Delta_r S^\circ(T_2 - T_1)$$
$$\Delta_r G^\circ_{T_2} - 2900 = -(-3.363)(100 - 25)$$
$$\Delta_r G^\circ_{T_2} = 3152.225 \text{ kJ mol}^{-1}$$

so, at $100\,^\circ\text{C}$,

$$0 - 3152.225 = -0.1881(P_2 - 1)$$
$$P_2 = 1 + \frac{3152.225}{0.1881}$$
$$= 16{,}759 \text{ bars}$$

E6.2 Nepheline + Albite = 2 Jadeite,

$$\Delta_r G^\circ = -14{,}600 \text{ J mol}^{-1}$$
$$\Delta_r S^\circ = -64.7 \text{ J mol}^{-1}\,\text{K}^{-1}$$
$$\Delta_r V^\circ = 3.343 \text{ J bar}^{-1}\,\text{mol}^{-1}$$

First get $\Delta_r G^\circ$ at $300\,^\circ\text{C}$,

$$\Delta_r G^\circ_{300} - \Delta_r G^\circ_{25} = -\Delta_r S^\circ(300 - 25)$$
$$\Delta_r G^\circ_{300} = 3192.5 \text{ J mol}^{-1}$$

then find P_2 for $\Delta_r G^\circ_{P_2} = 0$,

$$\Delta_r G^\circ_{P_2} - \Delta_r G^\circ_{P_1} = -\Delta_r V^\circ(P_2 - P_1)$$
$$0 - 3192.5 = -3.343(P_2 - 1)$$
$$P_2 = 956 \text{ bars}$$

E6.3

$$\text{Al}_2\text{O}_3 \cdot 3\text{H}_2\text{O} = \text{Al}_2\text{O}_3 + 3\text{H}_2\text{O},$$

$$\Delta_r G^\circ = -1582.3 + 3 \times -237.129 - (-2310.21)$$
$$= 16.523 \text{ kJ mol}^{-1}$$
$$\Delta_r S^\circ = 50.92 + 3 \times 69.91 - 136.90$$
$$= 123.75 \text{ J mol}^{-1}\,\text{K}^{-1}$$

$$\Delta_r G°_T - \Delta_r G_{298} = -\Delta_r S°(T - 298)$$

$$0 - 16.523 = -0.12375(T - 298)$$

$$T = 16.523/0.12375 + 298$$

$$= 431.5\,K$$

$$= 158.5\,°C$$

E6.4 $CaMg(CO_3)_2 = CaCO_3 + MgCO_3,$

$$\Delta_f G° = -1128.79 + (-1012.1) - (-2163.4)$$

$$= 22.51\,kJ\,mol^{-1}$$

which is positive. The reaction goes to the left, dolomite is more stable.

E6.5 (a) α is more stable under standard conditions because it has a lower value of $\Delta_f G°$.

(b) β is the high-T form because it has greater values of both $\Delta_f H°$ and $S°$.

(c) β is the high-P form because it has a smaller $V°$.

(d) Recall that 50 cm^3 = 5.0 J bar^{-1}, so

$$\frac{dP}{dT} = \frac{\Delta_r S°}{\Delta_r V°}$$

$$= \frac{S°_\beta - S°_\alpha}{V°_\beta - V°_\alpha}$$

$$= \frac{93.4 - 90.0}{4.9 - 5.0}$$

$$= \frac{3.4}{-0.1}$$

$$= -34.0\ \text{bars/degree}$$

(e)

$$\Delta_r G°_P - \Delta_r G°_{1\ bar} = \Delta_r V°(P - 1)$$

$$0 - 3000 = -0.1(P - 1)$$

$$P = 30,001\ \text{bars}$$

(f) See Figure C.3.

(g) See Figure C.4.

(h) The question is based on the example of silica solubility. Because no other data are given, you must assume that dissociation of α into other dissolved species (ions) is not important. For the $\alpha(s) = \alpha(aq)$ reaction,

$$\Delta_r G° = -2415 - (-2440)$$

$$= 25\ kJ$$

$$= -RT \ln K$$

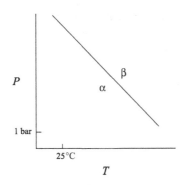

Figure C.3 *P–T* phase diagram of the α and β polymorphs.

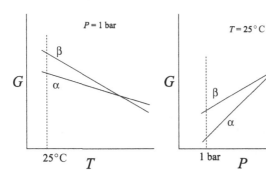

Figure C.4 *G–T* and *G–P* sections through the phase diagram.

so

$$\log K = \frac{-25{,}000}{2.303 \times 8.3145 \times 298.15}$$

$$= -4.380$$

Assuming that $a_{\alpha(s)} = 1$ and that $a_{\alpha(aq)} = m_\alpha$, the solubility of solid α is $10^{-4.38}$ molal.

(i) β is a metastable form and will have a greater solubility. You can explain it by invoking the effect of the larger $\Delta_f G°$ on $\log K$, or you can explain it to be the result of the fact that the metastable → stable reaction is spontaneous, so it can take place by dissolution of the metastable form and precipitation of the stable form. The dissolution → precipitation mechanism is one way to overcome the energy barrier which often prevents the spontaneous recrystallization of metastable phases.

Chapter 7

E7.1 $\Delta_{mix} G = -1668.4$ J.

E7.2 The chemical potential of A is 7979.5 J less in the solution than it would be in an ideal 1 molal solution.

E7.3 In an ideal gas solution, the volume percentage is proportional to the number of moles of gas, or to the mole fractions. Therefore the mole fraction of each gas is its volume percentage divided by 100. This also equals its fugacity and partial pressure in bars.

E7.4 At 100 bar, $f_{CH_4} = 1.7 \times 10^{-4}$ bar.

E7.5 $f_{CH_4} = 1.615 \times 10^{-4}$ bar.

Chapter 8

E8.1 (a) (i) 1.0
  (ii) 1.4928
  (iii) 1.4933
  (iv) $10^{-14.946}$

  (b) (i) 1.0
  (ii) 1.787
  (iii) 5.408
  (iv) Estimate a value for f_{ice} at T by any means, such as extrapolating fugacities from lower T.

  (c) (i) 0.001
  (ii) 0.000472

  (d) (i) At system equilibrium.
  (ii) When the same standard state is used for both; system equilibrium is irrelevant.

E8.2 $-11,533$ cal mol^{-1}.

E8.3 7.070 (Same as $a_{SiO_2}^{qtz}$ using this standard state).

E8.4 In Equation (8.17), the value of μ° varies with T. To use a constant-T standard state, you must include a term giving the difference between μ° in this equation and μ° in the state you have chosen. Nobody does this, because it is not useful.

Chapter 9

E9.1 (a) From Appendix B, $\Delta_r H^\circ = 22{,}100$ J mol^{-1} and $\Delta_r G^\circ = 39910$ J mol^{-1}.

  (b) See Figure C.5.

  (c) Using $\Delta_r G^\circ$ and $\Delta_r H^\circ$ calculated from Appendix B the check is exact. Using the Suleimenov and Seward $\Delta_r G^\circ$ in Table C.2 does not check exactly because the log K value is the experimental result at 25 °C and the $\Delta_r G^\circ$ value is from a theoretical fit to all the data.

  (d) Using the van 't Hoff equation assumes that $\Delta_r H^\circ$ is constant and therefore $\Delta_r C_P^\circ$ is zero. There are no commonly used equations showing the change of log K with T when $\Delta_r C_P^\circ$ is a constant because it doesn't work any better, especially with aqueous species (see Section 9.6.2).

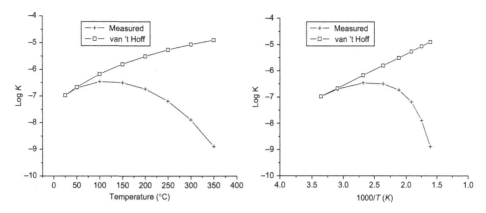

Figure C.5 The measured equilibrium constant for the reaction $H_2S(aq) = H^+ + HS^-$ compared with the result of a van 't Hoff extrapolation using a linear temperature axis (left) and a $1000/T$ (K) axis (right).

Table C.2. Data for Exercise E9.1.

T (°C)	Van 't Hoff	Suleimenov and Seward
25	−6.992	−6.99
50	−6.693	−6.68
100	−6.214	−6.49
150	−5.848	−6.49
200	−5.560	−6.73
250	−5.327	−7.19
300	−5.134	−7.89
350	−4.973	−8.89

It is interesting to note that Suleimenov and Seward give tables showing three different values of $\Delta G°$ and $\Delta H°$. One is for the ionization reaction, one for the liquid-vapor reaction, and one is formation from the elements. A bit confusing unless you read carefully.

It is also interesting to note that Suleimenov and Seward use the Clark and Glew (1966) method of determining the best-fit equation for their data.

E9.2 (a) $\Delta_r G° = -73{,}070\,\mathrm{J\,mol^{-1}}$. This shows that the reaction will indeed go in the direction of producing calcite in the cement blocks, at least if the CO_2 pressure is 1 bar, and the other phases are pure, and for CO_2 pressures greater than that calculated in part (b).

(b)

$$\log K = 73{,}070/(2.30259 \times 8.31451 \times 298.15)$$

$$= 12.801$$

$$K = 6.33 \times 10^{12}$$

$$= \frac{a_{H_2O} a_{CaCO_3}}{a_{Ca(OH)_2} a_{CO_2}}$$

$$= 1/f_{CO_2}$$

$$f_{CO_2} = 10^{-12.80}\,\text{bar}$$

$$= 1.58 \times 10^{-13}\,\text{bar}$$

This is the fugacity or partial pressure of CO_2 in equilibrium with both portlandite and calcite at 25 °C. Any greater CO_2 pressure will cause formation of calcite from portlandite. Any lesser pressure will cause formation of portlandite from calcite plus water. However, this is a very low value of fugacity, so virtually any CO_2 would react with portlandite.

(c) The reaction to write is

$$CaO(s) + H_2O(l) = Ca(OH)_2(s)$$

for which $\Delta_r G^\circ = -57.331\,\text{kJ}\,\text{mol}^{-1}$. This means, with no ambiguity, that CaO and water are not stable together. As water is present, CaO cannot be. Or, in other words, if CaO happened to be there, it would react with water to form portlandite until it was gone. Note that CaO and $Ca(OH)_2$ could theoretically coexist in the absence of water. However, as shown in part (d), it would have to be an extremely dry environment.

(d) The reaction for this question is the same as the last one, *except* that the water must be gaseous, not liquid, because we want to calculate a water fugacity, not a water mole fraction. Only gas activities are in the form of fugacities or partial pressures. Thus the reaction is

$$CaO(s) + H_2O(g) = Ca(OH)_2(s)$$

$$\log K = 65{,}888/(2.30259 \times 8.31451 \times 298.15)$$

$$= 11.543$$

$$K = 3.49 \times 10^{11}$$

$$= \frac{a_{Ca(OH)_2}}{a_{CaO} a_{H_2O(g)}}$$

$$= 1/f_{H_2O}$$

$$f_{H_2O} = 10^{-11.54}\,\text{bar, or}$$

$$= 2.86 \times 10^{-12}\,\text{bar}$$

This is the fugacity or partial pressure of water vapor in equilibrium with CaO and portlandite. As in part (b), it is a very low value. Most vacuum pumps cannot produce pressures this low. Therefore virtually any moisture in the air will cause lime to hydrate. Or you could use lime in a desiccator to produce extremely dry air.

(e) There are at least three ways to do this. They all give about the same answer.

$$\Delta_r H^\circ = -113{,}151 \, \text{J} \, \text{mol}^{-1}$$

$$\Delta_r S^\circ = -134.32 \, \text{J} \, \text{mol}^{-1}$$

$$\log K_{423} = \log K_{298} + \frac{-113{,}151}{2.30259 \times 8.31451} \left(\frac{1}{423.15} - \frac{1}{298.15} \right)$$

$$= 12.80 - 5.856$$

$$= 6.94$$

$$K = 10^{6.94}$$

$$= 8.71 \times 10^6$$

The reaction will be reversed when a temperature is reached at which $\Delta_r G^\circ = 0$, or $\log K_T = 0$. To find this T, write

$$0 = \log K_{298} + \frac{-113{,}151}{2.30259 \times 8.31451} \left(\frac{1}{T} - \frac{1}{298.15} \right)$$

$$T = 841.6 \, \text{K}$$

$$= 568 \, ^\circ\text{C}$$

Or

$$\Delta_r G_T^\circ = \Delta_r H_{298}^\circ - T \cdot \Delta_r S_{298}^\circ$$

$$= -113{,}151 - 423.15 \times (-134.32)$$

$$= -56{,}313.5 \, \text{J}$$

Then

$$\log K_{423} = \frac{-(-56{,}313.5)}{2.30259 \times 8.31451 \times 423.15}$$

$$= 6.95$$

and

$$\Delta_r G_T^\circ - \Delta_r G_{298}^\circ = -\Delta_r S_{298}^\circ (T - 298.15)$$

$$0 - (-73{,}070) = -(-134.32)(T - 298.15)$$

$$T = 842.2 \, \text{K}$$

$$= 569 \, ^\circ\text{C}$$

Note that in doing this calculation we have ignored the fact that liquid water cannot exist at 1 bar at this high temperature. If we were really interested in this reaction at high T, we would switch to $H_2O(g)$, which *can* exist at 1 bar at high T. Also, we would have to check whether in fact portlandite was stable at this T. It is possible that the lime–portlandite reaction would reverse at high T, making CaO more stable than $Ca(OH)_2$, even in the presence of water at 1 bar.

This raises the general point that there are at least two distinct aspects to doing calculations of this type. One is learning the mechanics – getting numbers from tables and correctly calculating a number for a *given* reaction. The second, and more difficult, is finding or writing the reaction most appropriate for a given problem. Not only must it be balanced, but also it must fairly closely match what would actually happen. In other words, your model must reflect reality. Thermodynamics allows you to calculate an infinite number of results which, though correct, mean virtually nothing, such as our answer to this question.

E9.3

$$CH_4(g) + 2\,O_2(g) = CO_2(g) + 2\,H_2O(l)$$

$$\log K = 143.29$$

$$f_{CH_4} = -145.4 \text{ bars}$$

Atmospheric CH_4 is continuously produced by life processes and is obviously not in equilibrium with atmospheric oxygen.

E9.4 The idea was to see that $\Delta_r G° = -168,600 + 75.729\,T$ could be considered equivalent to $\Delta_r G° = \Delta_r H° - T\,\Delta_r S°$, giving $\Delta_r H° = -168.60 \text{ kJ mol}^{-1}$ and $\Delta_r S° = -75.729 \text{ J mol}^{-1}\,\text{K}^{-1}$. $f_{O_2,\,600\,°C} = 10^{-12.26}$ bar.

E9.5 $f_{H_2O,\,25\,°C} = 10^{-1.499}$ bars, or 0.0317 bars. $f_{H_2O,\,100\,°C} = 1.123$ bars.

E9.6 $Al_2O_3 \cdot H_2O(s) = Al_2O_3(s) + H_2O(g)$. $\log K = -5.415$. Then $0.0317 > 10^{-5.4}$, so diaspore is stable. Or use

$$\Delta_r G° + RT \ln Q = 30,908 + 5708.04 \log(0.0317)$$

which is positive, so the reaction goes to the left; diaspore is stable.

E9.7

$$\Delta_f G°_{\text{amorphous silica}} = -850,700 \text{ J mol}^{-1}$$

$$\Delta_f G°_{H_4SiO_4} = -1,307,669 \text{ J mol}^{-1}$$

$$\Delta_f G°_{H_2O(l)} = -237,129 \text{ J mol}^{-1}$$

$$\Delta_r G° = -17,289 \text{ J mol}^{-1}$$

$$\log K = -3.029$$

$$K = 9.357 \times 10^{-4}$$

$$K = \frac{a_{H_4SiO_4}}{a_{SiO_2}\,a_{H_2O}^2}$$

$$9.357 \times 10^{-4} = \frac{m_{H_4SiO_4}}{a_{H_2O}^2}; \quad \gamma_{H\,H_4SiO_4} \approx 1.0$$

$$m_{H_4SiO_4} = 9.357 \times 10^{-4} \times 0.9^2$$

$$= 7.58 \times 10^{-4} m$$

$$\approx 73 \text{ ppm}$$

E9.8   2 S (orthorhombic) $= S_2(g)$. $\log K = \log f_{S_2} = -13.893$ at 298.15 K. $\log f_{S_2} = -12.153$ at 323.15 K, using Equation (9.21).

E9.9 $\log f_{O_2} = -211.64$ for equilibrium between Ca and CaO, which is so low that it never occurs on Earth.

E9.10   Equation (10.16) is

$$Al_2Si_2O_5(OH)_4 + 5 H_2O(l) = 2 Al(OH)_3 + 2 H_4SiO_4(aq)$$

Note that the formula for gibbsite can also be written as $Al(OH)_3$. The value of $\Delta_f G°$ for this gibbsite is one half the tabulated value (for $Al_2O_3 \cdot 3 H_2O$). $m_{H_4SiO_4} = 10^{-5.238}$ for gibbsite–kaolinite equilibrium at 298.15 K.

E9.11   100 ppm $= 10^{-2.78} m$ SiO_2 or H_4SiO_4. At equilibrium, $a_{H_4SiO_4} = 10^{-5.238}$; therefore kaolinite is expected.

E9.12 $CaMg(CO_3)_2(s) + 2 SiO_2(s) = CaMgSi_2O_6(s) + 2 CO_2(g)$. $\Delta_r G° = 55962$ J mol^{-1}; reaction does not proceed. $f_{CO_2} = 10^{-4.90}$ at 1 bar, 25 °C. $f_{CO_2} = 1$ bar at 194 °C.

E9.13 $m_{H_2S} = 0.0991$ at 25 °C, 0.020 at 100 °C.

E9.14   (a) $m_{SiO_2} = 10^{-3.029}$.
      (b) $\Delta_f G°_{H_4SiO_4(aq)} = -1,302,554$ J mol^{-1}; $\Delta_f G°_{SiO_2(aq)} = -828,246$ J mol^{-1}.

E9.15   (a) $H_2S(aq) = H^+ + HS^-$; $\log K_1 = -6.992$.
         $HS^- = H^+ + S^{2-}$; $\log K_2 = -12.915$. Overall, $K = K_1 K_2 = 10^{-19.907}$.
      (c) $PbS(s) = Pb^{2+} + S^{2-}$; $\log K_{sp} = -28.043$.
         $m_{Pb^{2+}} = 10^{-16.136}$.

E9.16 $CH_3OOH(aq) = CH_3OO^- + H^+$. $\log K_{25°C} = -4.757$. $\log K_{100°C} = -4.783$.

E9.17 $(f^2_{NO_2}/f_{N_2O_4}) = 0.148$; $f_{NO_2} + f_{N_2O_4} = 1.0$ bar; so $f_{NO_2} = x_{NO_2} = 0.318$; $f_{N_2O_4} = x_{N_2O_4} = 0.682$.

E9.18 $CaCO_3(calcite) + H^+ = Ca^{2+} + HCO_3^-$.

$$\Delta_r G = \Delta_r G° + RT \ln Q$$

$$= -11,560 + 5708.042 \log \left(\frac{10^{-6} \times 0.09}{10^{-8}} \right)$$

$$= -6113.14 \text{ J mol}^{-1}$$

which is negative, therefore calcite will not precipitate. Or calculate that

$$IAP = 10^{-9.38} < K_{sp}$$

$$Ca^{2+} + Mg^{2+} + 2 HCO_3^- = CaMg(CO_3)_2 + 2 H^+.$$

$\log K = -3.245$.

$a_{Mg^{2+}} = 10^{-4.66}$ at equilibrium, so $a_{Mg^{2+}} > 10^{-4.66}$ will precipitate dolomite. This is much less than the actual magnesium ion activity, so the oceans are supersaturated with dolomite. Petrologists call this the dolomite problem.

E9.19   (a) $a_{Ca^{2+}} = 0.249$.
      (b) $a_{Ca^{2+}} = 0.378$.

      With both minerals in the same solution at 1 bar, aragonite will always require a greater $a_{Ca^{2+}}$ for equilibrium than will calcite, no matter what the conditions.

Therefore, it should continue to dissolve while calcite precipitates until it is used up. This is consistent with the free energy relations, which say that aragonite should change to calcite spontaneously at 1 bar. The dissolution/precipitation simply provides a mechanism to do this.

E9.20 $K_1 = 10^{213.97}$; $K_2 = 10^{95.099}$; $K_3 = 10^{7.77}$. No amount of aqueous molecular oxygen can equilibrate with pyrite under these conditions.

E9.21 (a) The reaction is

$$FeS_2 = FeS + \tfrac{1}{2}S_2$$

for which

$$K = a_{FeS}(f_{S_2})^{0.5}$$

$$= 0.46 \times (10^{-1.95})^{0.5}$$

$$= 0.04872$$

$$\ln K = -3.0215$$

then

$$\Delta_r G° = -RT \ln K$$

$$= -8.31451 \times 875.15 \times (-3.0215)$$

$$= 21{,}986.1 \, J \, mol^{-1}$$

and for the opposite reaction

$$FeS + \tfrac{1}{2}S_2 = FeS_2 \qquad \Delta_r G° = -21{,}986.1 \, J \, mol^{-1}$$

(b) The pyrrhotite in this $\Delta_r G°$ term is stoichiometric FeS, *not* $Fe_{0.92}S$, because the calculated $\Delta_r G°$ is between standard states.

(c) K for the reaction $Fe + \tfrac{1}{2}S_2 = FeS$ is $1/f_{S_2}^{0.5}$, so $\log K = 6.25$, and $\Delta_r G° = 2.303RT \log K = -104{,}716 \, J \, mol^{-1}$.

Then, given

$$Fe + \tfrac{1}{2}S_2 = FeS \qquad \Delta_r G° = -104{,}716 \, J \, mol^{-1}$$

$$FeS + \tfrac{1}{2}S_2 = FeS_2 \qquad \Delta_r G° = -21{,}986 \, J \, mol^{-1}$$

upon adding the reactions,

$$Fe + S_2 = FeS_2 \qquad \Delta_f G° = -126{,}702 \, J \, mol^{-1}$$

at 602 °C.

E9.22 (a) $f_{CO_2} = 135$ bars.

(b) One way would be to use the fugacity of pure CO_2 at 749 K, 2 kbar, plus the Lewis fugacity rule, which is $f_i^{mixtrue} = x_i \cdot f_i^{pure}$.

E9.23 The value of $\log Ba^{2+}$ would need to be about 4.32, or over 10,000 molal to precipitate BaS if $a_{S^{2-}}$ were 0.1. With any less sulfide it would be even greater. Barite or witherite would precipitate if Ba^{2+} was appreciable.

E9.24

$$\frac{a_{H_2CO_3}}{a_{CO_2}a_{H_2O}} = 10^{-1.468}$$

$$a_{H_2CO_3} = 10^{-1.468} \times 10^{-1}$$

$$= 10^{-2.468}$$

$$m_{H_2CO_3} = 0.00341$$

To connect H_2CO_3 and CO_3^{2-}, we write

$$H_2CO_3 = 2\,H^+ + CO_3^{2-}$$

for which

$$\Delta_r G^\circ = 2(0) + (-527.81) - (-623.109)$$

$$= 95.299\,\text{kJ mol}^{-1}$$

$$\log K = -95,299/5708.042$$

$$= -16.696$$

$$\frac{a_{H^+}^2 a_{CO_3^{2-}}}{a_{H_2CO_3}} = 10^{-16.696}$$

$$a_{CO_3^{2-}} = \frac{10^{-16.696} \times 10^{-2.468}}{(10^{-5})^2}$$

$$= 10^{-9.164}$$

$$= 6.855 \times 10^{-10}$$

E9.25

$$\frac{a_{Zn^{2+}}a_{CO_3^{2-}}}{a_{ZnCO_3}} = 10^{-9.925}$$

$$a_{Zn^{2+}} = 10^{-9.925}/10^{-9.164}$$

$$= 10^{-0.761}$$

$$m_{Zn^{2+}} = 0.173$$

E9.26

$$\Delta_r G^\circ = -1523.16 + 4(0) - 2(-147.06) - (-856.64) - 2(-237.129)$$

$$= 101.858$$

$$= -RT \ln K$$

$$\log K = -101{,}858/5708.042$$

$$= -17.84$$

$$K = 1.445 \times 10^{-18}$$

$$= \frac{a_{H^+}^4}{a_{Zn^{2+}}^2}$$

$$= \frac{(10^{-5})^4}{a_{Zn^{2+}}^2}$$

$$a_{Zn^{2+}} = 10^{-1.078}$$

or

$$m_{Zn^{2+}} = 0.0836$$

This is the value of $a_{Zn^{2+}}$ for equilibrium. $a_{Zn^{2+}}$ or $m_{Zn^{2+}}$ is actually 0.173, so willemite should precipitate.

Or

$$\Delta_r G = \Delta_r G^\circ + RT \ln Q$$

$$= \frac{101{,}858 + 5708.042 \log (10^{-5})^4}{(10^{-0.762})^2}$$

$$= -3604 \,\mathrm{J\,mol^{-1}}$$

which is negative, reaction should go to the right, and willemite should precipitate.

E9.27 $H_2O(l) = H^+ + OH^-$, $\log K = -13.995$.

Absolutely pure water must have $a_{H^+} = a_{OH^-}$ for electrical neutrality, so with nothing else present both must be 10^{-7}, so pH $= -\log a_{H^+} = 7$. $f_1 = 10^{-12.30}$ bar. $f_{I_2} = 10^{-3.39}$ bar.

E9.28 $f_{O_2} = 10^{-83.09}$ bar. $f_{H_2} = 10^{-41.20}$ bar.

Chapter 10

E10.1 $Al(OH)_3$ (gibbsite) $= Al^{3+} + 3\,(OH)^-$. $\log K_{sp} = -34.75$. $a_{Al^{3+}} = 10^{-10.75}$. (Don't forget that $a_{H^+} \cdot a_{OH^-} = 10^{-14}$.)

E10.2 (a) The reaction given suggests $\log(a_{Na^+}^2/a_{Ca^{2+}})$ vs. $\log a_{SiO_2(aq)}$. The slope is -1.

(b) Only the ratio $(a_{Na^+}^2 a_{SiO_2(aq)}/a_{Ca^{2+}})$ is buffered. If quartz is present, then the ratio $(a_{Na^+}^2/a_{Ca^{2+}})$ is buffered.

E10.3 (b)

$$\log a_{Al^{3+}} = -3\,pH + 7.23$$

$$\log a_{Al(OH)^{2+}} = -2\,pH + 2.253$$

$$\log a_{Al(OH)_2^+} = -pH + 2.903$$

$$\log a_{Al(OH)_3^{\circ}} = -9.53$$

$$\log a_{Al(OH)_4^-} = pH - 14.932$$

E10.4 Your $\log(a_{K^+}/a_{H^+})$ vs. $\log a_{SiO_2(aq)}$ diagram represents equilibrium conditions among pure minerals. If the natural system (in this case altered granite and hot-spring water) is not near equilibrium, then the solution composition will not fall on the mineral field boundaries, and there may be too many minerals (the phase rule is not obeyed; see Chapter 12). In this case, with four components $(K_2O, Al_2O_3, SiO_2,$ and $H_2O)$ and two degrees of freedom (T and P), there should be no more than four phases coexisting, that is, three minerals plus water. Therefore the four-mineral granite mentioned, plus water, cannot fit the model. This is also shown by the diagram, in which no more than three minerals can coexist.

Nevertheless, if a solution having values of $\log(a_{K^+}/a_{H^+})$ and $\log a_{SiO_2}$ which fall in the kaolinite field is in contact with microcline, we can predict that microcline will react to form kaolinite and that the solution composition will move towards the microcline/kaolinite boundary. Most groundwaters do in fact plot in the kaolinite field, which is why feldspars are unstable in the weathering environment.

(a) (i) If quartz *and the other minerals* do not equilibrate with the solution, we can say little, except that the solution composition will not be on the quartz saturation line, except by accident. At low temperatures ($\lesssim 150\,°C$) solutions are quite often supersaturated and sometimes undersaturated with respect to quartz. At higher temperatures, saturation with quartz is the rule, and the silica content may even be used as a geothermometer. However, if we assume that quartz is the only recalcitrant mineral, then the solution composition should migrate toward the intersection of the microcline, muscovite, and kaolinite fields.

(ii) If quartz does equilibrate, then the solution composition lies on the quartz saturation line, but we can say little else. One of the other three minerals (either microcline or kaolinite) must disappear before equilibrium can be achieved. If microcline disappears, the solution could equilibrate at the muscovite–kaolinite–quartz intersection.

(iii) The answer depends on what stable or metastable assemblage you assume would exist. If, for example, you assume that K-feldspar changes metastably to kaolinite while at equilibrium with muscovite, $a_{K^+} \leq 10^{-2.45}$. Other "correct" answers are possible.

You also can say nothing about the mineral *proportions*, even at equilibrium. That is, it makes no sense to say "the granite would be mostly microcline, with some muscovite and kaolinite." If microcline is present in the system, its activity in the model is 1.0, irrespective of whether it makes up 1% or 99% of the mineral mass present.

E10.5 You would not expect to find much quartz in bauxite (Equation (10.16); Figure 10.8). Any you did see should be coated or armored by a layer or zone of kaolinite, which is common in bauxite.

E10.6 (a) Redox conditions have no effect.

(b) $a_{K^+} = 4.173$. A wrong answer might be obtained because of using data for $SiO_2(aq)$ instead of quartz.

(c) Measurement or calculation error, lack of equilibrium, possibly poor thermodynamic data.

E10.9 For

$$Al_2Si_4O_{10}(OH)_2 + H_2O(l) = Al_2Si_2O_7 \cdot 2H_2O + 2\,SiO_2(quartz)$$

$$\Delta_f G° = -3799.7 + 2(-856.64) - (-5268.14) - (-237.129)$$
$$= -7.711\,kJ\,mol^{-1}$$

which is negative, so kaolinite would be expected rather than pyrophyllite with quartz. If all three (kaolinite, quartz, and pyrophyllite) were found together, then either pyrophyllite is metastable in the presence of water, or water is not present, and the three phases can coexist at equilibrium. Think about this in terms of the phase rule, using three components (e.g., Al_2O_3, SiO_2, and H_2O).

E10.10 In this case the presence of the kaolinite is irrelevant. The silica content of the water is controlled by the quartz:

$$SiO_2(quartz) = SiO_2(aq)$$

$$\Delta_r G° = -833.411 - (-856.64)$$
$$= 23.229\,kJ$$

$$\log K = -23,229/5708.042$$
$$= -4.070$$

$$K = a_{SiO_2(aq)}$$
$$= 10^{-4.070}$$

A change in redox conditions would have no effect, as no elements are reduced or oxidized in this reaction.

E10.11 We have

$$Al_2Si_4O_{10}(OH)_2 + H_2O(l) = Al_2Si_2O_7 \cdot 2H_2O + 2\,SiO_2(aq)$$

This time we write the reaction using $SiO_2(aq)$ because we want to calculate $a_{SiO_2(aq)}$, and this time it is controlled by the assemblage kaolinite plus pyrophyllite. Thus

$$\Delta_r G° = -3799.7 + 2(-833.411) - (-5268.14) - (-237.129)$$

$$= 38.747\,kJ$$

$$\log K = -38,747/5708.042$$

$$= -6.788$$

$$K = 10^{-6.788}$$

$$= a_{SiO_2(aq)}^2$$

$$a_{SiO_2(aq)} = 10^{-3.394}$$

so the concentration of silica in this solution is about $10^{-3.394}$ molal.

Note that it is strictly speaking incorrect to say the activity of aqueous silica is $10^{-3.394}$ molal. Activities are dimensionless, so $a_{SiO_2(aq)} = 10^{-3.394}$, from which we conclude (from the fact that we are using the ideal 1 molal standard state for silica, and that the activity coefficient for $SiO_2(aq)$ is 1.0) that $m_{SiO_2(aq)} = 10^{-3.394}$, or that silica is $10^{-3.394}$ molal.

E10.12

$$\Delta_r H° = -4119.6 + 2(-877.699) - (-5642.04) - (-285.83)$$

$$= 52.872\,kJ\,mol^{-1}$$

$$\log K_{348} = \log K_{298} - \frac{\Delta_r H°}{2.30259 \times R}\left(\frac{1}{T} - \frac{1}{T_r}\right)$$

$$= -6.788 - \frac{52,872}{19.1449}\left(\frac{1}{348.15} - \frac{1}{298.15}\right)$$

$$= 6.788 - (-1.330)$$

$$= -5.458$$

$$a_{SiO_2(aq)} = 10^{-2.729}\ \text{at } 75\,°C$$

Or

$$\Delta_r S° = 205.0 + 2(75.312) = 239.41 - 69.91$$

$$= 46.304\,J\,mol^{-1}\,K^{-1}$$

$$\Delta_r G°_{348} = \Delta_r H°_{298} - T\,\Delta_r S°_{298}$$

$$= 52,872 - 348.15(46.304)$$

$$= 36,751\,J\,mol^{-1}$$

$$\log K_{348} = -36{,}751/(348.15 \times 19.1445)$$
$$= -5.514$$

$$a_{SiO_2(aq)} = 10^{-2.757}$$

Or

$$\Delta_r G_T^\circ = \Delta_r G_{T_r}^\circ - \Delta_r S_{T_r}^\circ (T - T_r)$$
$$\Delta_r G_{348}^\circ = 38{,}747 - 46.304(348.15 - 298.15)$$
$$= 36{,}432 \, \text{J mol}^{-1}$$

$$\log K_{348} = -36{,}432/(348.15 \times 19.1445)$$
$$= -5.466$$

$$a_{SiO_2(aq)} = 10^{-2.733}$$

Slight differences are due to the fact that thermodynamic data are derived in various places at various times by various laboratories, and, although they are accurate, they have various degrees of precision and are not necessarily perfectly consistent when used together in the standard equations. See Chapter 5.

Chapter 11

E11.1 See Figure C.6. Notice that there is a wide range of conditions under which galena, not anglesite, is the stable lead mineral in sulfate-dominant solutions.

E11.2 (a)

$$UO_2^{2+}(aq) + 2e = UO_2(s) \quad \mathcal{E}^\circ = 0.405 \, V$$

$$UO_2CO_3^\circ(aq) + 2H^+ + 2e = UO_2(s) + CO_2(g) + H_2O(l) \quad \mathcal{E}^\circ = 0.497 \, V$$

$$UO_2(CO_3)_2^{2-}(aq) + 4H^+ + 2e = UO_2(s) + 2CO_2(g) + 2H_2O(l) \quad \mathcal{E}^\circ = 0.677 \, V$$

(b)

$$UO_2CO_3^\circ(aq) + 2H^+ = UO_2^{2+}(aq) + CO_2(g) + H_2O(l); \quad pH = 2.06$$

$$UO_2(CO_3)_2^{2-}(aq) + 2H^+ = UO_2CO_3^\circ(aq) + CO_2(g) + H_2O(l); \quad pH = 3.53$$

(c) Boundary 1: $Eh = 0.405 \, V$.
Boundary 2: $Eh = 0.349 - 0.0592 \, pH$.
Boundary 3: $Eh = 0.558 - 0.1184 \, pH$.

E11.3 (a) $Eh = -0.141 \, V; f_{O_2} = 10^{-68.6}$ bar.

(b) Using $\Delta_r G_{573}^\circ = \Delta_r G_{298}^\circ - \Delta_r S^\circ (T - 298.15), f_{O_2} = 10^{-29.01}$ bars
Using $\Delta_r G_{573}^\circ = \Delta_r H_{298}^\circ - 573.15(\Delta_r S_{298}^\circ), f_{O_2} = 10^{-29.07}$ bars.

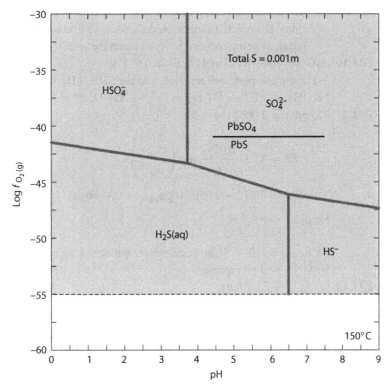

Figure C.6 The system H_2S–H_2O at 150 °C. Diagram provided by The Geochemist's Workbench (2015).

Using $\log K_{573} = \log K_{298} - (\Delta_r H^\circ_{298}/2.303R)[1/573.15 - 1/298.15], f_{O_2} = 10^{-28.97}$ bar.

E11.4 (a) I in I^- is -1; I in IO_3^- is $+5$.

(b) $IO_3^- + 6H^+ + 6e = I^- + 3H_2O(l)$. $\mathcal{E}^\circ = 1.097$ V; $Eh = 0.742$ V.

(c) $IO_3^- = I^- + 1.5O_2(g)$.

(d) $f_{O_2} = 10^{-8.93}$ bar; magnetite is stable only if $f_{O_2} \leq 10^{-68.6}$ bar, so expect iodide, I^-.

E11.5 The calculation follows that on page 266, just substituting $\Delta_f G^\circ$ for Zn^{2+} for that of Fe^{2+}. The zinc half-cell voltage (-0.762 V) is greater than that for iron (-0.409 V), and therefore gives a greater cell voltage, 1.101 V rather than 0.748 V, and so was favored.

E11.6 An Eh of zero just means that the redox conditions happen to be such that the pH and the activity ratio of reduced to oxidized species in the Nernst equation combine to give a voltage equal to the standard voltage $\mathcal{E}^\circ$. It has no particular theoretical significance. Naturally, however, in any particular system, solutions with a negative Eh will be more reduced than those with a positive Eh.

E11.7 $a_{Fe} = 10^{-6.20}$; $10^{-2.91}$; 1.0 for Fe_3O_4–Fe_2O_3, Fe_3O_4–FeO, Fe–FeO, respectively.

E11.8 $a_{Fe} = 10^{-5.95}$.

E11.9 (d) At this Eh and sulfate concentration, the equilibrium $H_2S(aq)$ activity is enormous. However, it cannot exceed about 0.1 (the solubility of H_2S in water), so sulfate is being reduced to try to reach the equilibrium concentration.

E11.10 $\Delta_rG° = -93.787\,kJ\,mol^{-1}$. $\mathcal{E}° = 0.972\,V$.

For the complete cell reaction, add the SHE, $\frac{1}{2}H_2 = H^+ + e$; $\mathcal{E}° = 0$, to get
$$Fe(OH)_3 + 2H^+ + \tfrac{1}{2}H_2(g) = Fe^{2+} + 3H_2O; \ \mathcal{E}° = 0.972\,V$$

E11.11 $475\,ppm = 0.00851\,m = 10^{-2.070}\,m$.

$$Eh = \mathcal{E}° - \frac{0.0592}{n}\log\left(\frac{a_{Fe^{2+}}}{a_{H^+}^3}\right)$$

$$0.533 = 0.972 - 0.0592\log a_{Fe^{2+}} - 0.1776\,pH$$

$$\log a_{Fe^{2+}} = -4.134$$

So $a_{Fe^{2+}} = 10^{-4.134}$ at equilibrium, the actual $a_{Fe^{2+}}$ is $10^{-2.070}$, and therefore $Fe(OH)_3$ will precipitate.

E11.12 $\log f_{O_2} = -31.7\,bars$.

E11.13

$$\Delta_rG° = -228.1 + 2(-237.129) - (-465.14)$$

$$= -237.218\,kJ$$

$$= -n\mathcal{F}\mathcal{E}°$$

$$\mathcal{E}° = 237,218/(96,485 \times 2)$$

$$= 1.229\,V$$

$$Eh = \mathcal{E}° - \frac{0.0592}{2}\log\left(\frac{a_{Mn^{2+}}}{a_{H^+}^4}\right)$$

$$0.6 = 1.229 - 0.0296\log\left(\frac{a_{Mn^{2+}}}{(10^{-7.5})^4}\right)$$

$$a_{Mn^{2+}} = 10^{-8.75} \ \text{at equilibrium.}$$

E11.14

$$Eh = 1.23 + 0.0148\log f_{O_2} - 0.0592\,pH$$

$$0.6 = 1.23 + 0.0148\log f_{O_2} - 0.0592(7.5)$$

$$\log f_{O_2} = -12.57$$

$$f_{O_2} = 10^{-12.57}\,bar.$$

$$Mn(s) + O_2(g) = MnO_2(s)$$

$$\Delta_r G° = -465.14 - 0 - 0$$

$$\log K = 465{,}140/5708.042$$

$$= 81.49$$

$$K = 1/f_{O_2}$$

so $f_{O_2} = 10^{-81.47}$ bar for Mn–MnO$_2$ equilibrium, but actually $f_{O_2} = 10^{-12}$ bar. Then $10^{-12} \gg 10^{-81.47}$, so yes, Mn should oxidize to MnO$_2$.

E11.15 $Eh = -0.141$ V; $f_{O_2} = 10^{-68.6}$ bar.

E11.16 $f_{O_2} = 10^{-29.0}$ bar.

E11.17 I in I$^-$ is -1; I in IO$_3^-$ is $+5$; $\mathcal{E}° = 1.097$ V; $Eh = 0.742$ V.

E11.18 $f_{O_2} = 10^{-8.93}$ bars; magnetite only stable if $f_{O_2} \leq 10^{-68.6}$ bars, so expect iodide, I$^-$.

Chapter 12

E12.1 $\Delta_f H_\beta° = -18{,}890$ cal mol^{-1}.

E12.2 $V_{H_2O(s)}° = 19.688$ cm^3 mol^{-1}.

E12.3 (a) Point 2 has four coexisting phases; the maximum is three for one component. Point 2 may actually be two points close together.

Point 5 has a triple point with one angle $> 180°$, which is not possible. Each metastable extension must lie between two stable curves. You can prove this to yourself by drawing G–T or G–P sections.

(d) $V_E° = 14.01$ cm^3 mol^{-1}.

(e) $S_C° = 21.26$ J mol^{-1}.

(f) $P_7 = 501.7$ bar.

(g) 15.0 cm^3 mol^{-1} $> V_B° > 11.62$ cm^3 mol^{-1}.

(h) Point 6 is a critical point. E is liquid. F is vapor or gas. Others are solids. Solid Λ floats. Solids B, C, D, and G sink.

(i) The slope dP/dT is not constant because $\Delta S/\Delta V$ is not constant, and this, in turn, is because the properties of phase F (a gas) change more rapidly with P, and to a lesser extent T, than do the properties of solids and liquids.

E12.4 T_m of A is 60 °C. T_m of B is 880 °C. Two polymorphs, the transition T is 275 °C. $S_\beta° = 6.35$ J mol^{-1} K^{-1}.

E12.7 Any way of looking at it that obeys the phase rule is right. Choose the one that suits you. The phase transition loop becomes detached from the temperature axis for component A when solution 1 is liquid, solution 2 is vapor, and the pressure of the T–X section is above the critical pressure of A, and similarly for B. Therefore, the loop may become detached from one or both axes.

E12.8 There are reasons why ice-IX does not exist, but they are not to be found within equilibrium thermodynamics, the subject of this book. Thermodynamics does not so much "explain" energetic relationships as define parameters and methods of measurement and calculation that bring order to our knowledge of existing substances. To understand why this order exists, and not some other kind of order, you must go to statistical and quantum chemistry. If ice-IX did exist, its melting temperature would have to be $> 0\,^\circ$C or $32\,^\circ$F, and $114.4\,^\circ$F is as good a number as any.

E12.9 (a) See Figure C.7.

     (b) Ice-V sinks in water.

     (c) See Figure C.8.

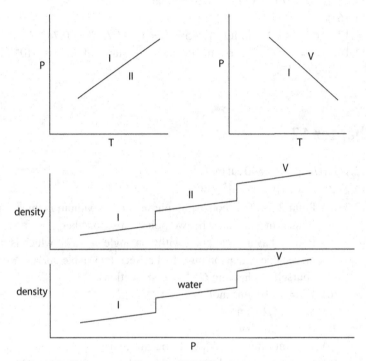

Figure C.7 The answer to Exercise E12.7(a).

Chapter 13

E13.1 (a)

$$dA = \left(\frac{\partial A}{\partial V}\right)_{T,\xi} dV + \left(\frac{\partial A}{\partial T}\right)_{V,\xi} dT + \left(\frac{\partial A}{\partial \xi}\right)_{T,V} d\xi$$

$$= -P\,dV - S\,dT - \mathcal{A}\,d\xi$$

$$dA_{T,V} = -\mathcal{A}\,d\xi$$

$$= 0 \quad \text{at equilibrium}$$

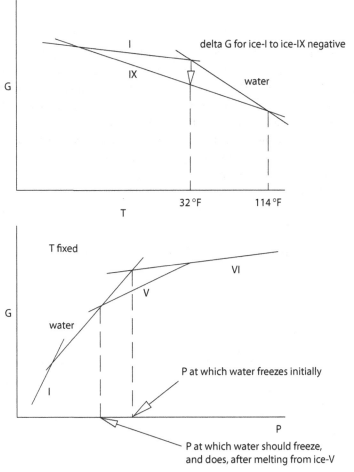

Figure C.8 The answer to Exercise E12.7(c).

(b) From Equation (4.30) we get

$$dH = \left(\frac{\partial H}{\partial S}\right)_{P,\xi} dS + \left(\frac{\partial H}{\partial P}\right)_{S,\xi} dP + \left(\frac{\partial H}{\partial \xi}\right)_{S,P} d\xi$$

$$= T\,dS + V\,dP - \mathcal{A}\,d\xi$$

$$dH_{S,P} = -\mathcal{A}\,d\xi$$

$$= 0 \quad \text{at equilibrium}$$

(c)

$$dV = \left(\frac{\partial V}{\partial U}\right)_{S,\xi} dU + \left(\frac{\partial V}{\partial S}\right)_{U,\xi} dS + \left(\frac{\partial V}{\partial \xi}\right)_{U,S} d\xi$$

$$= (1/P)dU - (T/P)dS - \mathcal{A}\,d\xi$$

$$dV_{U,S} = -\mathcal{A}\,d\xi$$

$$= 0 \quad \text{at equilibrium}$$

It is hard to imagine circumstances in which $V_{U,S}$ could be used as a thermodynamic potential, so it is never used as such.

E13.2 Both the internal energy per mole and the temperature are unchanged by ideal mixing. Equation (7.12) shows that the molar enthalpy does not change on ideal mixing, and neither does PV, so U does not change either. Of course, H and V do change on mixing. Equation (7.13) shows that changing the temperature does not change this fact. This is not to be confused with the change in enthalpy itself, which does change with temperature, as in Equation (3.33). It is the mixing which has no effect.

E13.3 After each increment of reaction, $n_{N_2} = 1 - \xi$, $n_{H_2} = 3(1 - \xi)$, and $n_{NH_3} = 2\xi$. The total number of moles in the system at any stage is

$$n_{N_2} + n_{H_2} + n_{NH_3} = (1 - \xi) + (3 - 3\xi) + (2\xi)$$

$$= 4 - 2\xi$$

and the mole fractions are

$$\left. \begin{aligned} x_{N_2} &= \frac{1 - \xi}{4 - 2\xi} \\[2mm] x_{H_2} &= \frac{3 - 3\xi}{4 - 2\xi} \\[2mm] x_{NH_3} &= \frac{2\xi}{4 - 2\xi} \end{aligned} \right\} \tag{C.1}$$

REFERENCES

Anderson, G. M. 2005. *Thermodynamics of Natural Systems*, 2nd edn. Cambridge: Cambridge University Press.

Anderson, G. M. 2015. The meaning of Δ. *Journal of Chemical Education*, **92**, 774–776.

Anderson, G. M., and Crerar, D. A. 1993. *Thermodynamics in Geochemistry*. Oxford: Oxford University Press.

Anderson, G. M., Castet, S., Mesmer, R. E., and Schott, J. 1991. The density model for estimation of thermodynamic parameters of reactions at high temperatures and pressures. *Geochimica et Cosmochimica Acta*, **55**, 1769–1779.

Atkins, P. 2010. Teaching thermodynamics: The challenge. *Pure and Applied Chemistry*, **83**(6), 1217–1220.

Badger, P. H. 1967. *Equilibrium Thermodynamics*. Boston, MA: Allyn and Bacon, Inc.

Barnes, I., Stuart, W. T., and Fisher, D. W. 1964. Field investigation of mine waters in the northern anthracite field, Pennsylvania. U.S. Geological Survey Professional Paper 473-B.

Barrett, T. J., and Anderson, G. M. 1982. The solubility of sphalerit and galena in NaCl brines. *Economic Geology*, **77**, 1923–1933.

Bent, H. A. 1965. *The Second Law. An Introduction to Classical and Statistical Thermodynamics*. New York: Oxford University Press.

Bethke, C. M. 2008. *Geochemical and Biogeochemical Reaction Modeling,* 2nd edn. Cambridge: Cambridge University Press.

Bowers, T. S., and Helgeson, H. C. 1983. Calculation of the thermodynamic and geochemical consequences of nonideal mixing in the system $H_2O–CO_2–NaCl$ on phase relations in geologic systems. *Geochimica et Cosmochimica Acta*, **47**, 1247–1275.

Bridgman, P. W. 1958. *The Physics of High Pressure*. London: G. W. Bell and Sons.

Bridgman, P. W. 1961. *The Nature of Thermodynamics*. New York: Harper Torchbooks. (First published in 1941 by Harvard University Press.)

Callen, H. B. 1985. *Thermodynamics and an Introduction to Thermostatistics*, 2nd edn. New York: John Wiley & Sons.

Canagaratna, S. G. 1969. A critique of the definitions of heat. *American Journal of Physics*, **37**, 679–683.

Carmichael, D. M. 1986. Induced stress and secondary mass transfer: thermodynamic basis for the tendency toward constant volume constraint in diffusion metasomatism, in Helgeson, H. C. (ed.), *Transport in Metasomatic Processes*. Dordrecht: D. Reidel Publishing Co., pp. 239–264.

Carnot, S. 1960. *Reflections on the Motive Power of Fire and Other Papers on the Second Law of Thermodynamics by E. Clapeyron and R. Clausius*. New York: Dover Publications Inc.

Carroll, S. 2010. *From Eternity to Here: The Quest for the Ultimate Theory of Time*. New York: Dutton.

Cartwright, N. 1983. *How the Laws of Physics Lie*. Oxford: Clarendon Press.

Clarke, E. C. W., and Glew, D. N. 1966. Evaluation of thermodynamic functions from equilibrium constants. *Transactions of the Faraday Society*, **62**, 539–547.

Clausius, R. 1867. *The Mechanical Theory of Heat with Its Applications to the Steam Engine and to the Physical Properties of Bodies*. London: John Van Voorst. (Translated and with an introduction by Professor John Tyndall. A collection of nine memoirs on the mechanical theory of heat published between 1850 and 1865.)

Clausius, R. 1879. *The Mechanical Theory of Heat*. London: MacMillan and Co. (Translated by Walter R. Browne. Probably the best summary by Clausius himself of his work in thermodynamics.)

Cohen, E. R., and Taylor, B. N. 1988. The 1986 CODATA recommended values of the fundamental physical constants. *Journal of Physical and Chemical Reference Data*, **17**, 1795–1803.

Craig, N. C. 1987. The chemist's delta. *Journal of Chemical Education*, **64**, 668–669.

De Donder, Th. 1920. *Leçons de thermodynamique et de chimie physique. Première partie: Théorie. Rédigées par F. H. van den Dungen and G. J. M. van Lerberghe*. Paris: Gauthiers-Villars.

De Donder, Th. 1922. L'affinité. Applications aux gaz parfaits. *Bullet in de l'Académie Royale de Belgique, Classe des Sciences*, **7**(5) 197–205.

De Donder, Th. 1927. *L'Affinité*. Brussels: M. Lamertin.

De Donder, Th., and van Rysselberghe, P. 1936. *Thermodynamic Theory of Affinity*. Stanford, CA: Stanford University Press.

de Heer, J. 1986. *Phenomenological Thermodynamics*. Englewood Cliffs, NJ: Prentice-Hall, Inc.

Denbigh, K. 1981. *The Principles of Chemical Equilibrium: with Applications in Chemistry and Chemical Engineering*, 4th edn. Cambridge: Cambridge University Press.

Dickerson, R. E. 1969. *Molecular Thermodynamics*. New York: W. A. Benjamin.

Douglas, T. B., and King, E. G. 1968. High temperature drop calorimetry, chapter in *Experimental Thermodynamics*. London: Butterworths, pp. 293–332.

Drever, J. I. 1988. *The Geochemistry of Natural Waters*, 2nd edn. Englewood Cliffs, NJ: Prentice-Hall.

Feynman, R. P., Leighton, R. B., and Sands, M. 1963. *The Feynman Lectures on Physics*, Vol. 1. New York: Addison Wesley Publishing Co.

Frigg, R., and Hartmann, S. 2012. Models in science, in Zalta, E. N. (ed.), *The Starford Encyclopedia of Philosophy*, fall 2012 edn. The Metaphysics Research Lab, Stanford University. (Available from http://plato.stanford.edu/archives/fall2012/entries/models-science/.)

Garrels, R. M., and Christ, C. L. 1965. *Solutions, Minerals, and Equilibria*. New York: Harper & Row.

Garrels, R. M., and Thompson, M. E. 1962. A chemical model for seawater at 25 °C and one atmosphere total pressure. *American Journal of Science*, **260**, 57–66.

Gibbs, J. W. 1961a. On the equilibrium of heterogeneous substances, in *The Scientific Papers of J. Willard Gibbs. Volume 1 Thermodynamics*. New York: New York: Dover Publications Inc., pp. 55–353. (Originally published 1875–1878.)

Gibbs, J. W. 1961b. A method of geometrical representation of the thermodynamic properties of substances by means of surfaces, in *The Scientific Papers of J. Willard Gibbs. Volume 1 Thermodynamics*. New York: Dover Publications Inc., pp. 33–54. (Originally published in 1873.) Chap. II. A Method of Geometrical Representation of the Thermodynamic Properties of Substances by Means of Surfaces, pages 33–54.

Guggenheim, E. A. 1959. *Thermodynamics. An Advanced Treatment for Chemists and Physicists*, 4th edn. Amsterdam: North-Holland Publishing Company.

Hahn, W. C., and Muan, A. 1962. Activity measurements in oxide solid solutions. The system "FeO"–MgO in the temperature interval 1100 to 1300 °C. *Transactions of the Metallurgical Society of the AIME*, **224**, 416–420.

Harvey, A. H., Peskin, A. P., and Klein, S. A. 2000. *NIST/ASME Steam Properties, NIST Standard Reference Database 10, Version 2.2*. Gaithersburg, MD: National Institute of Standards and Technology. (Available from the NIST website, www.nist.gov/srd/nist10.cfm.)

Helgeson, H. C. 1968. Evaluation of irreversible reactions in geochemical processes involving minerals and aqueous solutions. I. Theoretical relations. *Geochimica et Cosmochimica Acta*, **32**, 853–877.

Helgeson, H. C., Delaney, J. M., Nesbitt, H. W., and Bird, D. K. 1978. Summary and critique of the thermodynamic properties of rock-forming minerals. *American Journal of Science*, **278A**, 1–229.

Hemingway, B. S., Hass, J. L. Jr., and Robinson, G. R. Jr. 1982. *Thermodynamic Properties of Selected Minerals in the System $Al_2O_3-CaO-SiO_2-H_2O$ at 298.15 K and 1 bar (10^5 pascals) Pressure and at Higher Temperatures*. Washington, DC: US Geological Survey.

Herzfeld, C. M. 1962. The thermodynamic temperature scale, its definition and realization, chapter in *Temperature, Its Measurement and Control in Science and Industry, Volume 1, Basic Standards and Methods*. New York: Rheinhold Publishing Co., pp. 41–50.

Holland, T. J. B., and Powell, R. 2011. An improved and extended internally consistent thermodynamic dataset for phases of petrological interest, involving a new equation of state for solids. *Journal of Metamorphic Geology*, **29**, 333–383.

Holm, J. L., and Kleppa, O. J. 1966. The thermodynamic properties of the aluminum silicates. *American Mineralogist*, **51**, 1608–1622.

Johnson, J. W., Oelkers, E. H., and Helgeson, H. C. 1992. SUPCRT92: a software package for calculating the standard molal thermodynamic properties of minerals, gases, aqueous species and reactions from 1 to 5000 bars and 0 to 1000 °C. *Computers and Geosciences*, **18**, 899–947.

Jones, J. B., and Dugan, R. E. 1996. *Engineering Thermodynamics*. Englewood Cliffs, NJ: Prentice Hall, Inc.

Klotz, I. M. 1964. *Chemical Thermodynamics*. New York: W. A. Benjamin Inc. (With advice and suggestions from Thomas Fraser Young.)

Knapp, R. A. 1989. Spatial and temporal scales of local equilibrium in dynamic fluid–rock systems. *Geochimica et Cosmochimica Acta*, **53**, 1955–1964.

Krupp, R. E., and Seward, T. M. 1987. The Rotokawa geothermal system, New Zealand: An active epithermal gold-depositing environment. *Economic Geology*, **82**, 1109–1129.

Lambert, F. L. 1999. Shuffled cards, messy desks, and disorderly dorm rooms – examples of entropy increase? nonsense! *Journal of Chemical Education*, **76**, 1385–1387.

Lambert, F. L. 1998. Chemical kinetics: As important as the second law of thermodynamics? *The Chemical Educator*, **3**(2), 1–6.

Lewis, R. N., and Randell, M. 1923. *Thermodynamics and the Free Energy of Chemical Substances*, 1st edn. New York: McGraw Hill.

Lindberg, R. D., and Runnells, D. D. 1984. Groundwater redox reactions: An analysis of equilibrium state applied to *Eh* measurements and geochemical modeling. *Science*, **225**, 925–927.

Lupis, C. H. P. 1983. *Chemical Thermodynamics of Materials*. New York: North Holland.

Maier, C. G., and Kelly, K. K. 1932. An equation for the representation of high temperature heat content data. *Journal of the American Chemical Society*, **54**, 3243–3246.

Marsden, J. E., and Troma, A. J. 1988. *Vector Calculus*, 3rd edn. New York: W. H. Freeman and Company.

Maxima. 2013. *Maxima, a Computer Algebra System. Version 5.30.0*. (A free computer algebra program, http://maxima.sourceforge.net.)

McGlashan, M. L. 1979. *Chemical Thermodynamics*. London: Academic Press.

References

McQuarrie, D. A., and Simon, J. D. 1997. *Physical Chemistry, A Molecular Approach*. Sausalito, CA: University Science Books.

Melrose, M. P. 1970. Statistical mechanics and the third law. *Journal of Chemical Education*, **47**(4), 283–284.

Merino, E. 1975. Diagenesis in Tertiary sandstones from Kettleman North Dome, California – II. Interstitial solutions: Distribution of aqueous species at 100 °C and chemical relation to the diagenetic mineralogy. *Geochimica et Cosmochimica Acta*, **39**(12), 1629–1645.

Morey, G. W., Fournier, R. O., and Rowe, J. J. 1962. The solubility of quartz in the temperature interval from 25 to 300 °C. *Geochimica et Cosmochimica Acta*, **26**, 1029–1043.

Nahon, D., and Merino, E. 1987. Pseudomorphic replacement in tropical weathering: Evidence, geochemical consequences, and kinetic-rheological origin. *American Journal of Science*, **297**, 393–417.

Nash, L. K. 1974. *Elements of Statistical Thermodynamics*, 2nd edn. Reading, MA: Addison-Wesley Publishing Co.

Navrotsky, A., and Kasper, R. B. 1976. Spinel disproportionation at high pressure: calorimetric determination of enthalpy of formation of Mg_2SnO_4 and Co_2SnO_4 and some implications for silicates. *Earth and Planetary Science Letters*, **31**, 247–254.

Pankratz, L. B. 1964. *High-Temperature Heat Contents and Entropies of Muscovite and Dehydrated Muscovite*. Washington, D.C.: Government Printing Office.

Parkhurst, D. L., and Appelo, C. A. J. 2013. *PHREEQC Version 3. A Computer Program for Speciation, Batch-Reaction, One-Dimensional Transport, and Inverse Geochemical Calculations*. Reston, VA: U.S. Geological Survey. (Available from http://hydrochemistry.eu/downl.html.)

Pippard, A. B. 1966. *Elements of Classical Thermodynamics*. Cambridge: Cambridge University Press.

Pitzer, K. S., Peiper, J. C., and Busey, R. H. 1984. Thermodynamic properties of aqueous sodium chloride solutions. *Journal of Physical and Chemical Reference Data*, **13**, 1–102.

Poincaré, H. 1952. *Science and Hypothesis*. New York: Dover Publications. (Republication of the first English translation published in 1905.)

Prigogine, I., and Defay, R. 1954. *Chemical Thermodynamics*. London: Longmans Green. (Translation of the 1950 edition by D. H. Everett.)

Purrington, R. D. 1997. *Physics in the Nineteenth Century*. New Brunswick, NJ: Rutgers University Press.

Raff, L. M. 2014. Spontaneity and equilibrium II: Multireaction systems. *Journal of Chemical Education*, **91**, 839–847.

Robie, R. A. 1987. Calorimetry, chapter in *Hydrothermal Experimental Techniques*. New York: Wiley & Sons, pp. 389–422.

Robie, R. A., and Hemingway, B. S. 1972. *Calorimeters for Heat of Solution and Low-Temperature Heat Capacity Measurements*. Washington, DC: US Government Printing Office.

Robie, R. A., Bethke, P. M., and Beardsley, K. M. 1967. *Selected X-Ray Crystallographic Data, Molar Volumes, and Densities of Minerals and Related Substances*. Bulletin 1248. U.S. Geological Survey.

Robie, R. A., Hemingway, B. S., and Fisher, J. R. 1978. *Thermodynamic Properties of Minerals and Related Substances at 298.15 K and 1 bar (10^5 Pascals) Pressure and at Higher Temperatures*. Washington, DC: US Government Printing Office.

Schottky, W., Ulich, H., and Wagner., C. 1929. *Thermodynamik*. Berlin: Springer.

Shock, E. L., and Helgeson, H. C. 1990. Calculation of the thermodynamic and transport properties of aqueous species at high pressures and temperatures: Standard partial molal properties of organic species. *Geochimica et Cosmochimica Acta*, **54**, 915–945.

Sklar, L. 1993. *Physics and Chance: Philosophical Issues in the Foundations of Statistical Mechanics*. Cambridge: Cambridge University Press.

Smith, W. R., and Missen, R. W. 1982. *Chemical Reaction Equilibrium Analysis: Theory and Algorithms*. New York: John. Wiley and Sons.

Steinmann, P., Lichtner, P. C., and Shotyk, W. 1994. Reaction path approach to mineral weathering reactions. *Clays and Clay Minerals*, **42**, 197–206.

Stolper, E., and Asimow, P. 2007. Insights into partial melting from graphical analysis of one-component systems. *American Journal of Science*, **307**, 1051–1139.

Strang, G. 1991. *Calculus*. Wellesley, MA: Wellesley-Cambridge Press.

Suleimenov, O. M., and Seward, T. M. 1997. A spectrophotometric study of hydrogen sulphide ionisation in aqueous solutions to 350 °C. *Geochimica et Cosmochimica Acta*, **61**, 5187–5198.

The Geochemist's Workbench. 2015. A collection of speciation and reactive transport software, available from www.gwb.com.

Thompson, J. B. Jr. 1970. Geochemical reaction and open systems. *Geochimica et Cosmochimica Acta*, **34**, 529–551.

Tisza, L. 1966. *Generalized Thermodynamics*. Cambridge, MA: The M.I.T. Press.

Tuttle, O. F., and Bowen, N. L. 1958. *Origin of Granite in the Light of Experimental Studies in the System $NaAlSi_3O_8$–$KAlSi_3O_8$–SiO_2–H_2O*. Boulder, CO: Geological Society of America.

Ulmer, G. C., and Barnes, H. L. 1987. *Hydrothermal Experimental Techniques*. New York: John Wiley & Sons.

Van Zeggeren, F., and Storey, S. H. 1970. *The Computation of Chemical Equilibria*. Cambridge: Cambridge University Press.

Wagman, D. D., Evans, W. H., Parker, V. B., Schumm, R. H., Halow, I., and Bailey, S. M. 1982. The NBS Tables of Chemical Thermodynamic Properties. *Journal of Physical and Chemical Reference Data*, **II**, 392 pp. Supplement no. 2. (Published by the American Chemical Society for the National Bureau of Standards.)

Wellman, T. H. 1969. The vapor pressure of NaCl over decomposing sodalite. *Geochimica et Cosmochimica Acta*, **33**, 1302–1304.

Wilks, J. 1961. *The Third Law of Thermodynamics*. Oxford: Oxford University Press.

Zhu, C., and Anderson, G. M. 2002. *Environmental Applications of Geochemical Modeling*. Cambridge: Cambridge University Press.

Zimm, B. H., and Mayer, J. E. 1944. Vapor pressures, heats of vaporization, and entropies of some alkali halides. *The Journal Physics*, **12**, 362–369.

INDEX